Johannes Brügmann

Modelling and implementation of a microscopic traffic simulation system

Modelling and implementation of a microscopic traffic simulation system

Von der Fakultät für Ingenieurwissenschaften,
Abteilung Informatik und Angewandte Kognitionswissenschaft,
der Universität Duisburg-Essen
zur Erlangung des akademischen Grades eines

Doktors der Naturwissenschaften

genehmigte Dissertation

von

Johannes Brügmann

aus Stuttgart

1. Gutachter	Prof. Dr. Wolfram Luther
2. Gutachter	Prof. Dr. Michael Schreckenberg
Tag der mündlichen Prüfung	10. Juli 2015

Bibliografische Information der Deutschen Nationalbibliothek

Die Deutsche Nationalbibliothek verzeichnet diese Publikation in der
Deutschen Nationalbibliografie; detaillierte bibliografische Daten sind
im Internet über http://dnb.d-nb.de abrufbar.

ISBN 978-3-8325-4133-0

Logos Verlag Berlin GmbH
Comeniushof, Gubener Str. 47,
10243 Berlin
Tel.: +49 (0)30 42 85 10 90
Fax: +49 (0)30 42 85 10 92
INTERNET: http://www.logos-verlag.de

Contents

Introduction

This chapter introduces the microscopic traffic simulation systems. The online simulation system (OLSIM) that powers the traffic information platform `Autobahn.NRW.De` [146] employs such a microscopic traffic simulation. As part of the introduction, this chapter discusses several application domains as well as the motivation to use them. It explains their functioning and the basic principles on which they rely. The explanation also divides the functioning and the principles into subject-specific areas and associates the areas with their corresponding scientific domains. The chapter also highlights the most relevant parts of this thesis. These are several simulation models for real-world microscopic traffic simulations with various degrees of parallelism; the (semi-)formal specification for microscopic traffic simulations that smoothes formal verification especially for functional and declarative programming languages; classification of the simulation models into the systems theory by defining the corresponding state transition functions; a discussion of various implementation issues under the aspects of a productive system and of a verification and validation enhancement effort; and a presentation of the results of such a simulation systems as well as the benefits of this thesis.

The chapter starts with Section 1.1 that unveils the motivations as well as potential benefits for the application of traffic simulation systems. The section also explains the functioning and the basic principles of traffic simulation systems. Section 1.2 distinguishes the various traffic simulation systems under several different aspects and names their application domains. The section also names this thesis' subject-specific areas and associates them with their corresponding scientific domains. It also highlights the most relevant parts of this thesis. Section 1.3 concludes this chapter with an outline of this thesis.

1.1 Traffic simulation systems

In the last 30 years, the total number of vehicles, the total number of commuters, and the total number of kilometres driven on the highways of North-Rhine Westphalia (NRW) [145] increased more than the size of the road network did. Even worse, the network grew in size not only relatively slow. Undelayable reconstruction measures such as bridge reconstructions further reduce the network availability. Expressed in simple words, the traffic on the NRW highway network increased significantly. Supporters, maintainers, and administrators of the NRW highway network have to manage the traffic growth somehow. Maybe the most promising option to improve the situation exists, at least theoretically, with targeting a better capacity utilisation balance. While in the years before 2000 traffic information was mainly the domain of broadcasting stations and in the years before 1990 mainly fixed working times dominated the daily life [214], the means to achieve a better capacity utilisation balance have changed since then. Nowadays widespread flexible working times and the almost anywhere available internet support the distribution of traffic information and the commuters' flexible adaptation to spontaneously emerging traffic situations such as congested traffic and traffic accidents.

However, an increase in information does not automatically lead to a better highway capacity utilisation balance [188]. On the contrary, "advanced traffic information systems (ATIS) such as on-board navigation devices might make things even worse" [110]. When a highway user reaches a road that was in an uncongested traffic state at the time when the travel started, the road may be blocked when the user enters the road because traffic conditions change over time [110]. The factors that have an impact on the road user's decision range from monetary cost to loss in travel-time and also include non-causal behaviour. Among the factors that influence the highway users' route choice behaviour, the travel-time traffic information is the most important one as most highway users consider a loss in travel-time as 'lost time' [74] in [110]. The commuters' exact behaviour in response to the provided traffic information remains, however, an open question [188]. As a consequence, a warning information about a congested road segment "may act as a self-destroying prophecy; when all drivers choose alternative routes to avoid the critical road segment, the latter may be free of congestion, whereas the alternative formerly uncongested routes become congested" [110].

The approaches to achieve a better highway capacity utilisation balance include "adding more lanes to existing roads" (Schrank, Lomax and Eisele in [110]) and "constructing new ones" (Schrank, Lomax and Eisele in [110]). And they also include distributing the current traffic information to the highway users via internet and advanced traffic information systems. The approaches to heal the above mentioned consequences that arise from the increase in the amount of traffic information vary as described in the following. On the one side, they include changing the distribution of traffic information by making it available only for a fraction of the highway users. On the other side, the include even misinforming a fraction of the highway users about the traffic situation on particular road segments [110]. The final proof that an increase in information definitely leads to a better utilisation balance, is unfortunately not available. However, the need for traffic information is shared widely. Hall, Wakefield and Al-Kaisy assesses the travel time as the most important traffic information. They report "the results of focus group sessions [...] in which a group of commuters discussed their views about determinants of the freeway quality of service that they experienced. Total travel time is the most important determinant for them, but a number of other aspects of the trip also matter, including safety, traveler information, and maneuverability (density)" [74]. The question how much traffic information can help to relax the traffic situations on average and to better balance the highway capacity utilisation still lacks, however, a precise answer.

While the highway users' route choice behaviour has proven to be particularly difficult to influence and to direct into certain choices, delaying departure for several minutes may be possible for a relevant fraction of the highway users due to flexible working times. Traffic information distributed via internet may assist the highway users to make their decisions based on the current traffic situation. In contrast to the aforementioned advanced traffic information systems, an internet traffic information platform can assist the highway users to make their decision based on complete traffic information, to delay their trip, and to make their decision well-considered in contrast to the more spontaneous decision made during the travel and based on on-board devices. The development of the core of such a traffic information platform for the internet is the subject of this thesis.

Traffic information systems gather traffic data, optionally refine or combine them, and scatter the higher valued traffic information. As "packing the road infrastructure with a dense loop detector network is costly [...] data processing systems such as ASDA/FOTO [102, 203], VISUM [2] and microscopic traffic simulation systems such as VISSIM [3], SUMO [19], AIMSUN [197], OLSIMv3 [41] and now OLSIMv4 exist to complete the stationary data and provide valuable, spatio-temporal traffic information" [31]. The systems differ not only in how they refine and combine the traffic data into higher valued information, but also in how they gather and scatter their traffic information. Some systems collect their traffic data from movable devices such as mobile phones or on-board navigation devices while others use stationary devices such as cameras or inductive loop detectors. Some traffic information systems solely provide their traffic information after some time of data processing while others provide their information online, i.e. continuously and in real-time. Some systems refine their data by data processing while others employ simulations. The traffic information systems that employ a simulation system to refine and combine their traffic data additionally differ by their kind of simulation. The simulation systems distinguish between macroscopic, mesoscopic, and microscopic simulation systems.

This thesis presents a microscopic traffic simulation system that gathers its traffic data from inductive loop detectors and provides its simulation results as online traffic information. While macroscopic systems mostly model traffic by analogy to gas kinetic models, microscopic traffic simulations, in contrast, use "individual vehicles that move according to the rules of a microscopic traffic model in a virtual road traffic network to obtain traffic information" [30]. Consequently, a microscopic traffic simulation requires a microscopic traffic model that provides rules how the vehicles move inside the virtual network. The simulation requires also a network equipped with loop detectors and a virtual network, i.e. an instance of the (more or less) precise model of the network. The vehicles in the simulation enter the virtual network by a vehicle generation scheme. Some simulations use origin/destination matrices while others use tuning elements. The latter compares actual traffic data from the incoming loop detector data with should-be traffic values of the vehicles found in the simulation. The simulation system applies, then, the microscopic traffic model to the vehicles in the virtual network.

An analogy exists between the real-world road network and the virtual road network (cf. [180, Fig. 3]). In both networks it is possible to make experiments. The inductive loop detectors in the real-world road network measure the number of vehicles per single-lane and their velocities. Using an appropriate loop detector model, it is also possible to make the corresponding experiment in the simulation. In a perfect real world and a perfect simulation world, both data certainly would be identical. Until today, this is, obviously, not the case. However, the analogy as well as the differences between the two worlds facilitate to develop and exploit system theories for obtaining traffic information. The fundamental principle of obtaining the information acts on the assumption that if the measurements as the results from the experiments of both worlds match to a certain degree and for at least some time then the states of both worlds are assumed also to match. The assumption is, thus, a kind of a transitivity theorem in combination with a kind of an extrapolation.

However, to a certain limit, the analogy is well-suited enough, and the assumption is reasonable enough to compare the measurements from both worlds. Furthermore, the differences of both worlds facilitate to obtain higher valued traffic information. While the vehicles in the real world enter the road network due to the drivers' will, the vehicles in the virtual world enter due to a vehicle generation scheme. While the exact locations of the vehicles in the real world network are unknown for most of the time, they are well-known in the virtual world at any time. While calculating the density of vehicles on a certain road segment of the real-world is only a rough estimation by nature, calculating the density in the virtual network requires only to count the number of vehicles on the road segment and relate it to the total length of the road segment. In principle, such calculations represent, in fact, the making of experiments in both worlds (also called systems) where the experiments rely on system theories (also called models in the following). To obtain comparable measurements from both worlds, the models must correspond to each other. An inductive loop detector in the real-world relies on a traffic model that assumes all vehicles have velocities greater than zero when passing the detector (the real model is a technical standard and is much more complex). The corresponding model does not necessarily need to depend on the same assumption. In contrast, it should lead to comparable measurements that reflect the characteristics of the inductive loop detector model. Thus, conclusions that have been derived from the measurements of the experiments should always take into account that the model which is underlying the experiments may not be sufficiently well-suited to support a particular conclusion.

The comparison of the experiment results of both systems is not limited to the differences in the loop detector measurements. In principle, the systems support comparing any measurable phenomena such as road lengths, jam lengths, carbon emissions, vehicle masses and so on. The following section describes the application domains of a subset of such systems, namely the microscopic traffic simulations.

1.2 Application domains

The distinction of traffic simulation systems into macroscopic and microscopic systems is still not complete, and the distinction is not the only one. In addition to the already introduced simulation systems, mesoscopic traffic simulation systems co-exist and use an intermediate level of detail such as, for example, groups of vehicles. However, only the microscopic traffic simulation systems are in the focus of this thesis. Their space and time discretisation represent another important aspect for the distinction of the simulation systems. Some microscopic traffic simulations use space and time discretisation while others use only time discretisation to solve the underlying differential equation for the vehicle motion.

As the aforementioned distinctions are of technical nature, they remain secondary with respect to this chapter. The more important aspects for the classification of microscopic traffic simulations are the application domains, the kinds of traffic they simulate, as well as their current challenges.

Microscopic traffic simulations participate in various application domains such as traffic information systems, traffic control systems, and traffic planning systems. They assist the road users in choosing their best route with respect to the start and end time and also assist in finding alternative routes. They also assist the infrastructure planner in choosing among a couple of potential changes to the road network. And they assist the traffic control centres to base their dispositional actions on recent traffic information. All simulation systems have in common that they provide their traffic information to their users. The provided information refers to particular road segments and it ranges from traffic observables such as travel times, average velocities, and jam length over higher valued information such as level-of-service that indicates dense, viscous, or congested traffic for the segment to more abstract information such as road works messages and deviations. Composing the higher valued information from the more trivial ones is possible and a common practice in layered architectures. This thesis therefore focuses on gaining the trivial information and providing it to some traffic information system.

Short to medium road segments interrupted by traffic lights, low maximum velocities, a huge number of intersections, obstacles through parking vehicles, and bi-directional traffic all are common characteristics of urban traffic. In contrast thereto, highway traffic is uni-directional, and it consists of medium to long road segments most of the times without signalised intersections, multiple single-lanes per roadway, and relatively high maximum velocities. There also exist kinds of traffic without lane discipline but as is the case for urban traffic this is not in the focus of the thesis as the latter focuses on German highway traffic. In general, highways provide multiple single-lanes for the traffic. Depending on the respective traffic regulations, the vehicles can change in between the single-lanes and overtake on the left and on the right or either on the left or on the right. German traffic regulations demand the driver to overtake on the left lanes only. The aforementioned prerequisites and classifications delimit the subject of microscopic traffic simulation. The question emerges what kinds of sciences apply to microscopic traffic simulations.

From a physicist's point of view the question may arise "why physicists show so much interest in vehicular traffic. Motion, acceleration, and deceleration of an object such as a vehicle are topics, which are well understood since Isaac Newton. So, why still care about traffic?" [109, Chap. 1]. Knorr states in his answer to this question that vehicles as part of the traffic flow interact with each other and the interaction is the reason "which forces some vehicles to brake and allows others to accelerate" and "leads to collective effects". Thus, traffic belongs to the domain of the "dynamical systems with interacting particles". Helbing even entitles the many particle systems as "self-driven many-particle systems" [79] and relates the questions of why vehicles are sometimes stopped even though all drivers like to drive fast to the subject of physics [79]. The physicist's part of the subject covers, thus, studying and validating traffic system data, developing and deriving system theories, and explaining traffic phenomena.

From a mathematician's point of view, the mathematical systems theory "provides a fundamental, rigorous mathematical formalism for representing dynamical systems" [216, Chap. 1]. The "systems theory distinguishes between the structure (the inner constitution of a system) and behaviour (its outer manifestation)" [216, Chap. 1]. The theory also provides formalisms and structural patterns that have already certain properties such as closure under coupling granted. The formalisms help in modelling and understanding these complex systems. The complexity is frequently overlooked as large parts of these kind of dynamical systems look intuitive at the first glance. Applying the formalism to microscopic traffic simulations requires a formal specification of the latter. The specification is a milestone in mastering the complexity and requires the knowledge of semantic algebras to complete the formalism. Additionally, microscopic traffic simulations require numerical methods and stochastic methods such as random number generators.

From a computer scientist's point of view, simulation models have been used for a long time, and applying rules for vehicle motion to some set of vehicles in a road traffic network may sound like the state of the technology. However, "simulation models are increasingly being used to solve problems and to aid in decision-making" [180]. Therefore, a high degree of correctness, robustness, and reliability is of importance. As the simulations compute on machines, computer science has to fill the gap between the development of the traffic models and the application of general mathematical formalisms. This is what the thesis strives for. To fill the gap, this thesis provides a (semi-)formal specification for a family of microscopic traffic simulation systems. It builds a bridge from physics to mathematics using computer science. As the simulation systems from physics consist of a single road and a set of vehicles and as real-world scenarios involve complex road traffic networks, the step from physics to a real-world scenario without this thesis ends in an experiment. An experiment stands, however, in contrast to the aforementioned goals of importance. Additionally, the simulation models have to face conflicting goals, namely "information accuracy, increasing network size, network independency, and system scalability" [30]. The latter applies mainly to real-time systems such as the one described herein. Additionally, "the number and the complexity of the part-taking models pose the question of how to implement a simulation system with all these entities and still grant robustness and correctness of the computation" [30]. Physics and mathematics cannot answer these kind of questions, as the latter clearly belong to the subject of computer science. This thesis gives an answer to these questions from the perspective of a computer scientist.

This thesis presents the keys to a successful project and to a scalable, robust, and correct microscopic traffic simulation system. The keys involve a well-balanced amount of formal methods applied at the right places. As part of the keys, this thesis

- summarises the initial position as well as the state-of-the-art of science and technology. It thereby analyses the limitations of the current theories, systems, and models that participate in the context of microscopic traffic simulation systems. The analysis focuses on such formal aspects of the limitations that have the potential to contribute as determining factors when developing a microscopic traffic simulation system.

- introduces a (semi-)formal specification for multiple simulation models that consist of several subject-specific models with individual semantics, distinct complexities, several degrees of parallelism potential, and varying impacts and requirements on the network model. The subject–specific models include vehicle models, network models, microscopic traffic models, detector models, tuning models, and simulation (engine) models that apply single state transitions for entire road traffic networks. The involved models contribute in providing road network independency, traffic network portability across various kinds of traffic flow, information accuracy, robustness, correctness, and increase of parallelism and scalability.

- discusses the need for a software development process in a scientific research and development project, the choice of suitable programming languages for the development of a productive implementation, and the considered implementation goals.

- reflects the arising implementation issues by analysing the implementations of the real-world and real-time online microscopic traffic simulation system OLSIMv4 and the model checking implementation to enhance a future version of OLSIMv4. The reflection also identifies potential bottlenecks such as required library functionality, degree of parallelism, data storage, and data containers, and it also includes strategies for a future implementation aiming to exploit CPU thread-level parallelism.

- presents various kinds of traffic information as part of the benefits that result from this thesis and from validating the OLSIMv4 microscopic traffic simulation system. The presentation includes thoughts about the further refinement of the information and ideas about its further distribution. The presentation also covers some observations of the limitations of the current simulation approach.

- illustrates direct code verification options for implementations of the specified simulation models in a functional and a declarative programming language. Due to the analogy between the semantic algebras used by the specification and the means for the construction of abstract data types in functional and declarative programming languages, the direct code verification reduces to comparing the code with the specification.

- backs further verification options such as specification verification, model verification, and numerical solution verification that became available with the (semi-)formal specification. It demonstrates them exemplary as part of making the balance and presenting the benefits of this thesis.

- demonstrates numerical solution verification and model verification for the model checking implementation. The demonstration of the numerical solutions for a rather minimal simulation scenario with periodic boundary conditions shows the match with the solutions known from literature. The demonstration also verifies the model implementation for the minimal simulation scenario to be free from collisions. The demonstrations let validation and calibration activities for a future implementation of the OLSIMv4 system rely on verified models and a verified implementation.

- proposes implementation ideas for a future implementation of a productive implementation such as OLSIMv4. The implementation can rely on slim code, mature, well-structured, and orthogonal components. The proposed implementation results from the specification enhancement effort and the discussion of the implementation issues. The proposed specification and implementation ideas improve the transparency of the simulation model about which not much had been known previously. As a result from the specification and the implementation ideas, a future implementation will require a significantly reduced implementation effort only.

1.3 Outline

Chapter 2 recaptures the initial position that was present when the author joined the OLSIM team. At that time the third version of OLSIM software was already active and publicly available. As it was actively developed at that time, the project had to meet several requirements that ranged from the feature requests to re-architecture demands. Section 2.1 summarises the requirements and gives a short outline of the project history and its context. The chapter continues with an overview of the state of the art in Section 2.2. The overview comprises a system classification, a description of recent detector models, a discussion of the traffic models and its limitations, a sketch of current network models, a listing of vehicle generation schemes, and a criticism of common simulation systems. The chapter closes with a summary in Section 2.3.

Chapter 3 presents the conceptual design of the microscopic traffic simulation system in form of a (semi-)formal specification. The chapter starts with some general remarks about the methodology and the nomenclature of the specification in Subsection 3.1. The methodology employs semantic algebras to describe and to specify the participating models. The latter consist of the vehicle models, the network model, the traffic model, the detector models, the tuning element models, and the simulation models. The specification of the models starts with the vehicle models and the vehicle type models in Section 3.2. The specification continues with the network model in Section 3.3 that consists of cells, sections, roadways, a topological analysis, the occupation tables, and the switching areas. The roadways have a rectangular form and represent the longest possible segment of a road per direction of travel. The form and extent distinguish them from most other network models where they are partitioned into road segments. The non-partitioned form results in significant advantages for the parallelisation of the simulation. The presentation of the microscopic traffic model by Lee, Barlovic, Schreckenberg and Kim [124] in Section 3.4 forms the third model family of the specification. The model composes the resulting vehicle motion by a superposition of the longitudinal motion and the lateral motion part. The longitudinal motion part also presents an optimisation for the calculation of the next safe velocity. It closes with some considerations about configurations that lead to undefined model states. The lateral motion part presents the asymmetric lane changing rules as introduced by Habel and Schreckenberg [71, 70]. As the fourth participating model family, the specification defines next the detector model in Section 3.5. After defining the detector operations for each of the detector submodels, the section introduces several detector types that are part of the model family. The detector types distinguish between input and output detectors. Having comparable input and output detectors forms at the same time a basic modelling principle of the simulation model. The loop detectors, the prognosis detectors, and the invariant detectors fall into the category of the input detector types whereas the local and the global detectors belong to the output detector types. The tuning elements contain a of a pair of accumulators each and indirectly rely on the detectors through the accumulators. The accumulate group, average, and aggregate the traffic measurements of the detectors and provide the accumulated values to the tuning elements. The latter are part of the fifth specification in Section 3.6 and adjust their associated parts of the road according to the comparison of the actual and the should-be accumulator values. Section 3.7 completes the specification with the definitions for several simulation models. The simulation models are part of the systems theory formalism and describe the structure of the simulation systems and how the latter behave. The discrete time system specification represents the first of the simulation models. It is the basic model that is probably present in most microscopic traffic simulation systems. It provides the least system knowledge and has not much parallelism potential. Its semantics with respect to the traffic model depends on the implementation. The multicomponent model provides the next level of system knowledge and also some parallelism potential. For the scope of the microscopic traffic simulations, it models the intersections by means of the switching areas which make the components depend on each other during the vehicle update. The network of systems specifications removes this dependency for the traffic simulations case by modelling on- and off-ramps only implicitly through the detector measurements. The section continues with considerations about dynamic networks and closes with some thoughts about further decompositions. Section 3.8 summarises the contents of this chapter.

Chapter 4 discusses the implementation issues that emerged during the implementation of the fourth version of the productive simulation system OLSIM as well as the ones that occurred as part of the verification and validation enhancement effort. Section 4.1 opens the chapter with some general remarks about the development process, the programming languages used within the implementations, and the implementation goals. Section 4.2 introduces then the implementation of the simulation models. The introduction comprises the implementation of the simulation engine, the vehicles, the roadways and occupation tables, the detectors, the accumulators and tuning elements, and the remaining components of the simulation network. The section also discusses the difficulties, the pitfalls, the shortcomings, and the enhancements as well as further options. Section 4.3 continues the chapter by presenting the implementation of the first family of the model families, namely the vehicle models. Section 4.4 continues the presentation with the traffic models. The detector models follow in Section 4.5 and the tuning element models complete the presentation in Section 4.6. Section 4.7, finally, describes how to run the simulation. The chapter closes with a summary in Subsection 4.8.

Chapter 5 enumerates the benefits of the thesis that range from supporting to obtain and distribute traffic information over improving verification and validation activities to preparing the implementation of a next version. Section 5.1 presents the benefits related to the obtainment and distribution of traffic information. The latter compound road user and road operator traffic information as well as scientific traffic information. Section 5.2 details the enhanced verification and validation possibilities which exist, for example, with numerical solution verification possibilities. Section 5.3 gives advice based on the experiences of the previous implementations for a potential implementation of the next version. The chapter closes with a summary in Section 5.4.

Chapter 6 closes the thesis with an outlook on future tasks and some general thoughts about future work.

Initial position

OLSIM started as a dedicated research and development project with the goal of running a traffic information platform for the internet. The start of OLSIM was years before the author joined the OLSIM project team. Thus, the initial position refers to the point of time and the context of OLSIM that existed when the author started working for the OLSIM project team.

As part of the requirements description in Section 2.1, this chapter highlights several aspects of OLSIM that were present in the initial position. These aspects concern the project history, the project context, the project founders' wishes, and the features of OLSIMv3 and later. The chapter also describes several important architectural properties such as loop detector data, topology, traffic model, and user benefit of the initial position. Moreover, this chapter enumerates the shortcomings of the OLSIMv3 version of the initial position and derives the future advantages and the requirements of a potential revision. Finally, this chapter concludes with a summary of the resulting requirements and a listing of the necessary steps to revise from OLSIMv3 to OLSIMv4.

This chapter starts with the deriving of the requirements for a revision of OLSIM in Section 2.1. The requirements are derived from the project's context, the wishes of the founders, and the shortcomings that persisted in the initial position. Therefore, Subsection 2.1.1 reconstructs as a first aspect the milestones of the project history. Subsection 2.1.2 then characterises the project's infrastructural context. Thereafter, Subsection 2.1.3 lists the feature requests of the founder and of the users. Finally, Subsection 2.1.4 unveils the re-architecture demands.

This chapter continues with Section 2.2 that discusses the advantages and disadvantages of the various, state-of-the-art model kinds that OLSIM employs or that a revisioned OLSIM may employ possibly. The section thereby establishes the corresponding state of the art. The latter starts with a system classification for OLSIM in Subsection 2.2.1. Thereafter, Subsection 2.2.2 introduces the detector models. Subsection 2.2.3 then discusses the advantages and disadvantages of several microscopic traffic models in the context of OLSIM. Subsection 2.2.4 reviews the widely used network models. Subsection 2.2.4 presents the commonly used vehicle generation methods. Subsection 2.2.6 examines the background of some existing market solutions as simulation systems.

Finally, Section 2.3 concludes this chapter with a summary of the requirements for a revision of OLSIMv3 onto its next version.

2.1 Requirements

This section details the project history, the project context, the feature requests by the users and the project founder, and the revision requirements of OLSIMv3. It starts with a short outline of the project history in Section 2.1.1. The history reaches from the first publicly available version in 2002 to the starting time of the OLSIMv3 revision in 2005. As part of the project's context, Section 2.1.2 elaborates on the infrastructure conditions such as the dimension of the highway network, the traffic volume, and the kinds, quality, and amount of useful traffic information as well as available traffic data. Section 2.1.3 lists the feature requests that arose during the planning of OLSIMv4. The requested features involve the loop detector statistics, extension of the seven days long-term prognosis traffic information, a truck specialised traffic view, adaption to a dynamic road traffic network, and a catalogue system for the information around OLSIM. As the last part of this Section, Section 2.1.4 unveils several re-

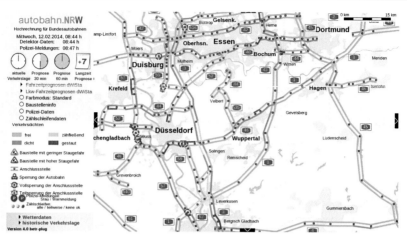

Figure 2.1: Restricted access area of the NRW traffic information platform `Autobahn.NRW.De` [146].

architecture demands that grew historically. This last Subsection examines at first the storage and access of loop detector data in Subsection 2.1.4.1. The topological data model and its representation follow in Subsection 2.1.4.2. Subsection 2.1.4.3 discusses the underlying data model and the data flow. This section completes with Subsection 2.1.4.4 that compares several traffic and simulation models.

2.1.1 Project history

The German Ministry of Economic Affairs, Technology, and Transport of North-Rhine Westphalia (NRW) in the role of a founder initiated OLSIM in 1999 as a research and development project [166]. The Physics of Transport and Traffic chair of the Department of Physics of the University of Duisburg-Essen joined the project in the role of the scientific contributor and the developer of the simulation software. The project aimed to develop a public traffic information platform for the internet. The intention of publishing the traffic information was to assist highway users in route planning in the hope that this results in a better capacity utilisation of the NRW highway network. As will be discussed later, that seems to be the most promising solution for the traffic problems in NRW. The first publicly available OLSIM version went online Thursday 19th September, 2002. Shortly after the startup, the second version of OLSIM went online Monday 31st March, 2003. It added a first short term prognosis to the features of the initial version. Two years later Thursday 20th October, 2005, OLSIMv3 went online. It offered the 30- and 60-minutes short-term prognoses and an integrated travel-time prognosis to further improve the route planning assistance. In the years following the start-up phase of the internet platform, the expectations of the project founder and all participants grew. As a next step, an additional restricted access platform that provided more detailed information such as traffic messages and road works messages evolved with OLSIMv3. Figure 2.1 shows the current view of the restricted area access platform. Publications that depict screenshots of previous versions of the traffic information platform are available for example in [42, Fig. 7] for 2002, in [186, Fig. 4] for 2003, in [165, Fig. 2] and [41, Fig. 2] for 2004, or in [139, Fig. 10] for 2005. The various Figures document the continuous feature extensions.

The author of this thesis came to the project team in 2005 starting with OLSIMv3. Since OLSIMv3, the founder of the research and development project defined the next few project goals periodically once a year. As a consequence of the experimental character of the project, project goals were re-defined on demand from time to time. The yearly project budget depended on the sum of tasks that derived from the project goals.

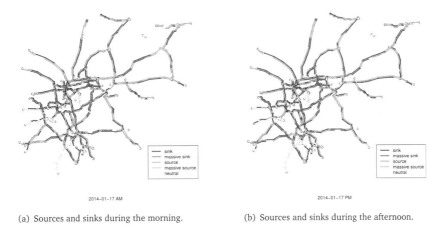

(a) Sources and sinks during the morning.

(b) Sources and sinks during the afternoon.

Figure 2.2: Density differences of subsequent NRW road segments for a.m./p.m. of 2014-01-17.

2.1.2 Project context

NRW contains several densely populated regions. The slightly increasing number of commuters [145, pp. 20-22] on their way to work and back home again cause a high traffic volume. Transients from the neighbouring countries that pass the NRW highway network further increase the traffic demand. Moreover, constantly growing heavy goods vehicle traffic [145, p. 83] degrades the situation. One of the secret expectations of the project was to improve the capacity utilisation balance of the very dense NRW highway network because of the following reasons. A lot of highway users might be able to schedule their routes more or less depending on the current traffic situation due to flexible work times. Transients might schedule their travels depending on the current traffic situation, too. Or, as a last option, road users may choose an alternative route.

However, not only information that assists the road users may be of use. Microscopic traffic simulations may also be part of complex information systems or support infrastructure planners [31, 131, 92]. As an example for how to assist the infrastructure planner, Figure 2.2 shows the density differences before and behind the entries and exits of the NRW highway network during the first and the second half of the day Friday 17th January, 2014. As the size of the time window is half a day, the traffic flow had only minor influence on the density differences. The sources and sinks in Figures 2.2(a) and 2.2(b) are similar to origin-destination matrices that are frequently used by infrastructure planners. More sophisticated tasks than the calculation of origin destination matrices include, for example, optimisation for a prospective location of a hub with respect to the travel times. Having a simulation system such as OLSIM with reliable travel times, solving such optimisation tasks becomes possible.

As the NRW vehicular traffic generally increases continuously and, in particular, traffic on highways increases the most significantly [145, p. 82], a better balancing of the highway network utilisation seems to be the best option to deal with this issue due to the following reasons. The NRW highway network is increasing constantly and further increase is only a last resort if at all. Between 1980 and 2010, the total NRW highway network length has already increased by 29 % [145, p. 27]. Similarly but disproportionately higher increased the passenger traffic volume by almost 50 % [145, p. 26] [1]. Currently, the "directional length of the NRW highway network is about 4600 km long. Up to 120 000 vehicles move in it simultaneously during rush-hours. Its main roadways consist of two to four lanes

---

[1]Assuming proportionality of traffic volume to the absolute number of licensed vehicles.

and vary in length from 50 km to 320 km " [31]. According to the OLSIMv4 network configuration, 31 highways interrupted by 563 exits and 70 interchanges form the network. The 31 highways consist of 2666 carriageways, on-ramps, and off-ramps that interconnect the main roadways. 1308 so-called tracks as part of the remaining 88 main roadways connect subsequent exits in the direction of travel. Each of the 2666 carriageways, on-ramps, and off-ramps forms a track, too. If at all possible, its extension would be very costly, extremely time consuming, and, even worse, potentially may not improve the overall network performance due to the traffic planning paradoxon known as the Braess paradoxon [26].

Useful traffic information that assists the highway user in route planning describes the traffic state by refering to a particular road segment. The description of the traffic state compounds traffic observables and road state information. The traffic observables can include some of the following. The density of a road segment counts the number of vehicles per length unit on it. It lacks velocity information, i.e. how fast the traffic flows on it. Therefore, the traffic flow expresses the number of vehicles per time unit. The level-of-service (LOS) combines the density, the velocity, and the traffic flow information into one discrete value that indicates free, dense, viscous, or congested traffic to the user. Finally, the travel time gives an information about how much time an average highway user would spent to travel through a specific road segment. The road state information provides the general information of road and single-lane closings and whether a particular road segment is affected by road works. The messages for the latter contain the active as well as scheduled and planned road works. Bridges over rivers and valleys are frequently subject of road works due to the high heavy goods vehicle traffic. As an example, starting with May 2014, the A1 NRW highway part that crosses the Rhine as the biggest river of NRW is closed for trucks for almost one year due to road works. Another consequence of the general road works includes that the highway users are urged to drive at slower velocities, in Germany generally 80 km/h or even 60 km/h . The traffic messages as another kind of traffic information originate from police observations and provide notifications about traffic events such as accidents, lane or even road closings, and traffic jam lengths. These kind of messages have informed highway users since the days of radio broadcasting stations. Lastly, the states of variable message signs could provide useful information, too, since they inform the highway user about speed limits and overtaking restrictions.

The NRW highway network is stocked "with more than 3000 loop detectors not including on- and off-ramps or intersection carriageways" [31]. The 3000 inductive loop detectors provide in each minute so called stationary traffic data that is classified into the vehicle types cars and trucks. The data consists of five fields of the size of 1 byte. The five fields represent the number of vehicles per measurement time interval–which is 1 min –the average velocity in kilometres per hour, and the occupation in percent of the minute wherein a detector was covered by a vehicle. For the loop detectors located in Germany in particular [199, 200, 201] and for loop detectors in general [105, 106] detailed technical specifications exist.

About 1300 measurement cross-sections bundle the 3000 loop detectors across the highway which leads to a statistical average distance of 3.5 km between two subsequent measurement cross-sections [31]. Taking into account that 5 % to 10 % of measured data is faulty or nonexistent per day the average distance between two subsequent measurement cross-sections prolongs to approximately 4 km [31]. Between subsequent detectors, the traffic situation is unknown due to the lack of detection. Even though there exist other, non-microscopic traffic simulation approaches [31] that are able to deal with these kind of distances [102, Fig. 7], the average distance of 4 km is greater than the critical distance of 3 km suggested, for example, by Treiber and Helbing [203]. Therefore, an online simulation system was chosen to complete the information gap between subsequent detectors. The following section describes the user's and operator's requests for the microscopic traffic simulation system that arised during the operation phase of OLSIMv3.

2.1.3 Feature requests

Since OLSIMv3, the microscopic traffic simulation calculated the following spatio-temporal traffic information in each minute and for each track of the 1308 main roadways – the level-of-service as a discrete value ranging from free flow to congested traffic, the average velocity in kilometres per hour, the density in vehicles per kilometre, and the travel time in minutes for some predefined set of routes. OLSIMv3 provided this traffic information for the current traffic situation and for the short-term 30 min and 60 min prognoses. The motivation for offering the prognoses services was to improve the assistance of user route planning. The prognoses were based on current and on predicted loop detector data and each prognosis instantiated its own simulation process [139, Fig. 1, Sec. 2]. However, mainly due to technical reasons the prognoses classified and extrapolated the historical loop detector data only for a subset of the network and not for each minute. Moreover, extreme traffic situations such as the start of the summer holidays lead to traffic breakdowns for large parts of the NRW highway network. Intended as a slight improvement to relax such kind of extreme situations, a seven-day long-term prognosis should assist the highway user in planning trips comfortably in advance. Additionally, the short term prognoses lacked availability due to unstable heuristics in prognosticating the loop detector data. Briefly summarised, these shortcomings required redesigning or even re-architecturing the whole prognosis system.

During the operation of OLSIMv3, several severe technical problems occured. Due to loss of connectivity, maintenance of traffic data centres, or road works, the loop detector data were from time to time completely or for individual locations not available to the simulation. As a consequence, the simulation results were inaccurate. Further investigations showed that the loop detector data was also unreliable. The loop detector data contained measurements with strange semantic errors such as trucks with extremely high velocities or zero traffic flows combined with velocities greater than zero. As the technical specifications allow errors to a certain degree and percentage [199, Part IV, 200, Part IV, 201, Part IV] such measurements may still be in accordance with the specification. To minimise the amount of questionable data, filtering incoming loop detector data was necessary. Finally, a loop detector traffic data statistics should classify the incoming data into available and erroneous traffic data and present it to the user as part of the expert view. Additionally, the context of a truck specialised traffic situation was requested to provide the spatio-temporal traffic information per vehicle type, i.e. all vehicles and trucks. Table 2.1.1 lists the domain and information type for the overall desired traffic information.

In addition to the needs of the group of the internet users, OLSIMv3 had to consider the special requirements of other user groups as well. One of them, the traffic data provider and traffic information consumer as a part of the project founder, the Landesbetrieb Strassen.NRW, upgraded his data distributing software to the next major version which is meanwhile available at [177]. The update involved to support the new platform as well. Another group of users requested the traffic information not only for the main roadway tracks but also for the on- and off-ramps and the carriageways. As the loop detection on the non-main roadway tracks is only available for a strictly limited subset, the questions whether and how to generate the traffic information for the non-main roadway tracks requires special attention. In addition to the limited data coverage that leads to questionable results, running the simulation on the non-main roadway tracks reduces, as a consequence, the set of potential parallel

Table 2.1.1: Domain and information type for desired traffic information.

Name	Information	Domain
j_ges	traffic flow of all vehicles	0 vehs/min to 254 vehs/min
j_lkw	traffic flow of trucks only	0 vehs/min to 254 vehs/min
v_pkw	average velocity of cars	0 km/h to 255 km/h
v_lkw	average velocity of trucks	0 km/h to 254 km/h
t_gap	average netto time headway	0 % to 100 %
p_occ	occupation time in percent	0 % to 100 %
density	density of all vehicles	0 vehs/km to 255 vehs/km
los	level of service	1 SL to 6 SL (SL = service level)
traveltime	travel time	0 s to 4095 s

interpretations significantly. It is, thus, questionable whether and how to run the simulation for the on- and off-ramps and carriageways. Section 3.7 discusses this aspect in more detail. Moreover, the so-called high flow states that will be discussed later in Sections 2.1.4.2, 2.1.4.3, and 3.4 only increase the doubt of whether simulating the on- and off-ramps and carriageways will produce accurate and valuable traffic information. OLSIMv4, thus, prefers using the traffic information of the corresponding downstream tracks for the on- and off-ramps and for the carriageways.

The NRW highway traffic network is a dynamic network. Regular road works affect about 10 % to 15 % of the network each day. Generally, road works have a speed limit of 80 km/h as a consequence. In many cases they lead to a reduced number of lanes that are available to the traffic participants. The information about road works is available to OLSIMv3 and OLSIMv4 through ALERT-C [86] encoded road works messages. Traffic messages that originate from police observations and police regulations also use the ALERT-C [86] encoding standard. In case of traffic accidents, the police sometimes closes a lane for the traffic. Evaluating and exploiting these kind of information into the simulation will result in more accurate traffic information especially for the travel times in the free flow traffic phase. Thus, OLSIMv4 needs to take over and exploit these kind of information. OLSIMv4 requires therefore a dynamic topology that supports runtime modification of the traffic properties of parts of the NRW highway network. A dynamic topology for OLSIMv4 was requested officially even though a concrete service that exploits this feature has not been instantiated until today.

The author was frequently astonished about the detailed expert knowledge of the NRW highway network that many traffic engineers provided in the project meetings. One of the project tasks was to provide expert answers concerning odd published traffic information and to trace these oddities. To solve these kind of tasks, knowledge is required about the particular highways including the parts of the network that are affected by road works. A tool to comfortably compare the measured traffic data with the published information would have been of great assistance. The lack of such a tool and the author's minor knowledge of the NRW highway network led to the development of an information system that enables the user to access a traffic catalogue for the highway road, the traffic network, the traffic data, and every evaluation capability of the microscopic traffic simulation system. Powered by a semantic traffic information system, the problem of giving expert answers such as the one from the beginning of this section reduces to a technical task.

2.1.4 Re-architecture demands

In general, scientific research and development projects do not aim at a professional software development process. A lot of software components have only experimental nature, i.e. implement something to get some information as a result and—in a lot of cases—forget about the software thereafter. The experimental development of parts of a software is symptomatically for computational science [35]. Needless to say that such projects lack a software quality engineering process [36]. As a result, most of the following re-architecture demands are—to some degree—in such a state due to the experimental nature of the research activities. However, most of the general criticisms presented in the following paragraph apply also for other and even enterprise traffic software systems such as [177]. While the data distribution software (DAV) [177] certainly is an excellent piece of software, it lacks, for example, any data research capability at all.

Future systems that want to benefit from some of the lessons learned that this section describes may consider the following. The topological data model impacts the simulation model and, consequently, the simulation algorithm as well as any potential speedup achievable by a parallel interpretation. Section 3.7 discusses this topic in more detail. Thus, any implementation should start with the design of the topological data model. The traffic data forms the second data source for any microscopic traffic simulation. The data originate from detector devices that refer to the road network as a representation of the topological data model. The data model for the traffic detector devices should be the same as for the virtual detector devices in the simulation. This allows to analyse the simulation state and input/output combinations to improve future versions, comfortable system debugging, and more advanced simulation setups such as the following example. By photographing vehicles on subsequent

Listing 2.1.1: An examplary data set for one minute of the incoming loop detector data.

```
# 13.12.2006 15:56:00 0100
# VRZ Leverkusen, 3305 DE data sets
# ID(=node-distance-channel) j_ges j_lkw v_pkw v_lkw t_gap p_occ sigma v_exp
04129-077-001 11 8 104 85 50 6 11 96
04129-077-002 20 4 124 98 28 4 13 118
04129-077-003 26 0 140 0 21 3 10 143
04129-077-033 12 3 109 83 47 4 14 109
04129-077-034 19 1 132 135 30 2 12 133
04129-077-035 9 0 170 0 65 0 15 172
...
```

bridges across a highway in multi-shot mode, the timestamps of the photos yield lead to travel times that specialised travel-time detectors can feed into a simulation. Comparing local detector output values from the simulation with real-world loop detector measurements provides another view on the simulation. The more established, general approach as exploited by many simulation systems such as SUMO [19] and OLSIMv3 simulates using the loop detector data and compares the virtual travel-times with the real-world ones. Such advanced simulation setups may be considerably useful for gaining new insights about the dynamics of the simulation. A last learned lesson that extends the concept of having comparable input/output data is that a simulation system should support the data combination between all the various data domains such as detector data, topological data, event calendar, and simulation results, i.e. virtual detector data. These kinds of simulation systems enable users then to explore complex data correlations as, for example, searching for high traffic flows and as the calculation of the density differences from Figure 2.2.

This section lists some architectural software improvements, the shortcomings that require further development, and the conceptual parts that have the necessity and the potential to increase the simulation software quality or its results significantly. As a first technical improvement, replacing the OLSIMv3 data model and, consequently, also its various representations will boost the comfort for storage, access, and maintenance of the traffic data and the various kinds of detectors. OLSIMv4 therefore employs such a data model for the loop detector data. Paragraph 2.1.4.1 discusses the shortcomings in more detail. As a second and probably the most significant improvement, OLSIMv4 eliminates the shortcomings of the topological data model. This model impacts the entire simulation system and is thus of great significance. For instance, dependencies between topological elements limit the speedup in a parallel interpretation. In OLSIMv3, the track structure dominated the topological data model and, consequently, the simulation algorithm, too. An improved topological data model has, thus, the potential to result in a qualitative improvement. Paragraph 2.1.4.2 discusses the shortcomings of the OLSIMv3 topological data model and its impacts on the simulation algorithm as well as on a possibly parallel interpretation. It discusses further the impacts on the mapping of traffic information between several and varying (i.e., dynamic) digital maps. Additionally, representing and integrating the topological data model in the database supports automated verification of the data integrity. Another qualitative improvement seeded by an appropriate topological data model frees it from any specific topological constraints such as tracks and intersections. Such a data model then can provide the additional benefit of a platform independent application programming interface for microscopic traffic models. OLSIMv4 employs such a topological data model and Paragraph 2.1.4.4 discusses this. As a last improvement, OLSIMv4 addresses the data combination and research facilities between the various participating domains such as topological data, loop detector data, time based event data, and simulation data. Paragraph 2.1.4.3 discusses the shortcomings of the existing OLSIMv3 approach.

2.1.4.1 Storage and access of loop detector data

As one of the benefits of OLSIM, the loop detector data became available for the research staff. The data amount of several years since the early days of the project had been stored on a filesystem database in a one-file-per-day format as described below. At the project start, the question that dominated the development of OLSIM was whether it would be possible to write a simulation using that data or not. It was unclear how much data to store and what else to do with it. Another irritating factor was the migration of the DAV data distribution centre software [177] that possibly would introduce a lot of changes. These circumstances led to the situation that several years after the start, there was a huge amount of data and a huge amount of software that stored and accessed the traffic data directly through low-level file input/output operations. No common data models and operations were available to access and re-store the available data. Set processing and direct interaction with the data as, for example, in accumulating traffic flows interactively was not possible [50]. As relational database management systems were widely available at that time, these shortcomings were too far behind the state of technology and, moreover, did not provide any benefit.

Additionally and as mentioned in the introduction of this Section, the data coupling between the several data domains was identified as a shortcoming of OLSIMv3 and, thus, intended to be enhanced in OLSIMv4. It was, however, unclear how the various data domains would look like because of the following reason. The project organisation was quite similar to modern agile software development processes. Once a feature idea was born and assigned for reasearch and development, it was added to the system immediately or at most half a year later to the next project meeting. Even though a lot of experimental research activities had been undertaken, no re-architecturing and re-factoring processes were existent or assigned. During these days the idea to re-factor or, to be more precise, re-architecture the so-called "simulation core" was born by the author as a feature request. A top down software design with specification of all data domains was not in the focus. Thus, the data coupling was intended while it was unclear when it would come to that point and how the data domains then would look like. A representation in a relational database management system promised the most flexible approach.

Listing 2.1.1 shows an example for the incoming loop detector data of one minute that has been converted into ASCII representation. The structure of the loop detector data consists of a 16 bytes header that contains the corresponding timestamp of the data and the number of data records following the header. The data usually came from two origins [209, Fig. 6], had 1500 to 5000 lines, and could contain duplicates. Each line that follows the header contained the loop detector identifier and the information that Table 2.1.2 summarizes. The three fields `t_gap`, `sigma`, `v_exp` of such a line turned out to contain unreliable data. The upgrade of the traffic data centre software to [177] brought technical improvements but did not touch the kind of data.

Accessing the data required to deal with file access details such as locking, error handling for missing data and parser errors. One year of loop detector data required about 30 GB of storage space which was a lot of space in the years 2000 to 2005. For scientific data research, the data persistency solution was inefficient, error-prone, and inconvenient. The refinement processes of the traffic data for the historical time series that build upon reliable loop detector data had to deal with these shortcomings. As the data volume grew in size during the years, searching for patterns became more inefficient. Data access ended up in calculating a position on the filesystem. However, "addressing the data by value, rather than by position, boosts the productivity of programmers as well as end users" [50, Sec. 3]. In order to speed-up searching, the addition of further indexes would have been required. Several applications would have required updating, too.

In the context of the scheduled feature request, the seven day long-term prognosis, re-architecturing the traffic data storage and access became unavoidable.

2.1.4.2 Topological data

OLSIMv3 and earlier versions modeled the highway network by dividing roadways into links and nodes [58, Sec. 4], [43, Sec. 4], [42, Sec. 4], [133, Sec. 3], [186, Sec. 5], [41, Sec. 4], and [139, Sec. 7]. OLSIMv3 used a configuration file for the network information. At the latest since 2003 [186], the

Table 2.1.2: Domain and information type for incoming loop detector data.

Name	Information	Domain
j_ges	traffic flow all vehicles	0 vehs/min to 255 vehs/min
j_lkw	traffic flow trucks only	0 vehs/min to 255 vehs/min
v_pkw	average velocity cars	0 km/h to 255 km/h
v_lkw	average velocity trucks	0 km/h to 255 km/h
t_gap	average netto time headway	0 % to 255 %
p_occ	occupation time in percent	0 % to 255 %
sigma	kind of variance for velocity	0 km/h to 255 km/h
v_exp	exponentially smoothened velocity	0 km/h to 255 km/h

OLSIMv3 configuration file used a special file format, called the OLSIM track data format (OTDF) [186] and exhibited exemplary in Listing 2.1.2.

OLSIMv3 defined tracks as such that they represent the area or part of a highway road-segment between the two imaginary middle points of two subsequent exits. This is a common technique that is found in several other digital maps until today, namely [211] and to some degree [86], too. Along a track the road geometry properties such as the number of lanes, lane closings, and merging areas were not allowed to vary. The traffic network properties such as the maximum velocity and overtaking restrictions were not allowed to change either. The concept is generally known as 'coordinate transformation' or—to be more precise—as a change of the orthonormal basis. Listing 2.1.2 contains an OTDF configuration entry with the two attributes LaneOffset and CellOffset.

Listing 2.1.2 shows an exemplary configuration for the 10100 m long track A001-SW-HF-098 on the A1 highway onto Leverkusen intersection into direction of travel south-west. The track has several exits each of them inside an Exit block structure. The coordinates LaneOffset and CellOffset describe the new basis for the destination track that is labelled as Destination. The Connection block structures inside a track configuration entry define routing possibilities and probabilities. The Destination attribute contains a destination main roadway connected by the Connection block structure and the appropriate Exit block structure. The Checkpoint block structure describes detector positions and checkpoint locations, i.e. a vehicle generation highway location where the simulation inserts and removes vehicles.

The track identifiers of the OLSIMv3 configuration represent so-called natural keys (in the terminology of relational database management systems). Each identifier consists roughly of four fields separated by the minus symbol. The first field denotes the highway number, the second represents the direction of travel. The third field provides the property whether the track belongs to a main roadway or is an on- or off-ramp or a carriageway. The last field specifies the destination exit number.

Since the network is a dynamic network, this procedure turned out to be impracticable. Some examples follow that contrast the highways of the road network and the naming procedure. Since from time to time administrative regulations bump up several roads to the state of a highway and degrade others back again, identifiers come and go. In general, these activities do not provoke identifier collisions rather than invalidate the previously stored data only. Unfortunately, however, administrations rename from time to time the exit numbers. Additionally, when road works introduce a new exit in-between of two existing subsequent ones, the new exit and one of the subsequent ones require new identifiers. The new identifier for the existing one results in a renaming procedure as well as changing the successor and predecessor neighbourship relationships. The renaming affects not only the OLSIMv3 digital map but also the mapping to other digital maps. Due to the renaming of the identifiers, several collisions occured in the past. As the OLSIMv3 configuration requires mappings between the OLSIMv3 highway network and several other digital maps, the collisions resulted in several workarounds that complicated the situation further.

As another issue, the definition of a track lacks an important point as it is unknown where exactly the

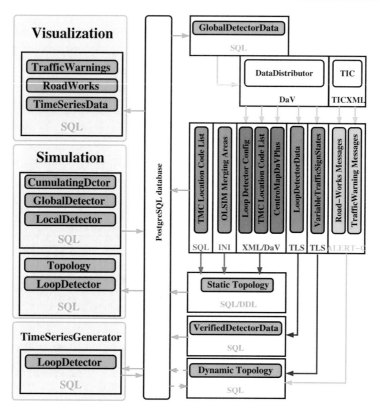

Figure 2.3: Kinds of data that participate in OLSIMv4. Dashed arrows represent planned services. TICXML represents XML data in a proprietary data format by the traffic information centre. The INI data format is a textual representation of the OLSIM track data format. The PostgreSQL database also stores the ALERT-C encoded road-works and traffic warning messages for the visualisation.

middle point of a given intersection or exit should be. This makes it difficult to identify corresponding highway positions between the OLSIMv3 highway network representation and other digital maps. As a consequence of the upgrade of the data distribution software (DAV) [177], all software versions after OLSIMv3 had to support the CentroMapDavPlus (CMDAV) proprietary digital map [191] that ships with DAV. This CMDAV digital map is based on the location code list (LCL) [86] which is a standard for location referencing of traffic information. The LCL, however, does not contain any length definitions for highway road-segments. The CMDAV is, thus, a more detailed version of the LCL that additionally provides "inner- and outer road-segments (i.e. driving connections)" [31]. The traffic information platform of OLSIMv3 visualizes ALERT-C encoded traffic messages [85]. Thus, support for the LCL digital map was required in all versions after OLSIMv3, too.

Furthermore to continue with the explanation to the exemplary configuration from Listing 2.1.2, one can see that except for the `StaticProbability` information, the simulation does not require the `Connection` block structure. The simulation can instead derive the information on demand by traversing along the neighbouring tracks of a particular track and starting with the exits that depart from it. The structure was designed to provide the routing information between main roadways. As it is generally more advisable to process explicit routing of vehicles in a control layer, such a layer could instead derive the information as provided by the `Connection` block structure through traversing over the exits. Deriving the information makes the `Connection` block structure superfluous again. However, additional issues with the explicit routing of vehicles exist. It is questionable whether a lateral merging process that requires the concept of changing the orthonormal basis is necessary at all due to the following reasons. Estimating from [115, Figs. 3 and 9] the maximum traffic flow for the number of vehicles that can enter a roadway via lateral merging for the brakelight traffic model is less than 16 vehs/min. A later publication [112, Tab. 2] supports this conclusion for the brakelight traffic model and extends it to further traffic models such as the intelligent driver model. To model on-ramps, inserting vehicles on the rightmost single-lane can achieve the same effect as lateral merging [112]. The German carriageway of the A1 Leverkusen intersection direction south-west may serve as a reference to compare the estimated model values for the maximum number of vehicles with the real-world ones. On Thursday 26th December, 2013, the detector at the end of the carriageway measured 1555 vehs/h of cars with an average velocity of about 90 km/h in the time between 1:30pm and 2:30pm. Even though this may be an exceptional situation, it is also a significant traffic situation because the Leverkusen intersection is of extreme importance for the traffic in the region around Cologne and the long distance transient highways A1, A3, and A4. Moreover, this high flow traffic pattern seems to appear regularly. Table 2.1.3 lists 10 further days where a similar traffic flow pattern occured. The maximum observed traffic flow was 1677 vehs/h.

Apart from the principle traffic model problems that probably will persist for some time, OLSIMv3 introduced the merging areas by OTDF. The length and position of the merging areas until today is, however, not present nor derivable in any other of the aforementioned digital map resources [191], [86], and [211] (which questions again the benefit of modelling the merging process in the simulation). A typical OTDF configuration file had about $100\,000$ lines of code. In case of configuration errors, run time exceptions occured.

2.1.4.3 Data model and data flow

OLSIMv3 integrated all aforementioned data sources for the purpose of visualisation. Road works messages, traffic messages, traffic information, loop detector data statistics, and—to some degree—road geometry data. Topologic data was hidden in an OTDF configuration file such as the one excerpted in Listing 2.1.2. Detailed traffic information from the simulation was available to some degree in the simulations' output files. Figure 2.3 shows all kinds of data that participate in the context of OLSIMv4.

To consider architectural and design options, expert knowledge about the requirements of a highway network is essential. As an example and as discussed in Section 2.1.4.2, the answer to the architectural question whether it might be necessary or even beneficial to model lateral merging or not may serve. When thinking about this question, it is of advantage to know the range and the circumstances of the

Table 2.1.3: High flow traffic patterns on carriageway `A001-SW-PF-098`.

day	start	end	total flow cars	average velocity
Sunday 30th September, 2012	1:00pm	2:00pm	1544 vehs/h	93 km/h
Saturday 17th November, 2012	5:00pm	6:00pm	1515 vehs/h	88 km/h
Sunday 14th April, 2013	6:00pm	7:00pm	1521 vehs/h	95 km/h
Saturday 27th April, 2013	2:00pm	3:00pm	1561 vehs/h	95 km/h
Saturday 24th August, 2013	2:00pm	3:00pm	1514 vehs/h	74 km/h
Sunday 13th October, 2013	3:00pm	4:00pm	1510 vehs/h	95 km/h
Sunday 20th October, 2013	2:00pm	3:00pm	1543 vehs/h	95 km/h
Sunday 1st December, 2013	2:00pm	3:00pm	1503 vehs/h	97 km/h
Sunday 8th December, 2013	5:00pm	6:00pm	1526 vehs/h	87 km/h
Sunday 19th January, 2014	4:00pm	5:00pm	1677 vehs/h	81 km/h

Table 2.1.4: High flow traffic patterns for carriageways in North-Rhine Westphalia on working days.

day	track	start	end	total flow
Tuesday 7th November, 2006	A001-SW-PF-086	7:00am	8:00am	2246 vehs/h
Tuesday 2nd May, 2006	A001-SW-PF-086	7:00am	8:00am	2232 vehs/h
Thursday 2nd May, 2013	A001-SW-PF-086	7:00am	8:00am	2230 vehs/h
Monday 13th February, 2006	A001-SW-PF-086	7:00am	8:00am	2227 vehs/h
Monday 27th May, 2013	A001-SW-PF-086	7:00am	8:00am	2227 vehs/h
Tuesday 11th September, 2012	A001-SW-PF-086	7:00am	8:00am	2205 vehs/h
Monday 27th August, 2012	A001-SW-PF-086	7:00am	8:00am	2197 vehs/h
Tuesday 14th November, 2006	A001-SW-PF-086	7:00am	8:00am	2192 vehs/h
...				
Thursday 7th January, 2010	A001-SW-PF-098	7:00am	8:00am	1807 vehs/h
Monday 10th December, 2012	A001-SW-PF-098	5:00pm	6:00pm	1626 vehs/h

measured and observed high flow traffic patterns in advance. Getting this information sounds easier than it actual was. How to get this information is described in the following.

At first, one had to extract a list with the identifiers and the properties of all carriageways of the NRW highway network from the topological configuration file. Then, one had to filter this list to contain only carriageways with loop detection. For each carriageway in the filtered list one had to isolate the loop detector identifiers of the appropriate detectors on the carriageways that are located next to the entrance of a main roadway. Using the identifier list one had to write a complex script that searches in thousands of files for a high flow traffic pattern. Finally, one had to sort the timestamps of the found patterns into classes such as working day, day before a holiday, weekend, or day in the holidays in order to exclude some potential explanations for singularities.

A frequent question in similar contexts is whether the affected traffic patterns were influenced by holidays, long weekends, road works, police actions, or traffic accidents. To answer questions like this, one had to find the location codes of the affected tracks which required a one-to-one correspondence between the OLSIMv3 tracks and the segments or points in the location code list. OLSIMv3 had a mapping between segments of the location code list to the tracks of the visualisation but not the other way round. To cut a long story short, the information was spread over several independent subsystems. Answering semantic queries such as the ones discussed caused an immense effort. Productivity was stuck due to technical problems. OLSIMv4 can limit the search to working days only. Table 2.1.4 shows some high flow states on carriageways on working days.

The scheduled software upgrade of the data provider and project member "Landesbetrieb Strassen.NRW" demanded to map the OLSIMv3 traffic information onto the CentroMapDavPlus digital map. Additionally, the loop detector key domain and the key identifiers expired. As a consequence, loop detector positions had to be remapped onto the CentroMapDavPlus digital map which depends on another version of the location code list. While the location code list versions for the road works and the traffic messages update once a year, the CentroMapDavPlus updates only occasionally. In addition to the loop detector key domain, the key identifiers, and their positions the traffic information that refers to the tracks had to be remapped onto the road segments of the CentroMapDavPlus digital map. Again both digital maps had different update cycles. Keeping track of several versions in parallel required a lot of effort.

The loop detector input data arrives from the traffic data centre each minute. OLSIMv4 returns the simulation result traffic information once a minute too. These time constraints identify OLSIMv4 as a real-time simulation system. The dynamic property of the road traffic network has already been mentioned. The ALERT-C [86] encoded traffic and road works messages contain information about lane, exit, or entrance closings and speed limits due to road works.

Listing 2.1.2: Configuration entry for track `A001-SW-HF-098` in OTDF.

```
[A001-SW-HF-098]
Length = 10.100
VMax = 160
Number Of Lanes = 3
Exit {
Destination = A001-SW-HF-099
LaneOffset = 0
CellOffset = -10.100
}
Exit {
Destination = A001-SW-PF-098
LaneOffset = 1
CellOffset = -9.900
TurnArea {
Begin = 9.900
End = 10.000
Lanes = -1
}
}
Connection {
Destination = A001-SW-HF-099
Routing = A001-SW-HF-099
Static Probability = 0.8
}
Connection {
Destination = A003-SW-HF-024
Routing = A001-SW-PF-098
Routing = A001-SW-LA-098
Routing = A003-SW-PF-023
Routing = A003-SW-HF-024
Static Probability = 0.1
}
...
Checkpoint {
ID = 4129-077-033
Position = 0.374
Lane = 0
}
Checkpoint {
ID = 4129-077-034
Position = 0.374
Lane = 1
}
...
```

2.1.4.4 Traffic and simulation model

OLSIMv3 used the brakelight traffic model as the underlying microscopic traffic model. This traffic model is an up to date model that is in good accordance with empirical findings [117, 112, 109, 113]. However, the implementation relied on the concepts of the underlying topology. Vehicles were associated with tracks. Both vehicle update procedures, a longitudinal drive update and a lateral lane change, require the positions and the velocities of the preceding or the neighbouring vehicles. These vehicles may be on the previous, on the succinctive, or even on other neighbouring tracks. In order to identify the neighbouring or the preceding vehicles, the update process contained some code to process searching for vehicles on neighbouring tracks. As these vehicles may not even exist in the neighbouring tracks, the implementation extended the search and the vehicle update to the further neighbours. Because nearly all microscopic traffic simulations also partition the roadways into segments, they likely uses this or a similar technique for searching the neighbours. Another similar technique of the same complexity is frequently used to implement traffic lights and lets vehicles search for topological elements in each time step. These techniques complicate adding further traffic models to an existing implementation such as OLSIMv3 in this case.

In addition to the aspect of the implementation complexity, the question of how portable an implementation is across several topological environments is of the same importance for examining the shortcomings of an implementation as they impact not only the implementation but also its verification. For the case of OLSIMv3, adding another traffic model or replacing the current one probably would consume the same effort as an implementation from scratch as the former beared too many uncertainties. As the vehicle update procedure had to search the topological structure anyway, it may have appeared at hand to the implementers to search also for so-called checkpoints and register vehicles to them. The checkpoints were responsible for measuring traffic observables and tuning the track. Due to the registering, the vehicle update procedure had not only to search for the neighbour tracks but also for the next positioned checkpoints. As the implementation of the traffic model depends on the concepts of the topological world, OLSIMv3 would have been hard to verify. Additionally, the vehicle update process was split into two separate cycles. In the first cycle, the velocities of all vehicles were updated. In the second cycle, the vehicle update process moved the vehicles according to their velocities. This technique is also used in SUMO [19]. It limits the family of possible traffic models to predefined two step models and state driven implementations. Moreover, the implementation had only limited parallelisation capabilities. Because of the splitted vehicle update and the registering of vehicles in the vicinity of a checkpoint a parallel implementation would have resulted in a high synchronisation effort. Lastly, the performance of the OLSIMv3 implementation was good but required almost a full real-time minute of CPU-time for a simulation minute during rush-hour. As a consequence, the computational capacity restrictions prevented the simulation from considering further topologic information such as overtaking restrictions, inner track lane closings and mergings, and per vehicle-type speed limits.

Due to the overlapping in the implementation of the topological concepts with the traffic model concepts extending the simulation model became difficult if not impossible. However, as the computing time was already at its limits adding new features were not allowed to consume computing power. Verification of the overall simulation model was not possible. The software had reached its end of life cycle.

2.2 State of the art

The following sections portrait the related work regarding the simulation system of OLSIMv4. At first, Section 2.2.1 classifies the simulation type. Thereafter, Section 2.2.2 introduces the input and output methods by discussing various existing detector models. Section 2.2.3 follows and details the state of the art of microscopic traffic simulations. Still continuing the related work, Section 2.2.5 describes some tuning models and strategies. Finally, Section 2.2.4 completes the state of the art Section and summarizes several existing road traffic network models.

2.2.1 System classification

The hierarchy of systems specifications [216, Sec. 1.3] provides a general concept for specifying dynamic systems at various levels of system knowledge. The specifications consist of five levels. At the lowest level, the specification describes the inputs and observables per time. At the second level the specification describes the behaviour of input and output data of a dynamic system. At the third level, the specification provides knowledge about the initial state and the functional relationship of input and output data. At the fourth level, the system specification can rely on the knowledge of state transitions. This specification requires again predictiveness of the simulation results which is not always given for the microscopic traffic simulations. At the highest level, knowledge about all components and their coupling is available. The conceptual design in Chapter 3 introduces a system specification for this level with its components and discusses this in more detail.

The general concept of the specifications requires a unique relation between an output and any input at the third level. This requirement seems to exclude stochastic simulations. However, as most stochastic simulations use pseudo-random number generators (PRNGs) the uniqueness requirement holds whenever the initial state of the PRNG is known because of the following facts. PRNGs produce numbers that have similar statistical properties as truly random number sequences. They are not unpredictive numbers because the sequence of a PRNG is unique to any given seed state. When using a hardware random number generator such as an audio or a random device provided by the operating system [53], any input from the random number device to the simulation system has to be recorded. The observed relationship between such input sequences and the output sequences will remain a functional relationship. However, when using large amounts of random numbers for a single simulation cycle—as is the case with many microscopic traffic simulations—the functional relationship may be practically unpredictable. As a consequence, the hierarchy of systems specifications is not fully applicable to microscopic traffic simulations at this level.

Dynamic systems vary their behaviour over time [216, Sec. 1.3]. As OLSIMv4 employs a microscopic traffic simulation system, it is thus a dynamic system. Apart from the aforementioned shortcomings, the hierarchy of systems specifications provides some reasonable specification levels that partly apply to OLSIMv4, too. A formal specification, however, relies on a prior accurate classification of the simulation system. According to [216], three main classifications for dynamic systems are widely used, namely the discrete event simulation systems, the differential equation simulation systems, and the discrete time simulation systems [216, Sec. 1.1, Figs. 4-6]. This kind of classification reflects, however, the "method of execution used on a model rather than the design structure" [62]. The latter publication proposes, thus, a "taxonomy whose categories are inspired from categories in programming language principles within the field of computer science". In addition to the focus on the execution method, for some simulation systems multiple classifications apply at the same time and transitions exist between the various classifications. Section 3.7.5 expresses such a transition for transforming the discrete time system specification of a microscopic traffic simulation into a discrete event modelling simulation. However, as an exact design structure is initially not available (it comes first available with the conceptual design from Chapter 3) and as the three classifications are wide spread in use, they are used herein also. Classifying a microscopic traffic simulation system into one of the three classes requires a first study about the details of the system. This is done in the following.

At the level of input and output observation, OLSIMv4 consumes the traffic data as already presented in Table 2.1.2. It produces the traffic information listed in Table 2.1.1. As another potential input data source, the OLSIMv4 simulation requires to adapt the topology in accordance to changes provided by road works or traffic messages. This property plays an important role for the classification of a microscopic traffic simulation that will be discussed later in more detail.

OLSIMv4 uses a microscopic traffic simulation with space and time discretisation, a variant of the so-called cellular automaton models. In addition to discrete space and time, these models have discrete and finite state sets as well as components, called cells, that are located on a one-, two-, or multi-dimensional grid [216, Sec. 3.2.2]. In cellular automaton models particles such as vehicles in the case of a microscopic traffic simulation can reside in the cells [183, Sec. 5.2.2] and move across the cell network. Multi-agent systems are very similar to the cellular automaton models. They "provide the particles in a cellular automation approach with more intelligence" [183, Sec. 5.2.2]. In these systems, agents "decide" about their actions to be taken depending on the constellation of the neighbouring vehicles as well as the current road and traffic context and their goals [183, Sec. 5.2.1].

In classical cellular automaton models such as the one by Nagel and Schreckenberg [147], each particle can reside on exactly one cell per time which corresponds to a spatial discretisation of 7.5 meters per cell. This discretisation turned out to be not fine-grained enough. In later models such as the brakelight model [118, 115] the discretisation is $1.5\,\mathrm{m}$ per cell. Consequently, a vehicle therein occupies several subsequent cells. Additionally, several vehicle types with different lengths and maximum velocities have been introduced [116, 164, 70, 112]. Moreover, models have been extended to let vehicles not only interact with each other but instead with the road traffic infrastructure, too [70]. Due to the lack of a formal definition for "intelligence" and the assimilation between both model approaches, the distinction between both microscopic traffic models and multi-agents is not very sharp.

The two distinct discrete event and discrete time system specifications have established for microscopic traffic simulations. In general, they do not lead to identical simulation results [183, Secs. 4.1.7, 4.4, 172, 57, 171, 117]. The subtle difference between the two system specifications lies in the vehicle update procedure that each system specification requires. The vehicle update procedure can be either a sequential or a parallel update, both explained in the following. In case of a sequential update, the vehicle update process chooses a vehicle at random or by its position in the sequence, updates it, and merges it back into the originating list of vehicles. In case of a parallel update, instead of merging the updated vehicle back into the originating list, the update process merges the updated vehicle into the list that contains the updated vehicles only. Additionally, the following variations exist that illustrate the assimilating behaviour of the two aforementioned. A random sequential update with very small time steps is almost identical to an ordered sequential update. The result of a sequential update for a single lane scenario with an update order in the direction of travel is identical to that of a parallel update as long as the two following conditions hold. The single vehicle update function must not depend on the state of the following vehicles and the single lane road must not have periodic boundary conditions, i.e. a ring structure. Additionally, in such a scenario the container that stores the vehicles has to tolerate that following vehicles may overlap.

For the family of simulations with sequential update the discrete event system specification [216, Sec. 1.3] applies. Discrete event simulation systems maintain an event queue in order to dispatch any incoming event such as change of time, distance, or state of a traffic light. After dispatching of the event, the simulation updates the affected vehicle. Multiple events may occur at the same time which leads to tie-breaking of the events or to parallel discrete event simulation. Several simulation models that use discrete event simulation systems exist [125, 17, 129]. Whenever the number of vehicles that do require updates is small compared to the total number of vehicles in a system, choosing a discrete event system specification may lead to several advantages over discrete time simulations such as better performance and better parallelisation possibilities. However, the real gain depends on the problem domain and without careful examination of the real traffic patterns that arise choosing the discrete event system may even lead to premature optimisations, incorrect results, and worse performance. In general, the random sequential update is unsuitable for microscopic traffic simulations as it loosens the coupling between vehicles and breaks intrinsic model characteristics [172, 171]. Consequently, the simulation results will lack accuracy. Thus, the parallel update is chosen from now on.

In a parallel update all vehicles anticipate the behaviour of their neighbouring vehicles. All vehicles operate on the same system state. The parallel update is therefore sometimes called a synchronous update [183, Sec. 5.2.2]. A simulation of a multi-lane road that updates all vehicles in parallel resembles best a Mealy network of flip-flops. In "Mealy networks the effects of an input can propagate throughout space in zero time – even cycling back [...]" [216, Sec. 3.2.5]. Thus, the family of simulations with parallel or synchronous update follow the discrete time systems specification for the Mealy-type system as proposed in [216, Sec. 6.4]. In contrast to discrete event simulation systems, time-discrete simulation systems advance the time in fixed increments. Furthermore, each vehicle in a discrete time system performs one update step called a »state transition« [216, Sec. 3.2.4]. In a microscopic traffic simulation at least the position variable and the velocity variable changes for each vehicle with a velocity greater than zero.

A microscopic traffic simulation such as OLSIMv4 can also be described as a multi-component system where the components consist of the cells, the lanes, the roadways, the occupation tables, the detectors, and the tuning elements. Multicomponent systems share a common state among the components. In a microscopic traffic simulation, the common states is required for modelling the traffic on the merging areas. For multicomponent systems, the hierarchy of systems specifications provides formalisms to couple these components into a complete system with a specification for each and for the overall system. Chapter 3 examines the relation of a formal specification for OLSIMv4 to the hierarchy of systems specifications in more detail. While it is certainly a good thing to have a formalism for microscopic traffic models, the presented formalism has also several shortcomings in the context of a possible implementation that Chapter 3 also discusses. The common states make the components depend on each other. In the case where the components depend only on the input and output of each other, the network specification applies. Chapter 3 also discusses the latter.

As OLSIMv4 uses stochastic microscopic traffic models, the dynamic behaviour of the simulation system is not necessarily deterministic. However, OLSIMv4 uses pseudo-random number generators which will result in a kind of a deterministic behaviour. One goal of an implementation for OLSIMv4 is to provide the choice between several microscopic traffic models. This choice should not be limited to cellular automaton models only but should cover so-called continuous models, too. The continuous models, however, fall into the same classification as the cellular automaton models because the vehicle update procedure discussion also applies to them. Additionally, most continuous models are time-discrete, and the time interval describes the discretisation of the numerical solution of the underlying differential equation. Section 2.2.3 discusses several microscopic traffic models in more detail.

2.2.2 Detector models

In the sense of the hierarchy of systems specifications, detectors process and provide input and output data. While traffic detector devices measure traffic observables in the real road network, detectors in a simulation calculate such traffic observables. Even though measuring traffic data involves some kind of calculations, too, a traffic detector device mostly deals with sensor technology [105, Chap. 2, 201, App. 3], classification of vehicle types [105, Figs. 2-29, 199, App. 2], connectivity details [106], and network stacks and protocols [199, Apps. 4, 5, 6, 200, 201].

Two classes of traffic detector devices exist, namely stationary traffic devices and moving devices. The stationary traffic devices in general use some kind of sensor technology to measure some traffic observables. Inductive loop detectors, magnetic detectors, video image processors, microwave radar sensors, infrared sensors, and laser radar sensors belong to this class. Excellent technical specifications exist in general [105, 106, 199, 200, 201]. The stationary traffic devices all have two properties in common. First, they observe only a fractional spatial extent of the network. Second, they rely on a concrete traffic model. For example, detectors in the NRW highway network measure the traffic data of Table 2.1.2. This data distinguishes vehicles into the two categories cars and trucks. Even though in Germany trucks are only allowed to drive at lower velocities of at maximum $80\,km/h$, the distinction or classification of the vehicles, however, is not based on the potential maximum velocity that a vehicle is able or allowed to drive. Instead it is based on the distance of the vehicle axles which occasionally leads to curious data such as trucks with velocities of $180\,km/h$ [114]. Additionally, the underlying traffic model of traffic devices such as loop detectors relies on the idealisation that traffic is always flowing. This weak assumption which is at the same time also a kind of a traffic model yields to the detector data ambiguity where loop detectors do not detect any vehicles either due to congested traffic or due to sparse free flow traffic. A different detector type technology that relies on stronger assumptions such as for instance a combination of a weighing machine and an inductive loop detector would have made such ambiguities impossible to occur. Such a detector device would then provide the semantically important measurement—detector blocked by one vehicle with zero velocity. As a side effect, the classification into cars and trucks would then correspond more accurately to the one that the German traffic administrative regulations contain.

The traffic devices of the second class [30]—the so-called moving devices [164, Sec. 2.1] — collect their traffic data in the vehicle and broadcast it to some provider. Several technologies exist that the devices can exploit. Namely, these are the global positioning system (GPS) and UMTS in case of cell phones or navigation devices. Public transport vehicles equipped with GPS devices such as taxis or busses could establish the provision of periodical data such as position and velocities. This kind of traffic data—also called floating car data—lacks some vehicle specific information such as length or vehicle class which is probably the biggest shortcoming of these group of moving devices. Another group of moving devices use LTE technology to submit so-called extended floating car data (xFCD) [77] to their fabricators. The vehicle specific traffic data can contain vehicle type, position, velocity, headway, and deceleration capability. While providing this data is certainly an excellent idea, it is currently a fabricator-specific technology. All moving devices share the difficulty of providing the total number of vehicles in a certain area. Therefore mathematical models exist and provide estimations for the total. They will be introduced in the following.

All aforementioned traffic detector devices can collect their data only for a locally limited part of a single road which is the most obvious difference to the virtual detectors in a simulation as the virtual detectors can operate on the whole road network. Several kinds of virtual detectors for the purpose of a simulation exist for various purposes. Constant detectors provide a constant measurement over all time and are especially useful for verification purposes [30]. Local detectors or virtual loop detectors simulate the behaviour of the inductive loop detectors. They aggregate the traffic observables of vehicles in the simulation that passed their cross-section. Their traffic data matches its real-world counterparts provided that the microscopic traffic simulation produces comparable traffic patterns. The publication [164, Sec. 2.1, 4.1] provides a mathematical model for the virtual loop detectors. The global detectors operate on complete sections or areas of a road. Their data conforms to the loop detector data only up to densities of about $40\,\text{vehs/km}$. The loop detectors "tend to underestimate the density in the congested regime" [183, Sec. 6.2] because the measuring method is event-driven and the only types of events rely on moving vehicles. The fundamental diagrams in [164, Figs. 5.5, 5.6] display the different effects of both approaches. However, there exist mathematical models for virtual local detectors that do reproduce even the traffic data of the global detectors.

Maybe the most important advantage of the global detectors is that they can determine the density on a road segment exactly in the following sense. In each second they count the number of vehicles on their associated part of the road. At the end of a measurement interval—one minute in the context of OLSIMv4—they determine the density by dividing the total number of vehicles with the total length of the road segment. As local detectors and inductive loop detectors measure locally they need to estimate the density through the hydro-dynamical relation $J(\rho) = \rho \cdot v$, where J denotes the traffic flow, ρ denotes the density, and v denotes the average velocity. This method has two disadvantages. First, it is based on events, i.e. only when a vehicle passes the loop detector a density value can be derived. In congested traffic however the vehicles may not be able to pass the detector. Second, for low velocities a vehicle may take several seconds to pass the detector which results in a greater uncertainty of the density value.

For the purpose of a simulation it is important that its traffic data can match the traffic data from the real-world detectors as this enables a formal verification for some parts of the simulation. Otherwise, the simulation even may not be able to reproduce the real-world scenarios adequately. Scenarios that rely on traffic data from moving devices require reformatting the floating car data into stationary traffic data [30]. In this case estimating the total number of vehicles from the fraction of vehicles equipped with a certain moving device technology is required. A recent publication [111] by Knorr and Schreckenberg provides a mathematical model for estimating the total number of vehicles in vehicular ad hoc networks [10] from the number of vehicles known to be equipped with the specific technology.

2.2.3 Traffic models

Microscopic traffic models describe traffic flow for individual vehicles in form of equations for vehicle motion. The time and spatial discretisation aspect of the vehicle motion is a key to distinguish so called continuous models and cellular automaton models. Continuous models solve the underlying differential equation for vehicle motion through discretisation, too. However, they only discretize the time but not the space. Additionally, the vehicles' state variables are continuous [39, Sec. 8]. Depending on the model, lane changing behaviour may be complementary to the vehicle motion, in particular for the so called, continuous car following models. The publications [100, 183, 117, 148, 79, 39] as well as the references contained therein provide a more detailed, general model discussion and share deep insights to vehicular traffic flow and microscopic traffic models.

Probably the most famous stochastic cellular automaton model is the Nagel-Schreckenberg-Model [147] even though it may have been not the first one [39, Sec. 8], [100, Sec. 5]. With having only two parameters [164, Sec. 3.1.1], [112, Tab. 1], it is rather minimalistic although able to reproduce the two traffic phases free flow and congestion quite well. The Nagel-Schreckenberg-Model uses a discretisation of $7.5\,m$ for vehicles and cells. However, it fails to reproduce synchronised traffic and other traffic phenomena [164, Sec. 3.1.1]. Several approaches that extend the Nagel-Schreckenberg-Model with lane changing rules exist [149]. The extended Nagel-Schreckenberg-Model has been validated for a scenario with German highway traffic in [112].

As a variation of the Nagel-Schreckenberg-Model, the brakelight model [118] introduces a finer discretisation of $1.5\,m$ for the cells only as well as a higher order synchronisation element between vehicles. A vehicle signalises its acceleration or deceleration behaviour to its follower by the state of a binary brakelight variable. In the sense of solving a vehicle's equation of motion, the brakelight variable represents a positive or negative slope for the derivative of the predecessor's velocity. A following vehicle takes its state into account. Thus, in contrast to a finer discretisation only that may lead to quantitative improvements, the brakelight model introduces a qualitative improvement over the Nagel-Schreckenberg-Model. With only five parameters, it is still quite minimalistic. The brakelight model reproduces synchronised traffic but may lack some accuracy in the single vehicle velocity data [164, Sec. 3.1.3]. In general, its results are in good if not excellent agreement with empirical findings [164, Sec. 3.1.3], [112, 113, 109]. The brakelight model has been extended with asymmetric lane change rules and varying vehicle types [115], e.g. cars and trucks. The model has been evaluated for time headway and distance headway distributions [117], for comparison with section control travel times [80] and in a scenario with German Highway traffic in comparison to the Nagel-Schreckenberg-Model and a car following model [112, 113, 109]. OLSIMv3 has used the extended brakelight model with asymmetric lane changing rules [138] and OLSIMv4 still supports it [31].

The model by Lee, Barlovic, Schreckenberg and Kim [124] is a more complex stochastic cellular automaton model that has six adjustable parameters [70, Sec. 4.3] for the longitudinal motion. Table 3.4.2 provides a more complete overview of the altogether sixteen model parameters. The model distinguishes vehicles into having an optimistic or a pessimistic driver behaviour that depends on the velocities and the distance headways of the next two predecessors. The method of regarding two predecessors for the distinction of optimistic or pessimistic behaviour results in a similar behaviour as the use of the brakelight variable in the brakelight model. The model has been defined by an inequality that compares the total covered distance headways of two succinctive vehicles in a full braking manoeuvre. The maximum velocity that still satisfies the inequality provides the next value for a vehicle's covered distance. Pottmeier has extended it with symmetric lane changing behaviour and multiple vehicle types in [164]. Habel and Schreckenberg [71, 70] introduced asymmetric lane changing rules for its use in OLSIMv4. The extended model has been validated for time and lane changing behaviour in [164] and for the comparison with section control travel times in [80]. The model reproduces empirical findings including synchronised traffic very well.

The car following models form another family of microscopic traffic models in addition to the stochastic cellular automaton models. While the parameters in the models of the car following model family represent and calibrate microscopic traffic relations such as safe distances and individual braking capabilities, the parameters in models of the cellular automaton model family calibrate, in contrast, macroscopic behaviour such as the outflow velocity of a wide moving jam which is calibrated by the dawdling parameter. The different impacts of the parameters among both model families make a direct comparison impossible. The knowledge of the relation between the microscopic parameters and the macroscopic behaviour eases turns out to be an advantage as it eases the calibration. Therefore and as a direct comparison is impossible, the model families are compared under the aspect of effort and benefit in the following.

The intelligent driver model [204] is a car following model. With six parameters [112] it is quite complex although its results are in good accordance with empirical findings [112, 109]. The authors of

[28, 27] list and examine calibration success in combination with the number of parameters of some cellular automaton models and further car following models such as the ones by Fritzsche, Wiedemann, and Gipps. Several full-featured traffic simulation systems of common market solutions such as AIMSUN [197], PARAMICS [51], and VISSIM [3] make use of the latter. The model by Fritzsche is the most notable among them and it provides and uses 13 calibratable parameters. The traffic simulation systems VISSIM [3] and PARAMICS [51] rely on the model by Fritzsche. Publication [156] lists 16 model parameters of the VISSIM model.

Brockfeld, Kühne, Skabardonis and Wagner state in their investigations [27] that "for this particular data set, the models that performed best were the ones with the smallest number of parameters." In their following work [28], they consolidate their initial statement as follows. "The calibration results agree with earlier studies as there are errors of 12% to 17% for all models and no model can be denoted to be the best. [...] But for special data sets with validation errors up to 60% the calibration process has reached what is known as 'overfitting': because of adaption to a particular situation, the models are not capable of generalizing to other situations." Similarly, concludes publication [183, Sec. 5.2.1] that "high fidelity models try to capture the complexity of decision making, actions, and so forth in a realistic way. In contrast, in the simplest models, agents are represented by particles without any intelligence. [...] Roughly speaking, the number of parameters in a model is a good measure for fidelity in the sense introduced here, but higher fidelity does not necessarily mean that empirical observations are reproduced in a better way!" According to [100, Sec. 5], the same criticism as for the Nagel-Schreckenberg-Model applies which is the lack of reproducing the synchronised traffic phase. When comparing the number of adjustable model parameters, any comparison should, thus, first confirm whether models belong to the same model class, i.e. whether they reproduce synchronised traffic or not. Kerner's publication [100, Sec. 5] may serve as a good starting point as it already provides a classification for most of today's traffic models.

The independent publications by Knorr and Schreckenberg [112] and by Brockfeld, Kühne, Skabardonis and Wagner [27, 28] compared empirical time series and the corresponding simulation results of two cellular automaton models as well as the intelligent driver model. The similarity of the simulation results in the publications confirmed the conclusion that the traffic models differ only little with respect to the quality of the simulation results. From that point of view, it is questionable whether more model parameters and finer discretisations generally and automatically lead to more accurate results. Knorr and Schreckenberg questioned it by means of an example using the intelligent driver model in [112] and Brockfeld and Wagner concluded the question in [28] from their comparison. However, it is clear that more model parameters and finer discretisations require more optimisation effort and more computational cost.

The so-called high flow states are also a critical challenge for microscopic traffic models, especially for the car following model family. The author of this thesis has contributed to a publication [114] that has examined the high flow states. Therein, the authors found loop detectors measurements with flow rates of more than 3000 vehs/h that lasted for a few minutes. As the high flow rates occur in reality, microscopic traffic models should be able to reproduce them. In a less extreme simulation scenario, unfortunately, none of the compared models [112, Tab. 2] was able to adapt to the measured flow rates on a German highway. The car following model turned out to have the largest difficulties in adapting to it. Knorr and Schreckenberg as the authors of publication [112] suppose that the realistic braking rules of the car following model are responsible for the lack of adapting to the high flow rates. To improve the results for the intelligent driver model, the author raised the vehicles' acceptable deceleration capability significantly to an unrealistic value. The latter, however, improved the situation only but did not make the model adapt to the high flow rates. The limited braking capability is common to all car following models but at the same time also present in the model by Lee, Barlovic, Schreckenberg and Kim [124]. Thus, their adaption to the high flow states will be similar to each other. However, it is questionable whether any microscopic traffic model is able to reproduce to the high flow states.

Another critical challenge in the design of microscopic traffic models exists with so-called accidents or dangerous situations [182]. This model characteristics is closely related to the high flow states by the safety distance that vehicles try to maintain. It is a challenge because, as of now, there are the following orthogonal design goals. A collision free model either uses unlimited braking capability or relies on completely safe distance headways. Cellular automaton models such as the Nagel-Schreckenberg-Model and the brakelight model provide unlimited braking capability. High fidelity models such as the ones by Wiedemann, Gipps, the intelligent driver model, and the Krauss model permit only limited braking capability but choose their next position and velocity as to be free of collisions [183, Sec. 9.5]. The freeness of collisions results in "unrealistic large decelerations" [182] as can be seen in [109, Fig. 4.8] for the brakelight model, in [112, Sec. 4.3] for the intelligent driver model, and in [61, Sec. 2.3.1] for the Wiedemann model. These "extreme braking manoeuvres might have a strong effect on the dynamics and are a possible origin of jams" [182].

Furthermore, accidents do occur in reality. For the most part, they do not occur due to technical problems of the vehicles but instead due to risky driving behaviour. According to [148], "in Germany cars are driving some 10^{11} km annually while generating about 10^6 crashes per year." Average reaction times range from $2/3$ s for simple decisions to not much less than 2 s for more complex decisions [137, Sec. 2.7.1]. This value range for the average reaction time corresponds with the value of 0.72 s that has been derived by the chosen distance which allows an estimation of the average assumed reaction time of the drivers in [148, Fig. 9][2]. The value range also corresponds to the values of 0.8 s to 1.8 s in [39, Fig. 8] that was taken from [151, Fig. 4]. However, in [148, Fig. 9] the data scattered largely and in both Figures a significant fraction of the vehicles or drivers chose a distance corresponding to considerably less than 0.5 s at velocities of about 100 km/h. Additionally, a significant fraction of vehicles on a single lane with a time headway below 0.5 s in France have been found [7]. These observations show that risky driving behaviour at least for the French and the German highways network is an empirical fact that ends in the observable accidents. Microscopic traffic models have to meet this phenomenon somehow. All models with realistic and limited braking capability, i.e. mostly the high fidelity models that also have collision freeness, do not reproduce these empirical observations.

In the models by Lee, Barlovic, Schreckenberg and Kim [124] with symmetric multi-lane extensions by Pottmeier [164] or asymmetric lane-changing rules by Habel [70, 71] collisions do occur [164, Sec. 5.5], [182, 167]. These models reproduce the short time and distance headways very well. Vehicles that change onto a lane can disturb subsequent vehicles in such a manner that the limited braking capability of the vehicles and the short distance headways lead to collisions. Apart from initialisation problems in periodic boundary condition scenarios, it is questionable whether collisions occur in the single lane model. The multi-lane model provides the collision feature and has been examined for it [164, Sec. 5.5]. The introduction of asymmetric lane-changing rules has reduced the frequency of collisions due to the density inversion phenomenon, i.e. on German highways overtaking is allowed for vehicles on a left single-lane. The phenomenon leads to the curious situation where more vehicles driving on the left lane than on the right. The publications [115, 208] provide more details. Even though Section 3.4.1.3 presents some vehicle constellations that have undefined next velocities and that the traffic model might produce, it remains, however, unclear what exactly causes the collisions and how to deal with them. As a consequence, the dynamics of this model remains to some degree unspecified [182].

2.2.4 Network models

A classical cellular automaton consists of a cell network wherein each cell can take on a defined state. In the case of the Nagel-Schreckenberg-Model the cell size was 7.5 m. However, as this discretisation was not fine-grained enough, the cell size in the simulation of OLSIMv3 decreased already to 1.5 m. Because a vehicle in this discretisation occupied several cells, the classical cellular automaton scheme in which cells adapt a certain state was no longer comfortably applicable. The decreased cell size results

---

[2]The note to [148, Fig. 9] determined the slope of the fitted piecewise linear function to be $1/\tau$ which yields to $\tau = 1.4$ s. However, concerning the value for the slope only, the correct value is $\tau = 1.4$ s^{-1}. The other value $1/\tau$ represents the average overall reaction time $1/\tau = 0.72$ s in [148, Fig. 9].

in the total number of cells $10\,882\,000$ for the NRW highway network. An explicit instantiation of each single cell as a component appears very inefficient in terms of memory usage as only starting with viscous traffic the average percentual density can rise up to more than $50\,\%$. In so-called continuous models, the discretisation of the network orientates at the floating precision of the computer model that is used. In general, simply using floating point numbers instead of discrete cell sizes does not lead to better information accuracy automatically.

A road traffic network model defines a view on the topological information of the real highway traffic network. In general, it considers the road geometry information such as the length, the number of lanes, the junctions, the intersections, the connections, lane closings or mergings, and the merging areas of all roadways in the network. In most contexts and apart from visualisation demands, microscopic traffic simulations do not require coordinates. As another type of information, a network model can consider traffic information such as speed limits, lane change, and overtaking restrictions all per vehicle type. However, as most models have a very similar road geometry model, only a few are able to represent traffic information adequately.

Most of the proposed road traffic network data models in the context of microscopic traffic simulations describe a road network by a directed and potentially cyclic graph of nodes and edges. In particular, these are "Visum, ArcView, FastLane, ARTEMIS, or Navteq-networks" [41], [120, 127, 198, 212], Location Code List [86], CentroMapDavPlus [191], and SUMO [19]. Even though VISSIM [61] and OLSIMv3 [41] do not explicitly instantiate a graph, they describe the network through links and connections between links which is very close to graphs. While a graph structure is certainly a great model to describe connections between road segments, it is not the best choice for a microscopic traffic simulation due to the following reasons. As the links are generally relatively short compared to the maximum roadway length in the network, this approach requires frequent sectioning of the roadways. Consequently, the vehicle update procedures have to search for neighbouring vehicles and for follow up positions across several connected links. As an example, in AIMSUN a "leaving vehicles entity process" [196, Figs. 275, 276] searches for positions on subsequent links or sections of updated vehicles that are in the first position of a link or section. Similar statements apply to OLSIMv3 [41], [64], and SUMO [19]. The road geometry and to some degree the traffic network may impact the parallelisation strategy considerably. Updating vehicles that may leave a link and enter to another link, require synchronisation between the two links.

A traffic network basically configures an underlying road geometry for the purpose of vehicular traffic. When choosing links and nodes as basic entities of a road network, the traffic configuration requires repartitioning of the links and nodes. As an example, lane changing may be possible according to the road geometry on a link but may be forbidden by the traffic rules on a roadway [30]. Except OLSIMv4 [30] and VISSIM [61], all "models do not reflect the important distinction between road geometry and traffic network" [30]. Another important distinction concerns the vehicle type. The OLSIMv3 traffic model, as presented in Section 2.1.4.2, contained already information about maximum allowed velocities in form of the topological configuration. However, the topological data model of OLSIMv3 does not support defining maximum velocities per vehicle type. It is not clear whether VISSIM distinguishes spatial objects for varying vehicle types [61, Sec. 2.2.2.2].

As is the case with most other road traffic networks, the NRW highway network is a dynamic network, too. Section 2.1.3 has already covered some aspects of this property. In addition to those aspects, variable message signs, daily road works due to tree trimming, and police actions in reaction to accidents or special traffic situations display the dynamic character of the network, too. Possible changes in the traffic configuration may include adapting the maximum velocity, closing of lanes, exits or entries, and overtaking restrictions. To the author's best knowledge, none of the discussed network models support this dynamic character of the network. From the dynamic network property follows for an online simulation that a network configuration is no longer constant and is subject of change. Either an event-driven approach or polling of configuration updates may be choices for an implementation.

The road traffic network does not only serve as a database for topologic information in the simulation but also enables detectors, tuning elements, and vehicles to reference to defined positions in the network. When vehicles interact with each other or with the topological infrastructure, the road traffic network data model serves as a specification basis for search algorithms that find neighbours and topological configurations. Additionally, the traffic information derived from the microscopic traffic simulation refers to parts of the road traffic network. Thus, a road traffic network data model that supports positioning of global detectors of any extent at any location in the network allows traffic information that refers to external digital maps to be derived. OLSIMv4, for example, has to deliver its traffic information to the DAV data distributor [177]. The DAV data distributor requires all information to refer to the CentroMapDavPlus [191]. The CentroMapDavPlus digital map models the NRW highway network with inner and outer road segments that in fact represent driving connections. The outer road segments connect nodes in the network. The inner road segments represent connections inside a node. Global detectors that refer to each of the inner and outer segments could easily collect the traffic information in the case where the network data model supports arbitrary positioning of the detectors.

2.2.5 Vehicle generation

Vehicle generation is common to all traffic simulations. The most common practice uses origin-destination matrices. In this case, vehicles enter the network through on-ramps or at the beginning of a road. The input data is then a finite matrix that schedules vehicles periodically onto the on-ramps. While this approach is mainly useful for infrastructure planners, Section 2.2.3 has already pointed out the shortcomings of this approach. Additionally, as already discussed in Section 2.1.4.3, most microscopic traffic models fail to insert vehicles from an on-ramp onto the main road when the number of vehicles to be inserted exceeds a critical value. As this critical value is much below measured values and as real-time data are available and are, in general, more up to date, in the context of an online simulation tuning elements are more appropriate.

In contrast to the origin-destination matrices, tuning elements can reside and operate on any location of a road. They compare an actual to a should-be value and, depending on that comparison and their implemented strategy, insert or remove vehicles directly into their associated area of the road. Unfortunately, except [94], there is a lack of studies investigating tuning elements and tuning strategy. One of the reasons might be their less widespread use. As the publication [94] is not publicly available, a short introduction of common difficulties in developing tuning strategies is given here. However, as the tuning strategy of OLSIMv4 is not the authors work, it is not in the scope of this thesis.

A first and very basic approach uses only the traffic flow values and tries to assimilate them. One of the pitfalls with this approach lies in the ambiguity of low traffic flow values. Whenever a negative difference of the actual and the should-be value indicates that the tuning element should insert vehicles, inserting vehicles could prove impossible due to congested traffic at the tuning elements insert area. The low traffic flow should-be value can, however, indicate free flow as well as congested traffic at the location of the should-be measuring device. A tuning element therefore first has to determine and track the traffic phases of the actual and the should-be devices. This makes the development of a successful strategy rather complex. Furthermore, tuning the road with vehicles can disturb the vehicle interactions that the traffic model produces which complicates the development of a successful tuning strategy. In any case, inserting and removing vehicles requires a strategy such that the impact on the vehicle structure is minimised [15]. The thesis [94] has proposed the traffic flow tuning strategy and the tuning of the mean gap strategy. However, OLSIMv4 uses its own tuning strategy which is not in this thesis' scope.

As a proof of concept of the tuning element approach, Knorr modeled the on-ramp effectively by tuning vehicles onto the lane instead of letting vehicles try to change the lane from the on-ramp to the main roadway [112]. Additionally, OLSIMv3 used tuning successfully for years at so-called checkpoints located at the measurement cross-sections [94] positions, too.

2.2.6 Simulation systems

Some of the common market solutions for microscopic traffic simulation systems are VISSIM [3], AIMSUN [197, 196], PARAMICS, and the open source software solution SUMO [19]. Their main application area focuses on assisting the infrastructure planner, traffic operations, facility design, planning studies, demand management, investment decisions, and policy analysis. [181, Fig. 1] provides a more detailed study on the application areas. Apart from AIMSUN that offers an optional online variant they provide—if at all—only limited online support in the sense of the discussion in [31]. Therein the (technical) aspects of real-time capability, parallelisation support, traffic generation through tuning elements using online traffic data, and parallel update have been chosen as they are closely related to OLSIMv4. As OLSIMv4 does not focus on assisting the infrastructure planner and as the market solutions do not focus on online simulation, a comparison of these systems remains vague.

However, the use of a microscopic traffic simulation is common to all these systems including OLSIM. As a matter of fact, designing and implementing a microscopic traffic simulation for a specific purpose will sooner or later end up in the validation the software system in a real-world scenario. According to [155] in [12], "validation is defined as the process of determining the degree to which a model is an accurate representation of the real-world from the perspective of its intended uses" (cf. [153, 154]). In general, the task of validation "is to compare the simulated outcomes of models to reality, to establish the degree of similarity using a (non-deterministic) validation metric, and to address model fidelity" [12]. Addressing model fidelity is possible and certainly worth studying across several simulation systems but left out here as it deals with model accuracy.

Because validation "takes place at a very late phase of the implementation process, differences from expected to measured results are hard to trace and may be caused by a lot of potential reasons – e.g. by software bugs, artefacts or measurement errors, unforeseen interaction effects of the models, poor calibrations, weak models, or combinations of any of the aforementioned" [30]. Even worse "programming errors in the computer code, deficiencies in the numerical algorithms, or inaccuracies in the numerical solution, for example, may cancel one another in specific validation calculations and give the illusion of an accurate representation of the experimental measurements" [153, 154]. Thus, "instead of expecting to detect the differences from expected to measured results posterior to the validation process model, code, and numerical verification in various degrees help to improve software quality during the design phase" [30] and also during the modelling phase.

Before the validation activities start, a microscopic traffic simulation system should, thus, first be verified at least numerically or, wherever possible, formally. The goal of the verification process is to provide "evidence, or substantiation, that the mathematical model, which is derived from the conceptual model, is solved correctly by the computer code that is being assessed" [153, 154]. Sargent distinguishes additionally between specification verification and implementation verification [180]. While the former assures "that the software design and the specification for programming and implementing the conceptual model on the specified computer system is satisfactory" [180] the latter assures "that the simulation model has been implemented according to the simulation model specification" [180]. As "mathematical error is as old as mathematics itself" [73] and traditional mathematical proof relies on intuitive arguments [73], formal proofs that involve checking of every logical inference [73] can help to detect logical errors. As specifications for simulation systems involve a lot of logical conclusions and inferences [73], specification verification which is a kind of a formal proof is essential before implementing a simulation system. Publication [73] gives a rather interesting motivation to the subject of formal proofs in various domains including mathematical proofs and computer software. Verification activities include code verification, formal verification, software quality and result verification, benchmarks, and accuracy assessment [12]. Auer, Luther and Cuypers distinguish the three categories static code verification, formal verification, and result verification for verification activities [12]. However, even though a wide range of literature deals with verification in general and with verification of simulation models in particular, few publications deal with the verification of microscopic traffic simulations, cf. [5, 82]. The general implementation cycle of microscopic traffic simulations seems to miss the verification stage entirely [169, App. G, 202, 59].

Specification verification and implementation verification require a written down description of the entire simulation model that is going to be assessed [180]. Most publications, however, deal with traffic models, detector models, and vehicle generation models only. As already mentioned, some systematic studies [172, 171] cover the influence of update schemes. The studies cover, however, only one-dimensional roads and not an entire road network. Consequently, the verification of an implementation with respect to the results known from the traffic model publications and the studies is only possible for identical networks, i.e. a single road in this case. Different execution models, network models, update schemes, and vehicle generation models produce, of course, different results in the traffic information. When extending the simulation scenario from a single road to an entire road network, the verification of the implementation is only possible for the case where the underlying simulation model traces the scenario for the entire network back to the single road case. Consequently, the simulation models that do not trace the scenario back to the single road case exclude result verification from the pool of verification activities. Additionally, Section 3.7 will show that for some execution models—probably the most common execution model—result verification depends on the concrete road traffic network. As the exact impact of the road network and the execution model on the simulation results is unknown, result verification is, thus, practically impossible for the combination of most networks and the affected execution model. Verification activities without any explicit written down specification is probably best compared with calibration activities without metrics—no reliable or sound conclusions are possible. The lack of complete specifications in combination with a complex execution model is probably the reason for verification efforts being rare. As a consequence, a calibration process without prior verification will never yield a systematic approach even when tried with the high effort as, for example, in [59].

An interesting part when comparing OLSIM with common market microscopic traffic simulations is the so-called calibration process. One of the goals of the calibration process is to find set ranges and default values for the parameters of the simulation model to use at simulation startup. Another goal is to find minimum durations within the system tunes itself to an initial state from whereon it can assimilate itself to the steady state. Both goals strive to reflect the reality as good as possible and strive to hold the mathematical model in numerical stable ranges. Calibration activities rely on verified and validated simulation models and, at least theoretically, involve sensitivity analysis and examination of the correlations among the parameters. Calibration activities include also to adjust the simulation results in form of traffic information quantitatively or to deform an existing model that—according to [173] in [154] (cf. [153])—has been validated successfully with a representative mixture of general, wide-spread field data or a reference model to better match local conditions. Ideally, the differences between local conditions and the representative field data or the assumed reference model are known in advance. Usually, the several parameters that a model provides change slightly. Otherwise, this process is more of a validation process. However, not everybody shares this understanding. Some examples to portray the more common understanding of the calibration processes for some of the common market solutions follow.

According to official "Guidelines for Applying Traffic Microsimulation Modeling Software" of the U.S. Department of Transportation from 2004 "calibration involves the review and adjustment of potentially hundreds of model parameters, each of which impacts the simulation results in a manner that is often highly correlated with that of the others. The analyst can easily get trapped in a never-ending circular process, fixing one problem only to find that a new one occurs somewhere else. Therefore, it is essential to break the calibration process into a series of logical, sequential steps—a strategy for calibration" [169, Sec. 5.2]. The "potentially hundreds of model parameters" are in a similar order of magnitude as the number of 192 parameters that a case study considered for calibrating VISSIM in modelling the Zurich network [66]. For the main part of the calibration, the sensitivity analysis regarded 14 microscopic traffic model parameters. In addition to [66], several guidelines of calibrating these complex software systems are available. They range from "engineering judgement" [202, App. H] to analytical considerations [65] and to non-linear optimisation strategies such as particle swarm optimisation [4]. Meanwhile the European Cooperation in Science and Technology funds a "Methods and tools for supporting the Use, caLibration and validaTIon of Traffic simUlations moDEls" (MULTITUDE) project that "is driven by the concern that, although modelling is now widespread, we are unsure how much we can trust our results and conclusions. Such issues force into question the trustworthiness of the results, and indeed how well we are using them" [59]. The project consists of four working groups with members that have a strong affinity to the common market solutions. While uncertainty is one of the project's and the report's main aspects, the report lacks a precise definition, i.e. whether aleatoric or epistemic uncertainty is targeted ([207] provides an interesting example and explains the differences). The classification as well as the characterisation of uncertainties and the propagation of the uncertainties through a system has a significant impact on the correct choice of the corresponding abstract data types, i.e. whether to use real numbers, integers, and so on.

Uncertainty analysis focuses mainly on the numerical aspects of the simulation and it is useless without prior numerical verification (not validation) of the simulation model. Numerical verification, in turn, relies on a correct (-ly implemented) algorithm and, thus, requires prior implementation verification. As the latter relies on a specification, the specification requires, first, a specification verification. Even though meanwhile there is significant progress in automatic theorem proving, e.g. Coq [25, 21], and model checking, e.g. Maude LTL model checker [56] and Real-Time Maude [158], it is needless to mention that most microscopic traffic simulation systems neither have an appropriate specification nor a specification verification. However, as microscopic traffic simulations have complex logic, they require specification verification and implementation verification in addition to the numerical verification activities. Subsection 3.4.1.3 will provide an example that questions the model's correctness and the robustness for the traffic model by Lee, Barlovic, Schreckenberg and Kim. As other commonly used microscopic traffic models are of comparable complexity, there are likely to be more of such modelling shortcomings in the specifications than only the presented one. Also needless to mention that the exact impacts of such shortcomings are unknown. As long as this is the case, calibration and uncertainty activities remain premature in nature.

Nevertheless, the complexity of the calibration process may not surprise as parameter calibration was proven to have a NP-complete complexity [81]. However, the majority of respondents in a survey to "establish a framework for identifying the pertinent information" [181, Fig. 4] found their calibration and validation results satisfactory and better. The survey [181, Tabs. 1 & 2] averaged the estimated cost of the calibration activities of their respondents to be about 40% of the total effort. The survey details 15 case studies. It summarizes the budget in combination with other details such as network size of each project in [181, Tab. 3]. Even though a straightforward comparison between the case study's projects requires comparable network sizes, the study, unfortunately, does not express the efforts in a comparable unit such as persons per years. However, a very rough estimation of the budgets average is $\$\,500\,000$. This results in an average amount of $\$\,200\,000$ for the calibration process. At the same time, only 4 projects from the 15 case studies had a network size greater than the NRW highway network. Additionally, five projects had the network data available from previous studies. Moreover, no study had online requirements in the sense that a continuous information stream had to be available as real-time output of the system. Most of the projects had simulation runs for a few hours of a day only. In any case, as the calibration of OLSIMv4 took significantly less than the estimated average budget, this first budget comparison of the calibration process allows the following concluding remark. OLSIMv4 as an absolute minimalistic simulation software solution requires probably not more calibration effort than the market solutions.

2.3 Summary

Section 2.1 details the project history, the project context, the feature requests by the users and the project founder, and the revision requirements of OLSIMv3. Handling the NRW traffic problems that originate from increased traffic by a better highway capacity utilisation was one of the goals when initiating the OLSIM project. Providing recent traffic information for the NRW highway network user through the internet and assisting the user with a better and flexible route planning is still the most promising way to achieve that goal.

Another motivation for using traffic simulation to complete the loop detector data was the cost argument. The average loop detector coverage distance is nowadays about $4\,\mathrm{km}$ and in the days when OLSIM started was significantly higher (maybe $5\,\mathrm{km}$ or even $6\,\mathrm{km}$). Thus, since the NRW highway network employs loop detectors to measure the traffic, an information gap between subsequent detectors exists. The online, microscopic traffic simulation system OLSIM has to complete this information gap. Viewed under the aspect of the project goals, the new features and the technical improvements for the next OLSIM version are intended to improve the information and assistence of the NRW highway user for route planning purposes. Three of the new features for the next OLSIM version focus on that subject. In particular, the seven day long-term prognosis, the truck specialised view, and the travel-times not only for a set of predefined routes but instead for individually chosen routes contribute to that goal. The remaining features focus on other categories, namely on enhancing the information accuracy, on combining the available information into more sophisticated traffic information, and on the new implementation of the simulation system.

The planned features for the next OLSIM version, namely the verification of incoming loop detector data, the loop detector data statistics, the dynamic topology, and the regarding of more topological information as well as the generalised road traffic network contribute to enhance the information accuracy. As another feature for OLSIMv4, it for the first time uses a database management system to store, access, and combine the available kinds of information namely the stationary loop detector data, the topological data, the event calendar, road-works messages, police traffic warning messages, and spatio-temporal traffic information into more sophisticated traffic information. Finally and maybe the most important change over OLSIMv3, is the new implementation of the simulation software which had reached its end of life cycle. The new implementation employs also a new topological data model, provides a platform independent application programming interface for microscopic traffic models, implements multiple microscopic traffic models, allows and provides a sequential as well as a parallel interpretation that exploits CPU thread-level parallelism, and introduces the new concepts of input/output detectors and tuning elements. All these new features required a fresh implementation as adding them to the previous version was not possible with reasonable effort.

In addition to the new features OLSIMv4 adapted some technical improvements over OLSIMv3, too. The filtering and semantic verifying of incoming loop detector data is a first one. As a second one, in OLSIMv4 the prognoses provide their data continuously and for the whole network in each minute. A third one replaces missing or faulty loop detector data on demand with prognosticated data. It uses the generated loop detector data time series in case of loss of connectivity or down time due to maintenance of the traffic data centre and serves as input for the simulation of the current traffic information. As a fourth technical enhancement, OLSIMv4 has adapted to the upgraded software of the traffic data centre. Additionally, it delivers simulation results for all road-segments to the traffic data centre. As a benefit of the feature that combines the various data kinds and the new simulation implementation, OLSIMv4 provides comparable input/output detector data per track.

The qualitative improvements for OLSIMv4 go much further than the technical enhancements. With the new implementation of the simulation software, it became possible to consider a lot of qualitative improvements for the implementation. At the ground, the new topological data model promotes to consider several, different simulation models for the purpose of a potential implementation. The simulation models in turn facilitate to examine their fitness for and degree of a potential, parallel interpretation. By means of the speed-up through hardware parallelism, that a given parallel interpretation ideally leads to on a multi-processor machine, they provide additionally available computing time. This allows to consider further algorithms that contribute in increasing the information accuracy and to implement a generic multi-component simulation system. The system consists of a platform independent application programming interface for microscopic traffic models, the various generic input/output detectors, and the tuning elements. As a benefit, it allows to choose advanced simulation setups. To increase information accuracy, OLSIMv4 takes into account additional and more detailed topological information such as per vehicle type speed limits, overtaking restrictions, and lane closings and mergings. Even though not each kind of information increases information accuracy automatically, the travel times in the free flow will benefit from the additionally available speed limits. Furthermore, the travel times and the densities may also benefit from the resistance induced by areas with relatively low speed limits such as in the case of road-works. Furthermore, OLSIMv4 takes into account a varying number of lanes per track and respects lane closings. They also lead to resistances or bottlenecks that provoke a qualitatively different and more realistic behaviour in the simulation.

Section 2.2 briefly reviewed the state of the art literature for various subjects. As the first of the subjects, the section classified the microscopic traffic simulation system as a dynamic system with space and time discretisation. In particular, the discrete time system specification applies and is in contrast to discrete event simulation systems. The classification also emphasised the need for a synchronous or parallel update scheme. The simulation system uses detectors to handle input into the simulation and output from the simulation. Additionally, the detectors provide the data as comparable input and output pairs which is an important property that the simulation model of the next chapter will exploit. The simulation employs two stochastic cellular automaton models which are in good if not excellent agreement with empirical findings. Both models reproduce the synchronised traffic phase which makes them distinct from most high fidelity models used nowadays in common market solutions. The section also discussed the number of parameters of the traffic models with respect to model calibration. OLSIM's network model distinguishes it from a classical cellular automaton model as it uses several cells per vehicle. For vehicle generation, OLSIM uses tuning elements which also makes it distinct from most other simulation models. Taken altogether, the simulation system contains a lot of properties that are uncommon to market solutions. The calibration process also reflects the differences in time and cost.

The conceptual design contributes in providing an explicit written-down specification of the entire simulation model. The design and the specification of the new topological data model, the simulation model, the multi-component simulation system and, in particular, its components are the subject of the next chapter.

Conceptual design

This chapter details the conceptual design for the next version of OLSIMv4. It reflects the results of the enhancement effort from the experiences of the OLSIMv4 implementation as well as its analysis and it also reflects the experiences from the model verification implementation of the author's publication [30]. The design strives for the following goals. It aims at increasing the information accuracy by providing high-quality and at least theoretically well-founded algorithms. It ambitions to improve the network platform portability by employing a portable network model and a network-independent microscopic traffic model. It targets on increasing the scalability and the efficiency of any possible implementation by reducing the synchronisation bottlenecks and by enabling a parallel interpretation at various degrees of parallelism. It intends to enable formal verification of an implementation and to prepare validation by sound theoretic specification of the simulation model as well as the subject-specific models. And finally, it ensures that a future implementation has a minimal yet orthogonal set of components.

Simulation is a tightly coupled and iterative three staged process that is composed of model design, model execution, and execution analysis [62]. This chapter's conceptual design is related to the model design stage whereas the next chapter deals with questions that belong to the model execution stage. The degree to which the components deal with low-level specifications such as formal semantics of a machine implementation distinguishes the two components model design and model execution from each other [62]. As computer code generally specifies a simulation model at a lower abstraction level, this chapter's conceptual design uses rules and equations to define the conceptual design of the presented simulation model. Alternative approaches use also graphs and visual representations in addition to the rules and equations. As the subject of microscopic traffic simulation requires the use of rules and equations for various components anyway, the author has chosen them as the primary specification means. In the sense of the hierarchy of system specifications, the conceptual design constitutes a significant enhancement over the simulation models of OLSIMv3 and OLSIMv4 as it provides knowledge about the functional relationship of input and output data. According to the specification hierarchy, the enhancement results in the highest specification level which is the state transition specification.

Each of this chapter's sections recaptures the initial position and the state of the art. The criticism on the existing solutions leads then to new solutions for OLSIMv4 and further developments. Moreover, each section provides a detailed specification of its contributing component to the simulation as described in the following. The simulation model intends to use orthogonal components that do not overlap or otherwise provide the same task by several concepts. As an explanation, detectors are designed to read and analyse the state of an occupation table. In contrast to them, tuning elements are designed to write or modify the state of an occupation table. Whenever users, simulators, and even implementers desire some information from the simulation, for example such as debugging information, they will use detectors. Whenever control layers, simulators, and even implementers want to manipulate an occupation table's state, as for example could be the case in tagging vehicles with routing information, they use tuning elements. The orthogonality grants simplicity and component-wise verification possibilities.

The sections start with general remarks in Section 3.1. The remarks compound some notes in Subsection 3.1.1 about this chapter's methodology and a nomenclature in Subsection 3.1.2 for the objects defined in this chapter.

Section 3.2 defines then the vehicle and vehicle type models. They include the basic vehicle type model definitions as presented in Section 3.2.1. Vehicle types are not only part of the vehicle model, but additionally are part of the network model, the microscopic traffic model, the detector models, and the tuning models. They play an important role for the internet user and the highway network user perspective. Section 3.2.1 discusses also the subtle differences between the several varying vehicle type classifications. Thereafter, Section 3.2.2 continues with the corresponding vehicle models for OLSIMv3 and OLSIMv4. As has been the case with the vehicles, the vehicle types also participate in other models, namely the microscopic traffic model, the detector model, and the tuning model.

Section 3.3 presents the network model. It consists of cells, sections, and roadways. Section 3.3.1 describes the cells. They compound all traffic rules that apply at a certain position of a single-lane section. Section 3.3.2 specifies the roadways' single-lane sections each employing a single prototype cell. Section 3.3.3 introduces the roadways itself. In addition to their structural components, the network model provides a topological analysis in Section 3.3.4. Section 3.3.5 then establishes the occupation tables that combine the roadway elements with the vehicles. As the final part of the network model, Section 3.3.6 introduces the switching areas that temporarily connect roadways with each other.

Section 3.4 describes the microscopic traffic models. As OLSIMv4 does not depend on any specific microscopic traffic model, it contributes in boosting the network platform portability. In addition to the brakelight model already present in OLSIMv3, it implements the model by Lee et al., extended for multi-lane traffic by Pottmeier and Habel. The traffic model composes the resulting motion from the longitudinal motion as described in Section 3.4.1 and the lateral motion as described in Section 3.4.2. An efficient calculation algorithm for the vehicle velocity in the next time step follows in Subsection 3.4.1.2. The section closes with some considerations about undefined configurations.

Section 3.5 expands the detector model for several kinds of detectors. Section 3.5.2 introduces the loop detector and Section 3.5.3 its variation as a prognosis detector. The so-called invariant detectors from Section 3.5.4 provide means to test the simulation model. The local detectors from Section 3.5.5 emulate the loop detectors and provide cross-section measurements from the simulation to the user or for the use in a feedback loop. Global detectors from Section 3.5.6 vary the measuring method of the local detectors. Instead of stationary traffic data they are able to provide spatio-temporal traffic information and, thus, represent the key detector type for OLSIMv4.

Section 3.6 presents the tuning model. It consists of the accumulators and the tuning elements. Section 3.6.1 introduces the former. They accumulate detector values depending on their accumulation algorithm such as cross-section-wise accumulation. Continuing with the tuning model, Section 3.6.2 presents the tuning elements. They remove and insert vehicles depending on their behaviour that results from the implemented tuning strategy. However, the OLSIMv4 tuning strategy is not subject of this thesis as the author did not contribute to it.

Section 3.7 discusses the simulation models for microscopic traffic simulations. Section 3.7.1 starts with the discussion and presents the most common simulation model for microscopic traffic simulations. It consists of a single discrete time system that updates the road traffic network sequentially or, in case of a parallel interpretation, with synchronisation in each time step. Section 3.7.2 decomposes such systems into a modular multicomponent discrete time system that requires less synchronisation and that provides more system knowledge. OLSIMv4 could have been implemented using the modular system approach. However, the author decided to employ a network of discrete time systems as introduced in Section 3.7.3. All components of such a network depend only on the output and not on the state of the connected components. Thus, they provide a high parallelism. Section 3.7.4 discusses the aspect of a dynamic network for OLSIMv4. Finally, Section 3.7.5 sketches further decompositions.

Section 3.8 concludes this chapter and summarises the most relevant concepts presented herein.

3.1 General remarks

This section presents some introductory thoughts in Subsection 3.1.1 about the methodology used throughout this chapter. It then introduces the nomenclature in Subsection 3.1.2 for the definitions and equations presented in this chapter.

3.1.1 Methodology

This chapter's conceptual design presents a semi-formal description of the physical foundations for a real-world microscopic traffic simulation and it aims at smoothing the way to a computer based model.

Section 2.2.6 has already discussed the need for an explicitly written-down specification of a microscopic traffic simulation system. As verification and validation are related to the model development process [180], neither specification verification nor implementation verification is possible without such a specification. The latter two are, however, milestones in making simulation system results trustworthy. This chapter aims at contributing to the explicitly written-down specification. Even though several parts have been implemented as part of OLSIMv4 and the Maude enhancement effort of the author's publication [30], the conceptual design has not yet been verified. However, it aims to have passed a conceptual model validation which is "determining that the theories and assumptions underlying the conceptual model are correct and that the model representation of the problem entity is 'reasonable' for the intended purpose of the model" according to [180].

This chapter describes the conceptual design using a (semi-)formal specification approach (see [185] for an introduction). The goal of using a formal specification approach is to enable formal verification thereafter. As part of a denotational semantics, the (semi-)formal specification uses semantic algebras that consist of structured sets, functions, and domains [185, Chap. 2]. The sets may have a global or a partial order. The structured sets additionally define constructors that combine values of already known sets into elements of the new set. Domains group sets and additional operations to have common properties. The function definitions consist of a signature part that describes the domain as well as the range of the function and an equational part that describes how to map elements of the domain into the range of the function. This chapter builds compound domains from primitive domains such as the natural numbers $\mathbb{N}$ using the product domain as in $\mathbb{N} \times \mathbb{N}$ and the function domain $\mathbb{N} \longrightarrow \mathbb{N}$.

Several circumstances have endorsed the decision to use semantic algebras to describe the conceptual design. Firstly, the semantic algebras enable a precise specification, i.e. each relevant domain kind and each range kind is properly defined. As set operations and the usual operations known from math can also be used, the specification takes place at a relatively high abstraction level. Secondly, the participating models such as the traffic models, the detector models, the tuning element model, and the execution model in general use an equational style, i.e. they define the calculation of variable values by means of equations. In general, it is possible to rewrite these rules into function representations without too much of an effort. Thirdly, the domain concept supports to integrate the variety of participating models in a way such that the variable names do not collide. Fourthly and lastly, the semantic algebra concept is closely related to the concept of abstract data types and, thus, formal verification in functional and declarative programming languages is eased.

The taxonomy in [62] distinguishes the simulation model categories into conceptual, declarative, functional, constraint, and spatial model types. Whereas the conceptual model type uses semantic networks, object oriented designs, and primitive model types, the declarative model type uses automata scripts and tracks, task checklists, logic programs, Petri nets, and space time diagrams. Still according to [62], the function model type employs block models, digital circuits, queuing networks, compartmental models, and system dynamics. The remaining model types use equations, analogue circuits, and bond graphs for the constraint model type and cellular automata and finite element modelling for the spatial model types. With the semantic algebra specification technique used herein this chapter, the classification as declarative model type applies for the simulation model of this chapter.

3.1.2 Nomenclature

This section introduces the Nomenclature for the conceptual design. It focuses on employing uniqueness and simplicity but it turns out that this is, unfortunately, not always possible. If not explicitly stated otherwise explicitly, the following rules apply.

Terminal typeset text represents symbols, i.e. single words of constant strings delimited by spaces, such as `car` and `truck` for the particular vehicle type names or `lc-safe?` for a more advanced symbol. Most of the time, type names make use of the symbols to associate a name with one or more values. The pair $(\text{car}, 150\,\text{km/h})$ as a combination of a vehicle type name 'car' and a velocity value $150\,\text{km/h}$ may serve as an example. This chapter, however, does not use string types except for the scope of symbols that represent type names. The vehicles, the cell rules, the detectors, and the tuning elements make use of the type names.

The symbols $\lfloor a \rfloor$, $\lceil a \rceil$, and $\lceil a \rfloor$ denote the value of a floored, ceiled, and rounded to the nearest integers. The term $\mathfrak{C} := \{b \in \mathfrak{B} \mid b \leqslant c\}$ reads as "$\mathfrak{C}$ is defined as the set of all b out of $\mathfrak{B}$ having $b \leqslant c$". The symbol ':=' introduces a definition, whereas the equal sign '=' denotes equality between the left hand and the right hand side. In general, the old german symbols $\mathfrak{B}$, $\mathfrak{C}$, ... name sets. However, certain sets have fixed names as they are already known in the mathematical context. Examples are the natural numbers $\mathbb{N}$ and the real numbers $\mathbb{R}$ as well as their variations such as $\mathbb{N}_0^{100}$ which denotes the subset of the natural numbers including zero with the maximum number 100. As the set definitions of this chapter sometimes overlap with the wide-spread known definitions, sometimes ambiguities occur. An example is the set of relative direction specifications $\mathbb{B}_{\text{cell}}^{\pm} := \{-1, 0, 1\}$. As the set in this form has no dimension, it reminds one on a subset of the integer numbers. The notation without the old german letters is, thus, preferred.

The italic typeset variable t denotes the time in seconds, i.e. $t \in \mathfrak{T}_{\text{secs}} := \{j\,\text{s} \mid j \in \mathbb{N}_0\}$. Ambiguities can occur whenever a variable $t_{\text{name}}^{\text{vehtypes}}$ refers to an element of a set of type names in order to denote a specific type name. This is the case, for example, in $t_{\text{name}}^{\text{vehtypes}} \in \mathfrak{M}_{\text{vehtype}} := \{\text{car}, \text{truck}\}$. Instead of the italic t a gothic typeset $\mathfrak{t}$ as in $\mathfrak{t}_{\text{name}}^{\text{vehtype}} \in \mathfrak{M}_{\text{vehtype}}$ will indicate the type name and its correspondence to the $\mathfrak{M}_{\text{vehtype}}$ set of type names. In general, the letters i, j, k, l, and n denote simple integers or indices, i.e. integer elements that refer to a certain position of an ordered sequence. A few exceptions to this rule occur mainly in this section where i sometimes refers to the complex number $i^2 = -1$. However, apart from this section, the complex numbers do not play a role in this chapter. The ambiguity is, thus, limited to the scope of this section. In general, functions are named with the capital letter $\mathbf{F}$ and a subscripted detailed definition such as $\mathbf{F}_{\text{rule}}$. The capital letter $\mathbf{F}$ is also typeset bold-face to enable a better distinguishing from the remaining symbols. This scheme is, however, not always applicable as some functions require to cite them literally as in their original publication. In any case the context will make clear whenever a function is defined. Similar to operator overloading in programming languages, distinct but equally named functions can operate on different domain kinds and/or ranges. To identify functions uniquely, the domain kinds and ranges in addition to its name are also required.

Function-like notations such as $v_n^{(t)}$ indicate a slight functional dependency of variables, which is in this case the velocity value v that depends on the time t. The superscripted functional dependency is intended to remind the reader that this variable may also be understood as a function of the super-scripted variable instead of the more widespread understanding as a pure variable. In the former case, the corresponding function selects a value from a series of values by the value of the superscripted argument. In the latter case, the corresponding variable denotes a name reference to a value that may change at any time. An example where the variable does not depend on the time but instead on the superscripted number value are the remainders and the quotients of the integer numbers such as in $c := \alpha^{(c)} D + \beta^{(c)}$. Here, the quotient $\alpha^{(c)}$ is not only strictly a variable, but has a unique functional dependency to the number c given by the Euclidean division theorem. Thus, a function-style notation may be more appropriate over the strict variable notation.

Terms such as (x, y) and (x, y, z) denote pairs and tuples. For these kind of values, the brackets in combination with separating the values by comma serve as a constructor syntax. To define a set of pairs that represent the set of the complex numbers as a domain kind, a definition such as $P_{\mathbb{C}} := \{(x, y) \mid \mathbf{z} = x + iy \wedge i^2 = -1 \wedge x, y \in \mathbb{R} \wedge \mathbf{z} \in \mathbb{C}\}$ could be used. Each element $p_k = (x_k, y_k)$ of the set of pairs $P_{\mathbb{C}}$ represents then also a complex number by the constraint $\mathbf{z} = x + iy$ that was provided in the definition. Concluding from the syntax only, is it not possible to distinguish a pair (x_k, y_k) that represents an element of the pairs $P_{\mathbb{C}}$ from another pair (x_n, y_n) that represents an element of the real number plane $\mathbb{R}^2 = \mathbb{R} \times \mathbb{R}$. As a consequence, there is no syntactic association between the elements of a set and the corresponding set definition. A special syntax to access the coordinates of a pair that, additionally, can be derived from the set definition implicitly is also not provided by a pair definition. Therefore, this chapter provides and makes use of a special variant of the pairs and tuples, namely the structures. They represent tuples with a named form and specifically defined domain kinds and ranges. They also introduce coordinate access through implicitly defined access functions. As an example, the definition for a set $S_{\mathbb{C}}$ of structure elements $\langle x, y \rangle^{\mathbb{C}}$ that represents the set of the complex numbers $\mathbb{C}$ by pairs will look like $S_{\mathbb{C}} := \{\langle x, y \rangle^{\mathbb{C}} \mid \mathbf{z} = x + iy \wedge i^2 = -1 \wedge x, y \in \mathbb{R}\}$. It consists of structure elements that represent complex numbers through the constraint $\mathbf{z}_= x + iy$. The capital letter S_{example} thereby is reserved for the sets of structure elements. Concluding from the syntax, a structure variable such as $s_{\mathbb{C}}$ always refers to an element of the corresponding set of structures $S_{\mathbb{C}}$. Additionally, if taken strictly, each structure definition would require set braces around the definition as in $S_{\text{example}} := \{\langle x, y \rangle^{\text{example}} \mid \ldots\}$. However, as the capital letter already makes clear that a definition of a set of structure elements will follow, the braces are unnecessary. Similarly follows from the context of the participating coordinates in a structure definition. Their context frequently makes their domain kind clear. To simplify structure definitions, the set braces and the definition of the domain kinds in the structure definition can be omitted whenever the coordinate types allow to derive that information from their corresponding type information and the context makes it clear that it is a structure definition. Additionally, variables that refer to structures as well as to vectors or tuples use the bold typeface. Thus, the structure variable $s_{\mathbb{C}}$ of the above example is instead correctly typeset as $\mathbf{s}_{\mathbb{C}}$. Structure names in the definition are optional whenever they are only used in the introduction.

Each set of structure elements definition also implicitly defines access functions–called accessor from now on. The $\#$-symbol followed by a coordinate name that matches the one from the structure definition denotes the coordinate accessor. As an example, the term $\mathbf{s}_{\mathbb{C}}\#x$ refers to the x-coordinate of the pair while $\mathbf{s}_{\mathbb{C}}\#y$ refers to the y-coordinate of the structure element $\mathbf{s}_{\mathbb{C}} \in S_{\mathbb{C}}$. The coordinate names of a structure definition are not global definitions. Their scope is limited to the structure only. For example, after the structure definition for the complex numbers from above $S_{\mathbb{C}}$, the variable x remains undefined. The variable x is only visible in the scope of a $S_{\mathbb{C}}$ structure as in $\mathbf{s}_{\mathbb{C}}\#x$. Accessing variables in nested structure elements requires, however, to use parentheses as in $(\mathbf{a}\#\mathbf{b})\#c$. Assigning values to structure elements or their coordinates in the sense of assigning a value to a variable is not in the scope of this chapter. Only instantiation of individual structure elements is used herein. As functions sometimes would like to construct structure elements with a lot of fields, i.e. coordinates, by only changing some field values from any already given original, the structures also define implicitly a clone function $\mathbf{F}_{\text{clone}}$ for each of the accessors. Each clone function takes three arguments and is defined as returning a fresh instance of the structure that was specified as the first argument and by replacing the value of the field named as the second argument with the value given as the third argument. In the case of the complex number example from above the clone function could be defined similar to $\mathbf{F}_{\text{clone}} : S_{\mathbb{C}} \times \mathbb{R} \times \mathbb{R} \longrightarrow S_{\mathbb{C}}$ with $\mathbf{F}_{\text{clone}}(\mathbf{s}, x, x^{'}) := \langle x^{'}, y \rangle^{\mathbb{C}}$ and $\mathbf{s} = \langle x, y \rangle^{\mathbb{C}}$. The second argument to the function is, however, not a number type but instead an expression that identifies the component of the structure. The signature of $\mathbf{F}_{\text{clone}}$ from above is, thus, somewhat misleading.

In each section, a summary of the relevant functions, structure definitions, variables, and constants is given. In general, it does not contain the names of variables, constants, definitions, and so on with only locally limited scope. It is intended as a quick lookup of definitions used by other sections. To ease the lookup of set definitions, most sets have subscript names that refer to the section where they are defined in. As an example, the set $\mathbb{B}^4_{\text{cell}}$ is defined in Section 3.3.1. Additionally, each section summarises the parameters in a separate table.

3.2 Vehicle model

Vehicles are the basic elements of a microscopic traffic simulation. In the original cellular automaton models such as the Nagel-Schreckenberg-Model [147] vehicles do not necessarily require explicit instantiation. However, a space discretisation where the cell size is smaller than the vehicle length requires explicit instantiation of the vehicles. Vehicles and vehicle types impact the results of a microscopic traffic model more than one might expect at first. As an example, introducing faster and slower vehicle types may lead to unrealistically long enduring overtaking manoeuvres that form a plug on a highway and prevent faster vehicles from driving at their possible maximum velocity [115]. The various vehicle

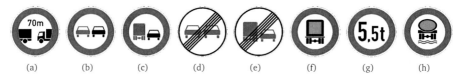

Figure 3.1: Traffic signs that discriminate vehicle types (traffic signs from source [194]).

types thus need to be chosen carefully. The vehicles themselves may introduce storage, maintainance, access, and dependency issues that result in poor model design. Detector, tuning elements, and road traffic network rules depend on vehicle types.

3.2.1 Vehicle types

Vehicle types occur in several varying contexts. Some of them are described in the following. As a first context, German traffic regulations discriminate vehicle types in speed limits, lane change possibilities, and overtaking restrictions. The discrimination depends on several vehicle aspects such as maximum weight, potentially dangerous cargo, number of axles, and passenger to cargo ratio. Figure 3.1 shows some of the vehicle type discriminating signs. The traffic signs 3.1 (a), (c), and (e-h) affect trucks while 3.1 (b) and (d) affect all vehicles regardless of their vehicle type. German traffic regulations discriminate another vehicle type class, namely the so-called privileged vehicles that consists of emergency services such as ambulance, fire engine, or police vehicles. The regulations grant privileged vehicles to pass and overtake non-privileged vehicles in between of two single lanes. Determining the fastest traffic routes for emergency services requires to take this special behaviour of the privileged vehicles into account [131].

As a second context, the inductive loop detectors distinguish between vehicle types when measuring traffic data. As the measurement method uses an electro-magnetic induction technique, the detectors distinguish the vehicle types depending on the axle distances [201, Apps. 2, 6, Secs. 6-3.2.8]. The loop detector control modules use a vehicle's measured axle distance to classify vehicles into the vehicle type classes cars and trucks. The classification depends on a configurable axle distance threshold value [201, Tab. 6-64]. While the loop detectors can distinguish up to 9 vehicle type classes, the incoming loop detector data for OLSIMv4 is split into cars and trucks only. The classifications in [201, App. 2] try to associate the measured axle distances to the corresponding car and truck vehicle types of the German traffic regulations. The association is a best practice method and by no means a bijective mapping. It contains a lot of classifications that correspond to those of the German traffic regulations but includes also non-corresponding associations as well as inmappable axle distances [201, App. 2].

As a third context, cars and trucks in reality behave different in traffic flow on German highways. The first classification of the German traffic regulations does not distinguish between slow and fast driving vehicles. Neither does the second one, the inductive loop detector classification. For instance, busses are allowed to drive with a maximum velocity of 100 km/h . In many cases, their real velocity is about 120 km/h which makes them difficult to distinguish from cars. However, they have almost the length of a truck and, consequently, in dense traffic they assimilate more to the behaviour of the trucks. Similar applies to vans. Loop detectors may have difficulties in classifying them as cars due to their relatively long axle distances. However, their behaviour in free to dense traffic flow is more like that of a fast car. In congested traffic their impact on the jam length is more similar to that of a truck. As a consequence, loop detector data frequently show relatively low traffic flow values for trucks with extremely high velocity values of about 140 km/h and even above.

Lastly, the user view on the classification of cars and trucks forms another context. The general user of the online information platform has its own classification of cars and trucks. A car driver might not want to distinguish between vans and trucks, and a freight forwarder might not classify vans as trucks as they can generally drive faster than trucks. This subjective classification might cause annoyance to some internet users as they are most probably not aware of the aforementioned subtle distinctions between the various vehicle type classes. Specifically, the published travel time information that distinguishes between the vehicle type classes 'all vehicle types' and 'trucks' might suggest to a car driver that the travel time of the former and unspecific class represents a travel time for passenger cars only which, consequently, might appear inappropriate or too pessimistic. However, the travel time for faster vehicles such as cars depends on the amount of disturbance by the slower vehicles but not the other way round. Thus, it makes sense to publish an independent travel time for slower vehicles in addition to the unspecific travel time.

All discussed contexts share and involve the common concept of vehicle types. A microscopic traffic simulation thus has to deal with different kinds of vehicle types. Despite all differences in the classifications of the cars and the trucks in the contexts of the German highway regulations and of the inductive loop detectors, OLSIMv4 uses the same classes for them and treats the two classifications as equivalent to each other. This is, of course, a rough simplification that, consequently, leads to several non-trivial difficulties such as the one described above and the one of poor and untrustworthy traffic data. However, the introduced concept or classification of vehicle types differs strongly from vehicle types such as fabricates [30]. From the user's perspective and for the simulation, the properties number and distance of axles, maximum weight, and potentially dangerous cargo are far too technical and do not provide any useful information.

Thus, the set of properties that characterises a vehicle type has to contain other properties than the ones provided by the various contexts. OLSIMv3 used the property pair that consists of the vehicle type and the vehicle type's maximum length. Equation 3.1 defines this structure S_{vehtype}^3 where the superscripted number 3 indicates that this structure is part of OLSIMv3. The OLSIMv3 trucks used the property pairs $s_{\text{vehtype}}^{\text{truck}} = \langle \texttt{truck}, v_i \text{ cells/s} \rangle^{\text{vehtype}}$ with $v_i \in \{14, 15, 16\}$ while the cars used the properties $s_{\text{vehtype}}^{\text{car}} = \langle \texttt{car}, v_i \text{ cells/s} \rangle^{\text{vehtype}}$ with $v_i \in \mathfrak{V}_{\text{vehtype}}^3 = \{20, 21, 22, 23, 24\}$. As these vehicle types do only cover the vehicle velocities, the lengths do not depend on the vehicle type. Under the aspect of unique vehicle lengths, this approach is very close to original cellular automata models wherein the cell's state indicates whether a vehicle occupies it or not. Due to the fact that in the original cellular automaton models the cell size corresponded to the vehicle size, the idea of having one vehicle length for all vehicle types is similar to the cellular automaton approach. However, the vehicle length for the trucks is unrealistic, results in too frequent and too short overtaking manoeuvres and leads to too high densities in congested traffic.

$$S_{\text{vehtype}}^3 := \langle t_{\text{name}}^{\text{vehtype}}, V_{\max} \rangle^{\text{vehtype}} \text{, where } t_{\text{name}}^{\text{vehtype}} \in \mathfrak{L}_{\text{vehtype}}^3 = \{\texttt{car}, \texttt{truck}\} \,. \tag{3.1}$$

The hyperrealistic approach where several varying vehicle types participate in the simulation hides some risks. In reality, truck overtaking manoeuvres endure for a long time, but they always end some time. A simulation is most likely not smart enough to cope with these situations adequately. Whenever a truck starts to overtake another truck, it will look forward at most for a few time steps. However, the short look ahead time horizon may result in an never ending overtaking manoeuvre. This also applies to the brakelight model [118, 115] and potentially other cellular automaton models as the authors of the brakelight model stated: "In order to implement an artificial asymmetry we introduced disorder by considering different types of vehicles, such as cars and trucks. Unfortunately, the introduction of disorder in cellular automaton models for two-lane traffic has some shortcomings: it is possible that two slow vehicles driving side by side on different lanes can form a plug which blocks the succeeding traffic" [115]. Especially in the context of the NRW highways that have an extent of up to 320 km such a scenario could invalidate the whole simulation results.

Due to the explained reasons from above, OLSIMv4 uses the classification of vehicle types into cars and trucks only but forbids trucks to overtake on the left on two lane roadways. Equation 3.2 displays the classification into cars and trucks as well as the properties of the OLSIMv4 vehicle type structure. The additional vehicle type class any has been adopted from the loop detectors that provide the data by the use of this class. The OLSIMv4 vehicle type properties extend those of OLSIMv3 by the maximum length $L_{\max}$ and a tolerance factor $V_{\text{mul}} \in \mathfrak{P}_{\text{vehtype}}^4 = \{i/100 \mid i \in \mathbb{N} \wedge i \leq 199\}$ for the strictness of how to obey to the topological speed limit constraints. The range for a vehicle type's maximum velocity $V_{\max}$

Table 3.2.1: Summary of relevant functions, structures, variables, and constants for Section 3.2.1.

definition	description	domain	range
$\mathfrak{L}^4_{\text{vehtype}}$	vehicle type names		$\{\texttt{any, car, truck}\}$
$\mathfrak{X}^4_{\text{vehtype}}$	maximum lengths		$\{i\,\text{cells} \mid i \in \mathbb{N}^{15}\}$
$\mathfrak{V}^4_{\text{vehtype}}$	maximum velocities		$\{i\,\text{cells/s} \mid i \in \mathbb{N}^{40}_0\}$
$\mathfrak{P}^4_{\text{vehtype}}$	speed limit tolerances		$\{i/100 \mid i \in \mathbb{N}^{199}\}$
$\langle a,b,c,d\rangle^{\text{vehtype}}$	vehicle type structure	$\mathfrak{L}^4_{\text{vehtype}} \times \mathfrak{V}^4_{\text{vehtype}} \times \mathfrak{X}^4_{\text{rway}} \times \mathfrak{P}^4_{\text{rway}}$	S^4_{vehtype}
$\text{s}_{\text{vehtype}}\#\text{t}^{\text{vehtype}}_{\text{name}}$	vehicle type name	$\text{s}_{\text{vehtype}} \in S^4_{\text{vehtype}}$	$\mathfrak{L}^4_{\text{vehtype}}$
$\text{s}_{\text{vehtype}}\#V_{\max}$	maximum velocity	$\text{s}_{\text{vehtype}} \in S^4_{\text{vehtype}}$	$\mathfrak{V}^4_{\text{vehtype}}$
$\text{s}_{\text{vehtype}}\#V_{\text{mul}}$	speed limit tolerance	$\text{s}_{\text{vehtype}} \in S^4_{\text{vehtype}}$	$\mathfrak{P}^4_{\text{vehtype}}$
$\text{s}_{\text{vehtype}}\#L_{\max}$	maximum length	$\text{s}_{\text{vehtype}} \in S^4_{\text{vehtype}}$	$\mathfrak{X}^4_{\text{vehtype}}$

matches the set $\mathfrak{V}^4_{\text{vehtype}} = \{i\,\text{cells/s} \mid i \in \mathbb{N}_0 \wedge i \leqslant 40\}$. However, the only defined two vehicle types $\text{s}^{\text{car}}_{\text{vehtype}}$ and $\text{s}^{\text{truck}}_{\text{vehtype}}$ make only use of the subset $\{22\,\text{cells/s}, 18\,\text{cells/s}\}$. Lastly, the maximum length $L_{\max}$ is an element of the set $\{5\,\text{cells}, 10\,\text{cells}\} \subset \mathfrak{X}^4_{\text{vehtype}} = \{i\,\text{cells} \mid i \leqslant 15\}$.

$$S^4_{\text{vehtype}} := \langle \text{t}^{\text{vehtype}}_{\text{name}}, V_{\max}, V_{\text{mul}}, L_{\max}\rangle^{\text{vehtype}}, \text{ where } \text{t}^{\text{vehtype}}_{\text{name}} \in \mathfrak{L}^4_{\text{vehtype}} = \{\texttt{any, car, truck}\}. \quad (3.2)$$

The OLSIMv4 vehicle types use the property combinations $\text{s}^{\text{car}}_{\text{vehtype}} = \langle \texttt{car}, 22\,\text{cells/s}, 1.15, 5\,\text{cells}\rangle^{\text{vehtype}}$ for the cars and $\text{s}^{\text{truck}}_{\text{vehtype}} = \langle \texttt{truck}, 18\,\text{cells/s}, 1.15, 10\,\text{cells}\rangle^{\text{vehtype}}$ for the trucks. The vehicle type structure provides the additional constant $\text{s}^{\text{any}}_{\text{vehtype}} = \langle \texttt{any}, 22\,\text{cells/s}, 1.15, 10\,\text{cells}\rangle^{\text{vehtype}}$.

The microscopic traffic model, the detector models, the tuning element models, and the road traffic network model depend on the classification into the various vehicle types. The microscopic traffic model applies a different maximum velocity depending on the vehicle type. The detector models distinguish vehicles based on a vehicle's type. The tuning element models add or remove a vehicle depending on the vehicle type that the detector data provided. The road traffic network data model distinguishes the topological constraints based on the vehicle type. And, finally, the detectors present traffic information to the user based on the vehicle type classification, too. Table 3.2.1 summarises the relevant functions, structure definitions, variables, and constants of this section.

3.2.2 Vehicles

OLSIMv3 used tracks as the basic road-segment structure. Roadways and their smaller units tracks can be viewed as a rectangular grid $\mathfrak{Y}_{\text{rway}} \times \mathfrak{X}_{\text{rway}}$ that consist of an ordered sequence $Y_{\min} \ldots Y_{\max} \wedge Y_{\min}, Y_{\max} \in \mathbb{Z}$ of lane numbers $\mathfrak{Y}_{\text{rway}} = \{i \in \mathbb{Z} \mid Y_{\min} \leqslant i \leqslant Y_{\max}\}$ where each lane with number $y \in \mathfrak{Y}_{\text{rway}}$ consists of an equally sized ordered sequence $X_{\min} \ldots X_{\max} \wedge X_{\min}, X_{\max} \in \mathbb{Z}$ of cell numbers $\mathfrak{X}_{\text{rway}} = \{i \in \mathbb{Z} \mid X_{\min} \leqslant i \leqslant X_{\max}\}$. Consequently, vehicles in OLSIMv3 that move on tracks represent an ordered sequence as well and they have a lane-wise index that in each time step uniquely identifies their relative position in the sequence of vehicles. In addition to the lane-wise index n, a vehicle in OLSIMv3 consisted of a constant vehicle type $\text{s}_{\text{vehtype}} \in S^3_{\text{vehtype}}$, a constant vehicle length $L \in \{5\,\text{cells}\}$, its current lane number $y^{(t)}_n \in \mathfrak{Y}_{\text{rway}}$, its current position $x^{(t)}_n \in \{i\,\text{cells} \mid i \in \mathfrak{X}_{\text{rway}}\}$ on that lane, its state of the brakelight $b^{(t)}_n \in \mathbb{B}^3_{\text{veh}} = \{0, 1\}$, and its longitudinal velocity $v^{(t)}_n \in \mathfrak{V}_{\text{vehtype},n} = \{i\,\text{cells/s} \mid i \in \mathbb{N}_0 \wedge i \leqslant \text{s}_{\text{vehtype}}\#V_{\max}\}$. Equation 3.3 displays the OLSIMv3 vehicle structure that despite all exactness still simplifies some aspects of the OLSIMv3 implementation.

$$S^3_{\text{veh}} := \langle \text{s}_{\text{vehtype}}, L, y^{(t)}_n, x^{(t)}_n, v^{(t)}_n, v^{(t-\Delta t)}_n, b^{(t)}_n, b^{(t-\Delta t)}_n\rangle^{\text{veh}}. \quad (3.3)$$

The S^3_{veh} vehicle structure provided the velocity and the state of the brakelight variable for the current time step t as well as for the previous time step $t - \Delta t$. OLSIMv4 generalised this concept of storing

Table 3.2.2: Summary of relevant functions, structures, variables, and constants for Section 3.2.2.

definition	description	domain	range
$\mathfrak{Y}_{\text{veh}}^4 = \mathfrak{Y}_{\text{rway}}^4$	lanes, lane numbers	$Y_{\min}, Y_{\max} \in \mathbb{Z}$	$\{i \in \mathbb{Z} \mid Y_{\min} \leqslant i \leqslant Y_{\max}\}$
$\mathfrak{X}_{\text{rway}}^4$	cell numbers	$X_{\min}, X_{\max} \in \mathbb{Z}$	$\{i \in \mathbb{Z} \mid X_{\min} \leqslant i \leqslant X_{\max}\}$
$\mathfrak{X}_{\text{veh}}^4$	cell position	$X_{\min}, X_{\max} \in \mathbb{Z}$	$\{i \text{ cells} \mid i \in \mathfrak{X}_{\text{rway}}\}$
$\mathfrak{Q}_{\text{coord}}^4$	set of trajectories		$(S_{\text{coord}})^{j+1} \subset \mathcal{P}_{\text{veh}}^4(S_{\text{coord}})$
$\langle l, p, v, b\rangle^{\text{coord}}$	coordinates structure	$\mathfrak{Y}_{\text{veh}} \times \mathfrak{X}_{\text{veh}} \times \mathfrak{V}_{\text{vehtype}} \times \mathbb{B}$	S_{coord}^4, see Equation 3.4
$\mathsf{s}_{\text{coord}}\#y$	vehicle's lane	$\mathsf{s}_{\text{coord}} \in S_{\text{coord}}^4$	$\mathfrak{Y}_{\text{veh}}^4$
$\mathsf{s}_{\text{coord}}\#x$	vehicle's position	$\mathsf{s}_{\text{coord}} \in S_{\text{coord}}^4$	$\mathfrak{X}_{\text{veh}}^4$
$\mathsf{s}_{\text{coord}}\#v$	vehicle's velocity	$\mathsf{s}_{\text{coord}} \in S_{\text{coord}}^4$	$\mathfrak{V}_{\text{vehtype},n}^4$
$\mathsf{s}_{\text{coord}}\#b$	vehicle's brakelight	$\mathsf{s}_{\text{coord}} \in S_{\text{coord}}^4$	$\mathbb{B}_{\text{veh}}^4$
$\langle a, \dots, q_0, \dots, q_j\rangle^{\text{veh}}$	vehicle structure	$S_{\text{vehtype}}^4 \times \cdots \times (S_{\text{coord}}^4)^{j+1}$	S_{veh}^4, see Equation 3.5
$\mathbf{k}_n^{(t)}\#\mathsf{s}_{\text{vehtype}}$	vehicle type	$\mathbf{k}_n^{(t)} \in S_{\text{veh}}^4$	S_{vehtype}^4
$\mathbf{k}_n^{(t)}\#L$	length	$\mathbf{k}_n^{(t)} \in S_{\text{veh}}^4$	$\mathfrak{X}_{\text{veh}}^4$
$\mathbf{k}_n^{(t)}\#V_{\max}$	maximum velocity	$\mathbf{k}_n^{(t)} \in S_{\text{veh}}^4$	$\mathfrak{V}_{\text{vehtype}}^4$
$\mathbf{k}_n^{(t)}\#A_{\max}$	maximum acceleration	$\mathbf{k}_n^{(t)} \in S_{\text{veh}}^4$	$\{1 \text{ cells/s}^2\}$
$\mathbf{k}_n^{(t)}\#D_{\max}$	maximum deceleration	$\mathbf{k}_n^{(t)} \in S_{\text{veh}}^4$	$\{2 \text{ cells/s}^2\}$
$\mathbf{k}_n^{(t)}\#V_{\text{mul}}$	speed limit tolerance	$\mathbf{k}_n^{(t)} \in S_{\text{veh}}^4$	$\mathfrak{P}_{\text{vehtype}}^4$
$\mathbf{k}_n^{(t)}\#\mathbf{q}^{(t)}, \mathbf{k}_n^{(t)}\#\mathbf{q}_0$	current coordinates	$\mathbf{k}_n^{(t)} \in S_{\text{veh}}^4$	S_{coord}^4
$\mathbf{k}_n^{(t)}\#\mathbf{q}^{(t-i\Delta t)}, \mathbf{k}_n^{(t)}\#\mathbf{q}_i$	trajectory coordinates	$\mathbf{k}_n^{(t)} \in S_{\text{veh}}^4 \wedge i \in \{1, \dots, j\}$	S_{coord}^4
$\mathbf{k}_n^{(t)}\#\mathbf{q}$	complete trajectory	$\mathbf{k}_n^{(t)} \in S_{\text{veh}}^4$	$\mathfrak{Q}_{\text{coord}}^4$

information from previous time steps into the storing of trajectories. Instead of storing the plain velocity and position for several time steps, OLSIMv4 uses a sequence of elements of the S_{coord}^4 coordinate structure that Equation 3.4 defines. The structure consists of the coordinates lane number $y \in \mathfrak{Y}_{\text{veh}}^4 = \mathfrak{Y}_{\text{rway}}$, position $x \in \mathfrak{X}_{\text{veh}}^4 = \{i \text{ cells} \mid i \in \mathfrak{X}_{\text{rway}}\}$, velocity $v \in \mathfrak{V}_{\text{veh},n} = \{i \text{ cells/s} \in \mathbb{N}_0 \mid i \leqslant \mathsf{s}_{\text{vehtype}}\#V_{\max}\}$ where $\mathsf{s}_{\text{vehtype}} \in S_{\text{vehtype}}^4$ denotes the vehicle's type, and state of the brakelight $b \in \mathbb{B}_{\text{veh}}^3 = \{0, 1\}$.

$$S_{\text{coord}}^4 := \langle y, x, v, b\rangle^{\text{coord}} . \tag{3.4}$$

With the help of the coordinate structure S_{coord}^4, Equation 3.5 specifies the OLSIMv4 vehicle structure S_{veh}^4 as described in the following. In a vehicle structure element $\mathbf{k}_n^{(t)} \in S_{\text{veh}}^4$, a vehicle stores its current coordinate $\mathbf{q}^{(t)} \in S_{\text{coord}}^4$ as well as a trajectory of the last j previous coordinates $\mathbf{q}^{(t-\Delta t)}, \dots, \mathbf{q}^{(t-j\Delta t)} \in S_{\text{coord}}^4$. As a special case, the trajectory of the last $j + 1$ previous coordinates including the current coordinate $\mathbf{q}^{(t)}, \mathbf{q}^{(t-\Delta t)}, \dots, \mathbf{q}^{(t-j\Delta t)} \in S_{\text{coord}}^4$ provides the abbreviated accessors $\mathbf{q}_0 := \mathbf{q}^{(t)}, \mathbf{q}_1 := \mathbf{q}^{(t-\Delta t)}, \dots, \mathbf{q}_j := \mathbf{q}^{(t-j\Delta t)}$. In addition to the constant vehicle parameters vehicle type $\mathsf{s}_{\text{vehtype}} \in S_{\text{vehtype}}^4$, length $L = \mathsf{s}_{\text{vehtype}}\#L_{\max}$, maximum velocity $V_{\max} \in \mathfrak{V}_{\text{vehtype}} = \{i \text{ cells/s} \mid i \in \mathbb{N}_0 \wedge i \leqslant \mathsf{s}_{\text{vehtype}}\#V_{\max}\}$, maximum acceleration $A_{\max} \in \{1 \text{ cells/s}^2\}$ capability, and maximum deceleration $D_{\max} \in \{2 \text{ cells/s}^2\}$ capability, it uses the tolerance factor $V_{\text{mul}} \in \mathfrak{P}_{\text{vehtype}}^4$ in percent that describes how strict a vehicle will obey to the topological speed limit constraints.

$$S_{\text{veh}}^4 := \langle \mathsf{s}_{\text{vehtype}}, L, V_{\max}, A_{\max}, D_{\max}, V_{\text{mul}}, \mathbf{q}^{(t)}, \mathbf{q}^{(t-\Delta t)}, \dots, \mathbf{q}^{(t-j\Delta t)}\rangle^{\text{veh}} . \tag{3.5}$$

The trajectories are useful not only for debugging purposes but additionally valuable for the measurements by the local detectors. The detectors measure some kind of traffic observables and thereby assume that each vehicle passes the detector continuously in one or more time steps, i.e. vehicles do not switch from or to another lane during the passing of the detector. With the help of the vehicle's position, its velocity, and the lane change velocity $\Delta y_n^{(t)} := 1/\Delta t \cdot ((\mathbf{k}_n^{(t)}\#\mathbf{q}^{(t)})\#y - (\mathbf{k}_n^{(t)}\#\mathbf{q}^{(t-\Delta t)})\#y)$, they can determine whether a vehicle $\mathbf{k}_n^{(t)} \in S_{\text{veh}}^4$ entered the lane and the detector in the last time step.

The accessor $\mathbf{k}_n^{(t)}\#\mathbf{q}$ provides the whole trajectory as a set $\mathfrak{Q}_{\text{coord}}^4 := \{\mathbf{q}^{(t)}, \mathbf{q}^{(t-\Delta t)}, \ldots, \mathbf{q}^{(t-j\Delta t)}\} \in (S_{\text{coord}})^{j+1} \subset \mathcal{P}_{\text{veh}}^4(S_{\text{coord}})$ with all $j+1$ coordinate structure elements and $\mathcal{P}_{\text{veh}}^4(S_{\text{coord}})$ as the power set over the coordinates.

Equation 3.6 defines a set of rather minimal vehicle structures S_{minveh}^4. In the strict sense, it is not a subset of the vehicle structure definition from above as it does not provide the maximum velocity V_{max}, maximum acceleration A_{max}, and maximum deceleration D_{max} vehicle parameters. However, it is quite similar to the one from above and has been used in [30]. It does not store coordinate trajectories. But it is sufficiently large enough to produce accurate results even though the local detectors may produce artefacts in scenarios with a lot of lane changes at low longitudinal velocities, i.e. velocities where vehicles take more than one time step to pass a detector. This combination is relatively seldom, as the most lane changes occur in free flow at relatively high velocities. Therefore the impact of such artefacts is small and, additionally, is also contained in the measured loop detector data.

$$S_{\text{minveh}}^4 := \langle \mathbf{s}_{\text{vehtype}}, L, V_{\text{mul}}, \mathbf{q}^{(t)} \rangle^{\text{minveh}} . \tag{3.6}$$

The OLSIMv4 vehicle structure individualised the parameters L, V_{max}, A_{max}, and D_{max} for each vehicle from the start on to have a flexible approach but that turned out to introduce non-trivial problems such as collisions and plugs of vehicles that drive side by side in an overtaking manoeuvre. For the cellular automaton based microscopic traffic simulations, the individualisation is, thus, superfluous. The per vehicle type constant parameterisation as introduced above delivers reliable results. Table 3.2.2 summarises the relevant functions, structure definitions, variables, and constants for this section.

3.3 Network model

OLSIMv3 used tracks as basic and atomic road segments. The OLSIMv3 tracks are rectangular areas with constant number of lanes and constant topologic constraints, namely maximum velocity and lane change restrictions. Theoretically, tracks reach from the centre of an arbitrary intersection to the centre of the next downstream intersection. The OLSIMv3 tracks did not support periodic boundary conditions. Section 2.1.4.2 has already discussed the OLSIMv3 tracks and its shortcomings. As it is unknown in OLSIMv3, where exactly the centre of the intersection is located, the mapping of topologic and traffic information between OLSIMv3 tracks and road-segments of other digital maps is very difficult, inexact, and error-prone. As highway intersections can extent up to 2 km and can include an arc-radius, it is debatable where exactly the centre of such an intersection should be. Even though connecting the road segments longitudinally is also common to many other network models, e.g. the SUMO [119] and the CMDAV digital map [191], it causes the most notable shortcoming of the track network model—the lack of a suitable parallel interpretation. Thus, OLSIMv4 eliminates the shortcomings of the longitudinally connected tracks by choosing the roadways with their full spatial extent as the atomic network elements. Simulating on the roadways as atomic units and thereby avoiding longitudinal and lateral connections results in a simulation model for which a significantly more parallel interpretation exists. Section 3.7 discusses this in more detail.

Starting with the presentation of the cells, Section 3.3.1 unveils the smallest units of this section's network model. They combine a bundle of topological rules into an atomic unit. As the next topological elements, the sections follow in Section 3.3.2. They describe an area on a single-lane where the topological properties do not vary. Consequently, they consist of a prototype cell and a length specifier. Section 3.3.3 then describes the roadway structure. It combines a specification of a roadway's spatial extent with a list of all the sections that are located on the roadway. Thereafter, Section 3.3.4 explains the details of the topological analysis that relies on the roadways. Section 3.3.5 introduces the occupation tables that combine a roadway with the vehicles that move on it. It also presents the filter method provided by the occupation tables that determines a vehicle's neighbouring vehicles. This section concludes with the specification of the switching areas that define temporary connections between roadways.

Table 3.3.1: Summary of relevant functions, structures, variables, and constants for Section 3.3.1.

definition	description	domain	range
$\mathfrak{L}_{\text{rule}}^4$	rule type names		$\{\texttt{maxvel}, \texttt{chlane}, \texttt{ovtake}\}$
$\mathbb{B}_{\text{cell}}^{\pm}$	direction specifications		$\{-1, 0, 1\}$
$\mathfrak{P}_{\text{cell}}^4$	lane change priorities		$\{1, 2, 3, 4\}$
$\langle r, t, y, v \rangle^{\text{rule}}$	rule structure $\texttt{maxvel}$	$\mathfrak{L}_{\text{rule}}^4 \times \mathfrak{L}_{\text{vehtype}}^4 \times \mathbb{B}_{\text{cell}}^{\pm} \times \mathfrak{V}_{\text{vehtype}}^4$	S_{rule}^4, see Equation 3.7
$\langle r, t, y, c \rangle^{\text{rule}}$	rule structure $\texttt{chlane}$	$\mathfrak{L}_{\text{rule}}^4 \times \mathfrak{L}_{\text{vehtype}}^4 \times \mathbb{B}_{\text{cell}}^{\pm} \times \mathfrak{P}_{\text{vehtype}}^4$	S_{rule}^4, see Equation 3.7
$\langle r, t, y, o \rangle^{\text{rule}}$	rule structure $\texttt{ovtake}$	$\mathfrak{L}_{\text{rule}}^4 \times \mathfrak{L}_{\text{vehtype}}^4 \times \mathbb{B}_{\text{cell}}^{\pm} \times \mathfrak{L}_{\text{vehtype}}^4$	S_{rule}^4, see Equation 3.7
$\mathbf{r}_i \# \mathfrak{t}_{\text{name}}^{\text{rule}}$	rule type name	$\mathbf{r}_i \in S_{\text{rule}}^4$	$\mathfrak{L}_{\text{rule}}^4$
$\mathbf{r}_i \# \mathfrak{t}_{\text{name}}^{\text{vehtype}}$	vehicle type name	$\mathbf{r}_i \in S_{\text{rule}}^4$	$\mathfrak{L}_{\text{vehtype}}^4$
$\mathbf{r}_i \# \Delta y_{\text{lc}}$	direction specification	$\mathbf{r}_i \in S_{\text{rule}}^4$	$\mathbb{B}_{\text{cell}}^{\pm}$
$\mathbf{r}_i \# v_{\text{max}}$	maximum velocity	$\mathbf{r}_i \in S_{\text{rule}}^4$	$\mathfrak{V}_{\text{cell}}^4$
$\mathbf{r}_i \# u_{\text{prio}}$	lane change priority	$\mathbf{r}_i \in S_{\text{rule}}^4$	$\mathfrak{P}_{\text{cell}}^{\pm}$
$\mathbf{r}_i \# \mathfrak{r}_{\text{name}}^{\text{vehtype}}$	vehtype not to ovtake	$\mathbf{r}_i \in S_{\text{rule}}^4$	$\mathfrak{L}_{\text{vehtype}}^4$
$\langle r_1, \dots, r_{\text{N}_i} \rangle^{\text{cell}}$	cell structure	$(S_{\text{rule}}^4)^{N_i} \wedge N_i \in \mathbb{N}$	S_{cell}^4, see Equation 3.8
$\mathbf{z}_i^{(t)} \# \mathbf{r}_i$	one rule	$\mathbf{z}_i^{(t)} \in S_{\text{cell}}^4 \wedge j \in \{1, \dots, N_i\}$	S_{rule}^4
$\mathbf{z}_i^{(t)} \# \mathbf{r}, N_i$	all rules, cardinality	$\mathbf{z}_i^{(t)} \in S_{\text{cell}}^4$	$\mathcal{P}_{\text{cell}}^4(S_{\text{rule}}^4), N_i := |\mathbf{z}_i^{(t)} \# \mathbf{r}|$
$\mathbf{F}_{\text{rule,dir}}(a, b, c)$	lane change rules	$S_{\text{cell}}^4 \times \mathfrak{L}_{\text{vehtype}}^4 \times \mathbb{B}_{\text{cell}}^{\pm}$	$\mathcal{P}_{\text{cell}}^4(S_{\text{rule}}^4)$
$\mathbf{F}_{\text{vehtype}}(a, b)$	rules for vehicle type	$S_{\text{cell}}^4 \times \mathfrak{L}_{\text{vehtype}}^4$	$\mathcal{P}_{\text{cell}}^4(S_{\text{rule}}^4)$
$\mathbf{F}_{\text{rule}}(a, b)$	rules for rule type	$S_{\text{cell}}^4 \times \mathfrak{L}_{\text{rule}}^4$	$\mathcal{P}_{\text{cell}}^4(S_{\text{rule}}^4)$
$\mathbf{F}_{\text{V}_{\text{max}}}(a, b)$	maximum velocity	$S_{\text{cell}}^4 \times \mathfrak{L}_{\text{vehtype}}^4$	$\mathfrak{V}_{\text{vehtype}}^4$, see Equation 3.12
$\mathbf{F}_{\text{Y}_{\text{max}}}(a, b, c)$	lane change priority	$S_{\text{cell}}^4 \times \mathfrak{L}_{\text{vehtype}}^4 \times \mathbb{B}_{\text{cell}}^{\pm}$	$\mathfrak{P}_{\text{cell}}^4$

3.3.1 Cells

A cell represents a collection of rules in combination with a cell size. As the cell size is a unique and globally defined parameter that represents the spatial discretisation, it is optimised out of the cell definition. The cells' rules model the topological constraints of the road traffic network. Equation 3.7 presents the rule structure S_{rule}^4 that distinguishes the three cases (a)-(c) for the use in the context of the author's publication [30] and this thesis. The property sets that have been used in the context of OLSIMv4 [70, Sec. 2.3.2] do not distinguish the vehicle types explicitly as they provide the distinction only as part of the property name. The rule based definition of Equation 3.7 removes this shortcoming. The cases of Equation 3.7 define in case (a) a speed limit rule, in case (b) a lane change restriction, and in case (c) an overtaking restriction. The definitions make use of the set of the rule names $\mathfrak{L}_{\text{rule}}^4 = \{\texttt{maxvel}, \texttt{chlane}, \texttt{ovtake}\}$, a set for the direction specifications $\mathbb{B}_{\text{cell}}^{\pm} = \{-1, 0, 1\}$, and a set for the lane change priorities $\mathfrak{P}_{\text{cell}}^4 = \{1, 2, 3, 4\}$.

In case (a), the rule definition combines a rule name $\mathfrak{t}_{\text{name}}^{\text{rule}} \in \mathfrak{L}_{\text{rule}}^4 \backslash \{\texttt{chlane}, \texttt{ovtake}\}$ with a vehicle type name $\mathfrak{t}_{\text{name}}^{\text{vehtype}} \in \mathfrak{L}_{\text{vehtype}}^4$, a direction specification $\Delta y_{\text{lc}} \in \mathbb{B}_{\text{cell}}^{\pm} \backslash \{-1, 1\}$, and a velocity restriction $v_{\text{max}} \in \mathfrak{V}_{\text{vehtype}}^4$. In case (b), the rule definition combines a rule name $\mathfrak{t}_{\text{name}}^{\text{rule}} \in \mathfrak{L}_{\text{rule}}^4 \backslash \{\texttt{maxvel}, \texttt{ovtake}\}$ with a vehicle type name $\mathfrak{t}_{\text{name}}^{\text{vehtype}} \in \mathfrak{L}_{\text{vehtype}}^4$, a direction specification $\Delta y_{\text{lc}} \in \mathbb{B}_{\text{cell}}^{\pm} \backslash \{0\}$, and a lane change priority specification $u_{\text{prio}} \in \mathfrak{P}_{\text{cell}}^4$. In case (c), the rule definition combines a rule name $\mathfrak{t}_{\text{name}}^{\text{rule}} \in \mathfrak{L}_{\text{rule}}^4 \backslash \{\texttt{maxvel}, \texttt{chlane}\}$ with a vehicle type name $\mathfrak{t}_{\text{name}}^{\text{vehtype}} \in \mathfrak{L}_{\text{vehtype}}^4$, a direction specification $\Delta y_{\text{lc}} \in \mathbb{B}_{\text{cell}}^{\pm} \backslash \{0\}$, and another vehicle type name $\mathfrak{r}_{\text{name}}^{\text{vehtype}} \in \mathfrak{L}_{\text{vehtype}}^4$.

$$S_{\text{rule}}^4 := \begin{cases} \langle \mathfrak{t}_{\text{name}}^{\text{rule}}, \mathfrak{t}_{\text{name}}^{\text{vehtype}}, \Delta y_{\text{lc}}, v_{\text{max}} \rangle^{\text{rule}} & \text{(a)} \\ \langle \mathfrak{t}_{\text{name}}^{\text{rule}}, \mathfrak{t}_{\text{name}}^{\text{vehtype}}, \Delta y_{\text{lc}}, u_{\text{prio}} \rangle^{\text{rule}} & \text{(b)} \\ \langle \mathfrak{t}_{\text{name}}^{\text{rule}}, \mathfrak{t}_{\text{name}}^{\text{vehtype}}, \Delta y_{\text{lc}}, \mathfrak{r}_{\text{name}}^{\text{vehtype}} \rangle^{\text{rule}} & \text{(c)} \end{cases} \tag{3.7}$$

Equation 3.8 defines the cell structure S_{cell}^4 that bundles a collection of rules into a cell. A cell

$\mathbf{z}_i^{(t)} \in S_{\text{cell}}^4$ in the context of OLSIMv4 is associated to a fixed length of $1.5\,\text{m}$ which is not part of the cell structure. A cell consists of a sequence of an arbitrary and cell position dependent number $N_i \in \mathbb{N}$ of rules $\mathbf{r}_1, \ldots, \mathbf{r}_{N_i} \in S_{\text{rule}}^4$ with $N_i := |\mathbf{r}|$ and $\mathbf{r} := \{\mathbf{r}_1, \ldots, \mathbf{r}_{N_i}\}$. Cells in OLSIMv4 do not maintain a state whether they are covered by a vehicle or not. The lack of maintaining a state distinguishes the OLSIMv4 approach from classical cellular automaton simulations. As the sequence and the number of rules of a cell may change over time, the cell identifiers $\mathbf{z}_i^{(t)} \in S_{\text{cell}}^4$ indicate the time dependence, too. As a repetition of the nomenclature from Section 3.1.2, the terms $\mathbf{z}_i^{(t)}\#\mathbf{r}_1, \ldots, \mathbf{z}_i^{(t)}\#\mathbf{r}_{N_i}$ provide access to the individual rules $\mathbf{r}_1, \ldots, \mathbf{r}_{N_i}$ of any given cell $\mathbf{z}_i^{(t)}$. The accessor $\mathbf{z}_i^{(t)}\#\mathbf{r}$ provides the corresponding set of all rules $= \{\mathbf{r}_1, \ldots, \mathbf{r}_{N_i} \in S_{\text{rule}}^4\} \subseteq \mathcal{P}_{\text{cell}}^4(S_{\text{rule}}^4)$ of a cell $\mathbf{z}_i^{(t)} \in S_{\text{cell}}^4$ where $\mathcal{P}_{\text{cell}}^4(S_{\text{rule}}^4)$ is the power set of the set of rules S_{rule}^4.

$$S_{\text{cell}}^4 := \langle \mathbf{r}_1, \ldots, \mathbf{r}_{N_i} \rangle^{\text{cell}} . \tag{3.8}$$

As an example to illustrate the construction of rules and cells, the term that constructs a rule for a closed cell for the car vehicle type has the form $\mathbf{r}_{\text{car}} := \langle \texttt{maxvel}, \texttt{car}, 0, 0\,\text{cells/s}\rangle^{\text{rule}}$. In combination with the similar rule $\mathbf{r}_{\text{truck}} := \langle \texttt{maxvel}, \texttt{truck}, 0, 0\,\text{cells/s}\rangle^{\text{rule}}$, the term to instantiate a closed cell for the vehicle types cars and trucks has the form $\langle \mathbf{r}_{\text{car}}, \mathbf{r}_{\text{truck}}\rangle^{\text{cell}}$. As soon as the information for the cells of a road traffic network are available, the focus can come to the main tasks of the cells. It consists of the storage and the access of the topological information and will be specified in the following.

The filter function $\mathbf{F}_{\text{rule}} : S_{\text{cell}}^4 \times \mathcal{L}_{\text{rule}}^4 \longrightarrow \mathcal{P}_{\text{cell}}^4(S_{\text{rule}}^4)$ filters the set of rules of a cell $\mathbf{z}_i^{(t)} \in S_{\text{cell}}^4$ for a certain rule type name $\mathfrak{t}_{\text{name}}^{\text{rule}} \in \mathcal{L}_{\text{rule}}^4$ and returns the sequence with the matching ones as a result. Equation 3.9 defines the result set of the function $\mathbf{F}_{\text{rule}}$. From the mathematical point of view, the function $\mathbf{F}_{\text{rule}}$ involves a projection on the rules of type name $\mathfrak{t}_{\text{name}}^{\text{rule}}$.

$$\mathbf{F}_{\text{rule}}(\mathbf{z}_i^{(t)}, \mathfrak{t}_{\text{name}}^{\text{rule}}) := \{\mathbf{q} \in \mathbf{z}_i^{(t)}\#\mathbf{r} \mid \mathfrak{t}_{\text{name}}^{\text{rule}} = \mathbf{q}\#\mathfrak{t}_{\text{name}}^{\text{rule}}\} . \tag{3.9}$$

As Equation 3.10 shows, the definition of the accessor function $\mathbf{F}_{\text{vehtype}} : S_{\text{cell}}^4 \times \mathcal{L}_{\text{vehtype}}^4 \longrightarrow \mathcal{P}_{\text{cell}}^4(S_{\text{rule}}^4)$ is quite similar to $\mathbf{F}_{\text{rule}}$. The function returns the filtered sequence of the set of rules of a cell that matched their vehicle type name to a supplied one $\mathfrak{t}_{\text{name}}^{\text{vehtype}} \in \mathcal{L}_{\text{vehtype}}^4$.

$$\mathbf{F}_{\text{vehtype}}(\mathbf{z}_i^{(t)}, \mathfrak{t}_{\text{name}}^{\text{vehtype}}) := \{\mathbf{q} \in \mathbf{z}_i^{(t)}\#\mathbf{r} \mid \mathfrak{t}_{\text{name}}^{\text{vehtype}} = \mathbf{q}\#\mathfrak{t}_{\text{name}}^{\text{vehtype}}\} . \tag{3.10}$$

As a last one of the filter functions, function $\mathbf{F}_{\text{rule,dir}} : S_{\text{cell}}^4 \times \mathcal{L}_{\text{rule}}^4 \times \mathbb{B}_{\text{cell}}^{\pm} \longrightarrow \mathcal{P}_{\text{cell}}^4(S_{\text{rule}}^4)$ filters the set of rules of a cell $\mathbf{z}_i^{(t)} \in S_{\text{cell}}^4$ for a certain rule type name $\mathfrak{t}_{\text{name}}^{\text{rule}} \in \mathcal{L}_{\text{rule}}^4$ and a specific direction Δy_{lc}. Equation 3.10 presents the definition for function $\mathbf{F}_{\text{rule,dir}}$.

$$\mathbf{F}_{\text{rule,dir}}(\mathbf{z}_i^{(t)}, \mathfrak{t}_{\text{name}}^{\text{rule}}, \Delta y_{\text{lc}}) := \{\mathbf{q} \in \mathbf{z}_i^{(t)}\#\mathbf{r} \mid \mathfrak{t}_{\text{name}}^{\text{rule}} = \mathbf{q}\#\mathfrak{t}_{\text{name}}^{\text{rule}} \wedge \Delta y_{\text{lc}} = \mathbf{q}\#\Delta y_{\text{lc}}\} . \tag{3.11}$$

The functions $\mathbf{F}_{\text{rule}}$, $\mathbf{F}_{\text{vehtype}}$, and $\mathbf{F}_{\text{rule,dir}}$ serve for the implementation of further and more comfortable functions. Function $\mathbf{F}_{\text{V}_{\max}} : S_{\text{cell}}^4 \times S_{\text{vehtype}}^4 \longrightarrow \mathfrak{V}_{\text{vehtype}}^4$ of them determines the maximum velocity for a given cell and vehicle type. As the other one of them, function $\mathbf{F}_{\text{Y}_{\max}} : S_{\text{cell}}^4 \times S_{\text{vehtype}}^4 \times \mathbb{B}_{\text{cell}}^{\pm} \backslash \{0\} \longrightarrow \mathfrak{P}_{\text{cell}}^4$ returns the lane change priority for a supplied vehicle type name and lane change direction. Equation 3.12 presents the result specification for one of the functions, namely $\mathbf{F}_{\text{V}_{\max}}$. It sequentially applies the functions $\mathbf{F}_{\text{vehtype}}$ and $\mathbf{F}_{\text{rule}}$ to the set of rules of a cell $\mathbf{z}_i^{(t)}$ and determines then the maximum of all velocity values of the remaining rules. The specification of the other function is analogous.

$$\mathbf{F}_{\text{V}_{\max}}(\mathbf{z}_i^{(t)}, \mathfrak{t}_{\text{name}}^{\text{vehtype}}) := \min \left\{ \mathbf{r}\#v_{\max} \mid \mathbf{r} \in \{\mathbf{F}_{\text{vehtype}}(\langle \mathbf{F}_{\text{rule}}(\mathbf{z}_i^{(t)}, \texttt{maxvel})\rangle^{\text{cell}}, \mathfrak{t}_{\text{name}}^{\text{vehtype}})\} \right\} . \tag{3.12}$$

Finally, Table 3.3.1 summarises all relevant functions, structure definitions, variables, and constants for Section 3.3.1.

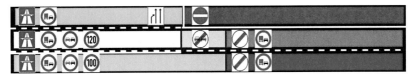

Figure 3.2: Sectioning of a roadway segment. The segment consists of six non-closed sections (traffic sign pictures taken from [194]). Figure 4.3 presents another roadway segment with its sectioning.

3.3.2 Sections

Sections define a rectangular area on a single lane along which the topological constraints do not vary. They summarise the topological information that rules on the section's spatial extent of the single lane in one single prototype cell. As another aspect of the same property, they group a roadway into areas of identical topological information. Figure 3.2 depicts a roadway with its sectioning. Identical sections therein have the same colour. Figure 4.3 presents another roadway segment with its sectioning. The section's set of German traffic signs in combination with the road and lane marking therein indicate a section's topological traffic information. As a consequence of the sections' properties, their structure definition combines the topological information with the information about the dimensions of their spatial extent. The sections structure consists thus of a prototype cell $\mathbf{z}_i^{(t)} \in S_{\text{cell}}^4$ and the specification of the section's spatial extent $\mathbf{s}_{\text{area}} \in S_{\text{area}}^4$ where S_{area}^4 represents the area structure set as defined by Equation 3.13. The Equation defines the area structure S_{area}^4 as a rectangular segment of a roadway. Even though the next paragraph will present and discuss the roadways as part of the network model in more detail, Section 3.2.2 has already introduced the roadways as a rectangular grid $\mathfrak{Y}_{\text{veh}}^4 \times \mathfrak{X}_{\text{veh}}^4 = \{i \mid i \in \mathbb{Z} \wedge Y_{\min} \leqslant i \leqslant Y_{\max}\} \times \{i\,\text{cells} \mid i \in \mathbb{Z} \wedge X_{\min} \leqslant i \leqslant X_{\max}\}$ detailed enough for the help in defining the area structure. That consists of the specification of the lane number range $[y_{\min}; y_{\max}]$ with $Y_{\min} \leqslant y_{\min} \leqslant y_{\max} \leqslant Y_{\max}$ as well as $y_{\min}\,\text{cells}, y_{\max}\,\text{cells} \in \mathfrak{Y}_{\text{veh}}^4$ and the cell range $[x_{\min}\,\text{cells}; x_{\max}\,\text{cells}]$ where $X_{\min} \leqslant x_{\min} \leqslant x_{\max} \leqslant X_{\max}$ and $x_{\min}\,\text{cells}, x_{\max}\,\text{cells} \in \mathfrak{X}_{\text{veh}}^4$.

$$S_{\text{area}}^4 := \langle y_{\min}, y_{\max}, x_{\min}, x_{\max} \rangle^{\text{area}} . \tag{3.13}$$

As the area structure is defined in the context of the roadway sections and the sections cover only single lanes, a specialised constructor for area structure elements may be handy. The constructor $\langle y_{\min}, x_{\min}, x_{\max} \rangle^{\text{area}} := \langle y_{\min}, y_{\min}, x_{\min}, x_{\max} \rangle^{\text{area}}$ takes only one lane range argument $y_{\min}$ in addition to the two regular cell range arguments $x_{\min}, x_{\max}$. It uses the same lane range argument for the lower and the upper bound specification of the regular area structure lane range. In addition to the specialised constructor for area structure elements, areas provide functions to calculate the length and the number of lanes of any area. Equation 3.14 defines the function $\mathbf{F}_{|Y|} : S_{\text{area}}^4 \longrightarrow \mathfrak{Y}_{\text{lanes}}^4$ with $\mathfrak{Y}_{\text{lanes}}^4 := \{i \in \mathbb{N} \mid i \leqslant 1 - Y_{\min} + Y_{\max}\}$ to determine the number of lanes for any given area structure element $\mathbf{s}_{\text{area}} \in S_{\text{area}}^4$.

$$\mathbf{F}_{|Y|}(\mathbf{s}_{\text{area}}) := 1 - \mathbf{s}_{\text{area}}\#y_{\min} + \mathbf{s}_{\text{area}}\#y_{\max} . \tag{3.14}$$

Equation 3.15 function $\mathbf{F}_{|X|} : S_{\text{area}}^4 \longrightarrow \mathfrak{X}_{\text{cells}}^4$ with $\mathfrak{X}_{\text{length}}^4 := \{i\,\text{cells} \mid i \in \mathbb{N} \wedge i \leqslant 1 - X_{\min} + X_{\max}\}$ calculates the length in cells for any given area structure element $\mathbf{s}_{\text{area}} \in S_{\text{area}}^4$. As sections do only cover areas on single lanes, the extent has the dimension of a length in cells.

$$\mathbf{F}_{|X|}(\mathbf{s}_{\text{area}}) := 1\,\text{cells} - \mathbf{s}_{\text{area}}\#x_{\min}\,\text{cells} + \mathbf{s}_{\text{area}}\#x_{\max}\,\text{cells} . \tag{3.15}$$

With the area structure definitions and specifications, the prerequisites are complete to define the set of section structure elements and its constructor. A section structure element $\mathbf{s}_{\text{section}} \in S_{\text{section}}^4$ combines an area structure element $\mathbf{s}_{\text{area}} \in S_{\text{area}}^4$ with a prototype cell $\mathbf{z}_i^{(t)} \in S_{\text{cell}}^4$. The index i of the prototype cell is only a reminder that the cell refers to a certain position in the ordered sequence of a roadway's cells. Equation 3.16 presents the definition for the set of the section structure elements. While a section indicates that the cells may change over time by the time dependent notation for the cells, the spatial

Table 3.3.2: Summary of relevant functions, structures, variables, and constants for Section 3.3.2.

definition	description	domain	range
$\mathfrak{Y}^4_{\text{lanes}}$	multiple lanes		$\{i \in \mathbb{N} \mid i \leqslant 1 + |Y_{\min}| + |Y_{\max}|\}$
$\mathfrak{X}^4_{\text{cells}}$	multiple cells		$\{i\,\text{cells} \mid i \in \mathbb{N} \wedge i \leqslant 1 + |X_{\min}| + |X_{\max}|\}$
$\langle y_0, y_1, x_0, x_1 \rangle^{\text{area}}$	area structure	$\mathfrak{Y}^4_{\text{veh}} \times \mathfrak{X}^4_{\text{veh}}$	S^4_{area}, see Equation 3.13
$\langle y_0, x_0, x_1 \rangle^{\text{area}}$	area structure	$\mathfrak{Y}^4_{\text{veh}} \times \mathfrak{X}^4_{\text{veh}}$	S^4_{area}, see Equation 3.13
$\text{s}_{\text{area}} \# y_{\min}$	lower lane range	$\text{s}_{\text{area}} \in S^4_{\text{area}}$	$\mathfrak{Y}^4_{\text{veh}}$
$\text{s}_{\text{area}} \# y_{\max}$	upper lane range	$\text{s}_{\text{area}} \in S^4_{\text{area}}$	$\mathfrak{Y}^4_{\text{veh}}$
$\text{s}_{\text{area}} \# x_{\min}$	lower cell range	$\text{s}_{\text{area}} \in S^4_{\text{area}}$	$\mathfrak{X}^4_{\text{veh}}$
$\text{s}_{\text{area}} \# x_{\max}$	upper cell range	$\text{s}_{\text{area}} \in S^4_{\text{area}}$	$\mathfrak{X}^4_{\text{veh}}$
$\mathbf{F}_{|Y|}(\text{s}_{\text{area}})$	lanes of area	$\text{s}_{\text{area}} \in S^4_{\text{area}}$	$\mathfrak{Y}^4_{\text{lanes}}$, see Equation 3.14
$\mathbf{F}_{|X|}(\text{s}_{\text{area}})$	cells of area	$\text{s}_{\text{area}} \in S^4_{\text{area}}$	$\mathfrak{X}^4_{\text{cells}}$, see Equation 3.15
$\langle a, b \rangle^{\text{section}}$	section structure	$S^4_{\text{area}} \times S^4_{\text{cell}}$	S^4_{section}, see Equation 3.16
$\mathbf{q} \# \text{s}_{\text{area}}$	spatial extent	$\mathbf{q} \in S^4_{\text{section}}$	S^4_{area}
$\mathbf{q} \# \mathbf{z}^{(t)}$	prototype cell	$\mathbf{q} \in S^4_{\text{section}}$	S^4_{cell}

extents as specified by the area structure do not change over time. The sectioning of a roadway remains thus constant.

$$S^4_{\text{section}} := \langle \text{s}_{\text{area}}, \mathbf{z}^{(t)} \rangle^{\text{section}} . \tag{3.16}$$

The regular structure accessors $\mathbf{q} \# \text{s}_{\text{area}}$ and $\mathbf{q} \# \mathbf{z}^{(t)}$ for any section structure elements $\mathbf{q} \in S^4_{\text{section}}$ apply here also. Table 3.3.2 summarises the relevant functions, structure definitions, variables, and constants for Section 3.3.2.

3.3.3   Roadways

Roadways represent the longest possible segment of a road per direction of travel, i.e. roadways in the context of OLSIMv4 evolve from connecting straight ahead in the direction of travel all road-segments of a highway virtually at each intersection. Consequently, they have the maximum possible spatial extent and the sections of a roadway cover the whole spatial extent. Figure 3.3 illustrates some of the OLSIMv4 roadways that stand in contrast to the tracks that used to dominate in OLSIMv3. The blue colored roadway named A003-NO-HF-034//A002-NO-HF-035 is the longest one. It covers the whole highway that is spanned by the two connected highways A3 and A2. It starts at the NRW border in the south as the A3 highway and, as the A3 and the A2 highways yield into another, it connects all road-segments up to the north-east frontier of the A2 highway. Its directional length is about $320\,\text{km}$. The tracks in Figure 3.3 can be distinguished from the roadways as road segments that are interrupted by intersections as represented by the black circles. The roadway names in Figure 3.3 are OLSIMv4 intern identifiers $\mathfrak{t}^{\text{roadway}}_{\text{name}} \in \{$ A003-NO-HF-034//A002-NO-HF-035, A061-NO-HF-029//A043-NO-HF-001, A046-NO-HF-021//A001-NO-HF-071, A040-SW-HF-044//A040-SW-HF-001, ... $\} \subset \mathfrak{L}^4_{\text{roadway}}$ that encode the highway numbers, the abbreviated direction, the roadway type, and the starting and ending exit numbers. The lengths of the other roadways are $204\,\text{km}$ for the A061-NO-HF-029-//A043-NO-HF-001, $182\,\text{km}$ for the A046-NO-HF-021//A001-NO-HF-071, and $93\,\text{km}$ for the A040-SW-HF-044//A040-SW-HF-001 roadway.

The roadways bunch the sections of all their lanes into a single collection and provide access by lane and position to each of the sections. By means of the latter, they also provide access to the prototype cell that contains the topological information for the particular lane and position. The sectioning of a roadway does not need to be unique across neighbouring lanes (even though it is in the context of OLSIMv4) and, thus, the sections on neighbouring lanes may vary in length and topological information. Figure 3.2 sketches an example roadway segment with such a sectioning.

The roadway structure S^4_{rway} contains a unique roadway name $\mathfrak{t}^{\text{roadway}}_{\text{name}} \in \mathfrak{L}^4_{\text{roadway}}$ where $\mathfrak{L}^4_{\text{roadway}}$ represents the set of all roadway names. The unique constraint for the roadway names implies that each

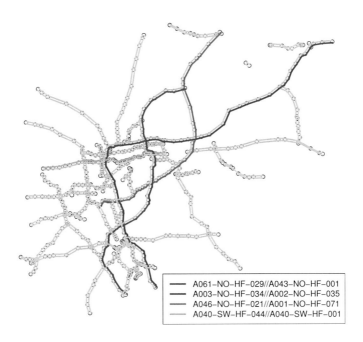

Figure 3.3: Roadways as the longest possible road segments in OLSIMv4 (source [30, Fig. 3]).

element of the set of all roadway names $\mathfrak{L}_{\text{roadway}}^4$ is used in a roadway at most once. As the second element of the roadway structure definition, an area structure element specifies $s_{\text{area}} \in S_{\text{area}}^4$ the roadway's total spatial extent. A sequence of elements with all N_i non-overlapping sections $\mathbf{q}_1, \ldots, \mathbf{q}_{N_i} \in S_{\text{section}}^4$ of the roadway with index i in the set of all roadways completes the definition of the roadway structure. Equation 3.17 presents this definition.

$$S_{\text{roadway}}^4 := \langle \mathbf{t}_{\text{name}}^{\text{roadway}}, \mathbf{s}_{\text{area}}, \mathbf{q}_1, \ldots, \mathbf{q}_{N_i} \rangle^{\text{roadway}} . \tag{3.17}$$

Similar to the accessor for the set of rules $\mathbf{z}^{(t)} \# \mathbf{r}$ of a cell $\mathbf{z}^{(t)} \in S_{\text{cell}}^4$, the roadway structure supplies the accessor for the set of all sections $\mathbf{w}^{(t)} \# \mathbf{q} := \{\mathbf{q}_1, \ldots, \mathbf{q}_{N_i} \in S_{\text{section}}^4\} \subseteq \mathcal{P}_{\text{roadway}}^4(S_{\text{section}}^4)$ of the roadway where $\mathcal{P}_{\text{roadway}}^4(S_{\text{section}}^4)$ is the power set of the set of sections S_{section}^4. Additionally, the roadway structure provides the function $\mathbf{F}_{\text{Y}}$ as shown by Equation 3.18 that filters the sections of a roadway for a given lane number y. With the help of this function, the lane-wise minimum and maximum position of a roadway $\mathbf{w}_i \in S_{\text{roadway}}^4$ become available as $\mathbf{F}_{\text{Y},\text{X}_{\text{min}}}(\mathbf{w}_i, y) := \min \{(\mathbf{q}_j \# s_{\text{area}}) \# x_{\text{min}} \mid \mathbf{q}_j \in \mathbf{F}_{\text{Y}}(\mathbf{w}_i, y)\}$ and as $\mathbf{F}_{\text{Y},\text{X}_{\text{max}}}(\mathbf{w}_i, y) := \max \{(\mathbf{q}_j \# s_{\text{area}}) \# x_{\text{max}} \mid \mathbf{q}_j \in \mathbf{F}_{\text{Y}}(\mathbf{w}_i, y)\}$ for any given lane number y. The functions $\mathbf{F}_{\text{Y}_{\text{min}}}(\mathbf{w}_i) := \{(\mathbf{q}_j \# s_{\text{area}}) \# y_{\text{min}} \mid \mathbf{q}_j \in \mathbf{w}_i \# \mathbf{q}\}$ and $\mathbf{F}_{\text{Y}_{\text{max}}}(\mathbf{w}_i) := \{(\mathbf{q}_j \# s_{\text{area}}) \# y_{\text{max}} \mid \mathbf{q}_j \in \mathbf{w}_i \# \mathbf{q}\}$ determine the minimum and maximum lane number analogously.

$$\mathbf{F}_{\text{Y}}(\mathbf{w}_i, y) := \{\mathbf{q}_j \in \mathbf{w}_i \# \mathbf{q} \mid (\mathbf{q}_j \# \mathbf{s}_{\text{area}}) \# y_{\text{min}} = y\} . \tag{3.18}$$

Table 3.3.3: Summary of relevant functions, structures, variables, and constants for Section 3.3.3.

definition	description	domain	range
$\mathfrak{L}^4_{\text{roadway}}, \mathfrak{L}^4_{\text{rway}}$	roadway names		$\{\texttt{A001}, \ldots, \texttt{A565}\}$
N_i	number of sections	$i \in \mathbb{N}, i^{\text{th}}$-roadway	$N_i \in \mathbb{N}$
$\langle a, b, \mathbf{q}_1, \ldots, \mathbf{q}_{N_i}\rangle^{\text{roadway}}$	roadway structure	$\mathfrak{L}^4_{\text{roadway}} \times S^4_{\text{area}} \times (S^4_{\text{section}})^{N_i}$	S^4_{roadway}, see Eq. 3.17
$\langle a, \mathbf{q}_1, \ldots, \mathbf{q}_{N_i}\rangle^{\text{roadway}}$	roadway structure	$\mathfrak{L}^4_{\text{roadway}} \times (S^4_{\text{section}})^{N_i}$	S^4_{roadway}, see Eq. 3.21
$\mathbf{w}_i \# \mathbf{t}^{\text{roadway}}_{\text{name}}$	roadway name	$\mathbf{w}_i \in S^4_{\text{roadway}}$	$\mathfrak{L}^4_{\text{roadway}}$
$\mathbf{w}_i \# \mathbf{s}_{\text{area}}$	spatial extent	$\mathbf{w}_i \in S^4_{\text{roadway}}$	S^4_{area}
$\mathbf{w}_i \# \mathbf{q}_1, \ldots, \mathbf{w}_i \# \mathbf{q}_{N_i}$	section access	$\mathbf{w}_i \in S^4_{\text{roadway}}$	S^4_{section}
$\mathbf{w}_i \# \mathbf{q}$	all sections	$\mathbf{w}_i \in S^4_{\text{roadway}}$	$\mathcal{P}^4_{\text{roadway}}(S^4_{\text{section}})$
$\mathbf{F}^{\text{section}}_{\text{Y} \times \text{X}}(\mathbf{w}_i, y, x)$	section by lane/pos.	$S^4_{\text{roadway}} \times \mathfrak{Y}^4_{\text{veh}} \times \mathfrak{X}^4_{\text{veh}}$	S^4_{section}, see Eq. 3.20
$\mathbf{F}^{\text{cell}}_{\text{Y} \times \text{X}}(\mathbf{w}_i, y, x)$	cell by lane/pos.	$S^4_{\text{roadway}} \times \mathfrak{Y}^4_{\text{veh}} \times \mathfrak{X}^4_{\text{veh}}$	S^4_{cell}, see Eq. 3.22

Similarly, the roadway structure also provides the function $\mathbf{F}_{\text{X}}$ that filters the sections of a roadway to match with the cell range a given cell position x. Equation 3.19 presents the function.

$$\mathbf{F}_{\text{X}}(\mathbf{w}_i, x) := \{\mathbf{q}_j \in \mathbf{w}_i \# \mathbf{q} \mid (\mathbf{q}_j \# \mathbf{s}_{\text{area}}) \# x_{\min} \leqslant x \leqslant (\mathbf{q}_j \# \mathbf{s}_{\text{area}}) \# x_{\max}\} . \qquad (3.19)$$

Moreover, the roadway structure provides also a combined lane and cell position accessor $\mathbf{F}^{\text{section}}_{\text{Y} \times \text{X}}$ that filters the sections of a roadway to match a given lane number and a given cell position x. Equation 3.20 presents this function.

$$\mathbf{F}^{\text{section}}_{\text{Y} \times \text{X}}(\mathbf{w}_i, y, x) := \{\mathbf{q}_j \in \mathbf{F}_{\text{Y}}(\mathbf{w}_i, y) \mid (\mathbf{q}_j \# \mathbf{s}_{\text{area}}) \# x_{\min} \leqslant x \leqslant (\mathbf{q}_j \# \mathbf{s}_{\text{area}}) \# x_{\max}\} . \qquad (3.20)$$

With the help of the introduced functions $\mathbf{F}_{\text{Y}_{\min}}$, $\mathbf{F}_{\text{Y}_{\max}}$, $\mathbf{F}_{\text{Y}, \text{X}_{\min}}$, and $\mathbf{F}_{\text{Y}, \text{X}_{\max}}$, the roadway structure can determine its maximum spatial extent automatically. Equation 3.21 demonstrates this by using $\langle y_{\min}, y_{\max}, x_{\min}, x_{\max}\rangle^{\text{area}}$ as the default value for the spatial extent argument. Thereby, the arguments $y_{\min}, y_{\max}, x_{\min}, x_{\max}$ to the area structure constructor are derived as follows. Starting with the initial term $\mathbf{w}_i = \langle \mathbf{t}^{\text{roadway}}_{\text{name}}, \langle 0, 0, 0, 0\rangle^{\text{area}}, \mathbf{q}_1, \ldots, \mathbf{q}_{N_i}\rangle^{\text{roadway}}$ that employs an initial area structure element, the minimum lane number $y_{\min} = \mathbf{F}_{\text{Y}_{\min}}(\mathbf{w}_i)$ and maximum lane number $y_{\max} = \mathbf{F}_{\text{Y}_{\max}}(\mathbf{w}_i)$ form the lane range $[y_{\min}; y_{\max}]$ of the roadway's spatial extent. Similarly, the cell position range matches the range between the minimum and the maximum positions, $x_{\min} = \min\{\mathbf{F}_{\text{Y}, \text{X}_{\min}}(\mathbf{w}_i, y) \mid \forall y \in [y_{\min}; y_{\max}]\}$ and $x_{\max} = \max\{\mathbf{F}_{\text{Y}, \text{X}_{\max}}(\mathbf{w}_i, y) \mid \forall y \in [y_{\min}; y_{\max}]\}$. The values for the minimum and maximum lane numbers and cell positions allow then to define a roadway constructor variant that does not require the spatial extent argument. Instead, the constructor derives the maximum spatial extent automatically. Equation 3.21 presents the constructor.

$$\langle \mathbf{t}^{\text{roadway}}_{\text{name}}, \mathbf{q}_1, \ldots, \mathbf{q}_{N_i}\rangle^{\text{roadway}} := \langle \mathbf{t}^{\text{roadway}}_{\text{name}}, \langle y_{\min}, y_{\max}, x_{\min}, x_{\max}\rangle^{\text{area}}, \mathbf{q}_1, \ldots, \mathbf{q}_{N_i}\rangle^{\text{roadway}} . \qquad (3.21)$$

Function $\mathbf{F}^{\text{section}}_{\text{Y} \times \text{X}}(\mathbf{w}_i, y, x)$ from Equation 3.20 enables the access of individual cells of a roadway based on lane and position. The definition as shown by Equation 3.22 extends it with a default cell $\mathbf{z}_{\text{default}} \in S^4_{\text{cell}}$ and defines it for the access of individual cells based on lane and position. Table 3.3.3 summarises all relevant functions, structure definitions, variables, and constants for Section 3.3.3.

$$\mathbf{F}^{\text{cell}}_{\text{Y} \times \text{X}}(\mathbf{w}_i, y, x) := \begin{cases} \mathbf{z}_{\text{default}} & , \text{if } \mathbf{F}^{\text{section}}_{\text{Y} \times \text{X}}(\mathbf{w}_i, y, x) = \{\}, \\ \mathbf{F}^{\text{section}}_{\text{Y} \times \text{X}}(\mathbf{w}_i, y, x) \# \mathbf{z}^{(t)} & , \text{otherwise}. \end{cases} \qquad (3.22)$$

3.3.4 Topological analysis

The topological analysis determines and summarises the topological information that apply on a certain lane and position for a given vehicle. Its main goal is to provide the topological information in a platform

Table 3.3.4: Summary of relevant functions, structures, variables, and constants for Section 3.3.4.

definition	description	domain	range
$\langle a, b, c, d \rangle^{\text{topo}}$	topology structure	$\mathfrak{P}^4_{\text{cell}} \times (\mathfrak{V}^4_{\text{vehtype}})^3$	S^4_{topo}, see Equation 3.23
$\mathbf{p} \# u_{\text{prio}}$	lane change priority	$\mathbf{p} \in S^4_{\text{topo}}$	$\mathfrak{P}^4_{\text{cell}}$
$\mathbf{p} \# v_{+1,\text{max}}$	left maximum velocity	$\mathbf{p} \in S^4_{\text{topo}}$	$\mathfrak{V}^4_{\text{vehtype}}$
$\mathbf{p} \# v_{\pm 0,\text{max}}$	ahead maximum velocity	$\mathbf{p} \in S^4_{\text{topo}}$	$\mathfrak{V}^4_{\text{vehtype}}$
$\mathbf{p} \# v_{-1,\text{max}}$	right maximum velocity	$\mathbf{p} \in S^4_{\text{topo}}$	$\mathfrak{V}^4_{\text{vehtype}}$
$\mathbf{F}_{\text{topo}}(\mathbf{w}_i, \mathbf{k}_n^{(t)})$	topologic analysis	$S^4_{\text{roadway}} \times S^4_{\text{veh}}$	S^4_{topo}, see Equation 3.24

independent way for the use by the microscopic traffic model. The analysis therefore collects the topological information of the three cells $\mathbf{z}_0^{(t)}, \mathbf{z}_{\pm 1}^{(t)} \in S^4_{\text{cell}}$ that apply to a vehicle $\mathbf{k}_n^{(t)} \in S^4_{\text{veh}}$ with vehicle type $\mathbf{t}_{\text{name}}^{\text{vehtype}} := (\mathbf{k}_n^{(t)} \# \mathbf{s}_{\text{vehtype}}) \# \mathbf{t}_{\text{name}}^{\text{vehtype}} \in \mathcal{L}^4_{\text{vehtype}}$ on its current lane $y = (\mathbf{k}_n^{(t)} \# \mathbf{q}^{(t)}) \# y \in \mathfrak{Y}^4_{\text{veh}}$ and position $x = (\mathbf{k}_n^{(t)} \# \mathbf{q}^{(t)}) \# x \in \mathfrak{X}^4_{\text{veh}}$ of a given roadway $\mathbf{w}_i \in S^4_{\text{roadway}}$. As a part of the collection process, it acquires first the cell $\mathbf{z}_0^{(t)}$ where the vehicle position refers to on the vehicle's current lane. Thereafter, it acquires the cells on the neighbouring lanes $\mathbf{z}_{\pm 1}^{(t)}$ that have the same position. To acquire the cells, it evaluates the cell access function from Equation 3.22 for the current lane $\mathbf{z}_0^{(t)} := \mathbf{F}_{Y \times X}^{\text{cell}}(\mathbf{w}_i, y, x)$ and for the neighbouring lanes of the current lane $\mathbf{z}_{\pm 1}^{(t)} := \mathbf{F}_{Y \times X}^{\text{cell}}(\mathbf{w}_i, y \pm 1, x)$. Thereafter, it evaluates the cell maximum velocity function $v_{\text{max}}^j := \mathbf{F}_{V_{\text{max}}}(\mathbf{z}_j^{(t)}, \mathbf{t}_{\text{name}}^{\text{vehtype}})$ for each of the cells $\mathbf{z}_j^{(t)} \in S^4_{\text{cell}} \wedge j \in \{-1, 0, 1\}$. To complete the topological analysis, it evaluates then the cell lane change priority function $u_{\text{prio}}^j := \mathbf{F}_{Y_{\text{max}}}(\mathbf{z}^{(t)}, \mathbf{t}_{\text{name}}^{\text{vehtype}}, \pm 1)$ from Table 3.3.1 for each of the cells $\mathbf{z}_j^{(t)} \in S^4_{\text{cell}} \wedge j \in \{-1, 1\}$. Finally, the topology structure S^4_{topo} can then summarise the topological information in the form of the definition as shown by Equation 3.23. The structure contains the priority $u_{\text{prio}} := \min \{u_{\text{prio}}^{-1}, u_{\text{prio}}^1\}$ for a lane change to take place and the maximum velocities of a vehicles current lane $v_{\pm 0,\text{max}}$ as well as for its neighbouring lanes $v_{-1,\text{max}}, v_{+1,\text{max}}$.

$$S^4_{\text{topo}} := \langle u_{\text{prio}}, v_{-1,\text{max}}, v_{\pm 0,\text{max}}, v_{+1,\text{max}} \rangle^{\text{topo}} . \qquad (3.23)$$

The topological analysis procedure determines the summarised information for each vehicle individually from the roadway structure. It therefore employs the topological analysis function $\mathbf{F}_{\text{topo}}$: $S^4_{\text{roadway}} \times S^4_{\text{veh}} \longrightarrow S^4_{\text{topo}}$. The function processes all the steps as described above and wraps the above four values for a roadway $\mathbf{w}_i \in S^4_{\text{roadway}}$ and a given vehicle $\mathbf{k}_n^{(t)} \in S^4_{\text{veh}}$ into a resulting topology structure element. Equation 3.24 presents function $\mathbf{F}_{\text{topo}}$. Finally, Table 3.3.4 completes this section with a summary of all relevant functions, structure definitions, variables, and constants for Section 3.3.4.

$$\mathbf{F}_{\text{topo}}(\mathbf{w}_i, \mathbf{k}_n^{(t)}) := \langle u_{\text{prio}}, v_{-1,\text{max}}, v_{\pm 0,\text{max}}, v_{+1,\text{max}} \rangle^{\text{topo}} . \qquad (3.24)$$

3.3.5 Occupation tables

Strictly speaking, the storage and access of the vehicles by lane and index is not part of a network model. However, as the vehicles drive on road-segments, they refer to road-segments. Additionally, the microscopic traffic models update procedures require not only a vehicle's predecessors and its neighbouring vehicles but also require the summarised topological constraints in the form of S^4_{topo} as determined by the topological analysis that the previous section has introduced. As the topological analysis depends on a vehicle's vehicle type, its lane, and its position, it also depends on the vehicles. For this reason, the occupation tables are treated in the network section. In the case when a container combines or associates the list of vehicles with the road segment where the vehicles drive on, the vehicles do not need to maintain a reference to their roadway. This storage method may look like a premature optimisation, but, as a matter of fact, is used by many microscopic traffic simulations, among them SUMO [19], OLSIMv3, and, when applied to the roadways, OLSIMv4 [31], too. However, the storage and access methods differ over the various implementations in several small but nevertheless important details. They are described in the following.

OLSIMv3 stored and maintained all vehicles of the simulation in a global list and maintained per vehicle references to the leading vehicle and the neighbouring vehicles. The vehicles in OLSIMv3 also maintained per vehicle references to the track where the vehicle drives on as the vehicle update procedures depend on them. More precisely, the value of a vehicle's safe velocity for the next time step in the traffic model depends on the position and the velocity of the leading vehicle. Similarly, the lane change update depends on the positions and velocities of the neighbouring vehicles. Each overtaking manoeuvre, each lane change, and each track change required thus reassigning the references.

As the OLSIMv3 vehicle update procedures search for a vehicle's leading vehicle and the neighbouring vehicles in each update, the search for the neighbours involves also searching the vehicles' list of the neighbouring tracks—in all possible directions. Even though the search cancelled whenever it reached a maximum distance between a vehicle and a possible neighbour track, there are several other downsides with this approach. At first, the algorithm implies that the network structure, i.e. the tracks and their connections, remains constant. Second, the concepts of the tracks topological structure, the vehicle container, and the vehicle update process are coupled. This coupling is not only poor software technique but more than that requires re-implementation for scenarios that do not rely on the concept of tracks. Platform independence is thus hard to achieve.

Another storage method that is used by SUMO [19] and probably by several market solutions, too, stores the vehicles in one container per track. The vehicle update procedures ensure that the vehicles remain sorted in the track container. However, before a vehicle can leave one track and enter the next one, a search such as the one from above has to be processed. According to [31] for the case of AIMSUN, this requires to redirect section leader vehicles in a separate "leaving vehicles entity process" [196, Figs. 275, 276]. This concept relies on the state of the neighboured track and, in the worst case, may require two synchronous write updates [30, Sec. 3] of the associated container structures for each time step within which vehicles are redirected. Taking into account that the number of vehicles that change the track in the direction of travel is roughly proportional to the flow through the track, this concept requires a lot more synchronisation than the track changing by lateral motion only. As will be seen later in this section, the situation worsens further when random number generators are involved because they require storing, maintaining, and updating a state thereby requiring additional synchronisation in case of a non-parallel random number generator [123]. sections 4.2.2.1 and 4.2.5 will discuss this further. As the vehicle container does not necessarily need to rely on the topological concepts, OLSIMv4 stores the vehicles lane-wise per roadway.

The occupation table structure S_{occ}^4 as defined in Equation 3.25 represents occupation tables. Its elements combine a random number generator's (RNG) state $s_{\mathrm{rand}} \in \mathfrak{R}_{\mathrm{occ}}^4 := \{i \in \mathbb{N}_0 \mid i < R_{\max}\} \subset \mathbb{N}_0$ with $R_{\max} \in \mathbb{N}$, a periodic boundary condition flag $p_i \in \mathfrak{B}_{\mathrm{occ}}^4 := \{\mathtt{true}, \mathtt{false}\}$, a roadway $\mathbf{w}_i \in S_{\mathrm{roadway}}^4$ with index i in the list of all roadways, and all the K_i vehicles $\mathbf{k}_1^{(t)}, \ldots, \mathbf{k}_{K_i}^{(t)} \in S_{\mathrm{veh}}^4$ positioned on it. In addition to the vehicle access functions $\mathbf{o} \# \mathbf{k}_j^{(t)} \in S_{\mathrm{veh}}^4$ with $j \in 1, \ldots, K_i$, the occupation table structure provides the function $\mathbf{o} \# \mathbf{k}^{(t)} \in \mathcal{P}_{\mathrm{veh}}^4(S_{\mathrm{veh}}^4)$ to access the set of all the K_i vehicles positioned on it. Whenever the periodic boundary condition flag of an occupation table is set, the occupation table maps vehicles with positions beyond the spatial extent of the roadway back. The final position is then the remainder of the division by the total length of the roadway.

$$S_{\mathrm{occ}}^4 := \langle s_{\mathrm{rand}}, p_i, \mathbf{w}_i, \mathbf{k}_1^{(t)}, \ldots, \mathbf{k}_{K_i}^{(t)} \rangle^{\mathrm{occ}} . \tag{3.25}$$

Similar to the sections access by lane and position that the roadway structure provides from Equation 3.20, the occupation table structure provides the two vehicle access functions, $\mathbf{F}_{\mathrm{Y}} : S_{\mathrm{occ}}^4 \times \mathfrak{Y}_{\mathrm{veh}}^4 \longrightarrow \mathcal{P}_{\mathrm{veh}}^4(S_{\mathrm{veh}}^4)$ which filters all vehicles of an occupation table $\mathbf{o} \in S_{\mathrm{occ}}^4$ to match a lane $y \in \mathfrak{Y}_{\mathrm{veh}}^4$ and $\mathbf{F}_{\mathrm{Y} \times \mathrm{X}} : S_{\mathrm{occ}}^4 \times \mathfrak{Y}_{\mathrm{veh}}^4 \times \mathfrak{X}_{\mathrm{veh}}^4 \longrightarrow S_{\mathrm{veh}}^4$ which accesses by lane $y \in \mathfrak{Y}_{\mathrm{veh}}^4$ and position $x \in \mathfrak{X}_{\mathrm{veh}}^4$. Equation

Table 3.3.5: Summary of relevant functions, structures, variables, and constants for Section 3.3.5.

definition	description	domain	range
$\mathfrak{B}_{occ}^4$	boolean values		$\{\texttt{true},\texttt{false}\}$
$\mathfrak{R}_{occ}^4$	random numbers		$\{i \in \mathbb{N}_0 \mid i < R_{max}\}$
$\langle a, b, c, \mathbf{k}_1^{(t)}, \ldots, \mathbf{k}_{K_i}^{(t)}\rangle^{occ}$	occupation tables	$\mathfrak{R}_{occ}^4 \times S_{roadway}^4 \times (S_{veh}^4)^{K_i}$	S_{occ}^4, Eq. 3.25
$\mathbf{F}_{insert}(\mathbf{o}, \mathbf{k}_r^{(t)})$	insert vehicle	$\mathfrak{R}_{occ}^4 \times S_{roadway}^4 \times (S_{veh}^4)^{K_i+1}$	S_{occ}^4, Eq. 3.28
$\mathbf{F}_{remove}(\mathbf{o}, \mathbf{k}_r^{(t)})$	remove vehicle	$\mathfrak{R}_{occ}^4 \times S_{roadway}^4 \times (S_{veh}^4)^{K_i-1}$	S_{occ}^4, Eq. 3.29
$\mathbf{o}\#s_{rand}$	state of RNG	$\mathbf{o} \in S_{occ}^4$	$\mathfrak{R}_{occ}^4$
$\mathbf{o}\#p_i$	periodic boundaries	$\mathbf{o} \in S_{occ}^4$	$\mathfrak{B}_{occ}^4$
$\mathbf{o}\#\mathbf{w}_i$	roadway access	$\mathbf{o} \in S_{occ}^4$	S_{occ}^4
$\mathbf{o}\#\mathbf{k}_1^{(t)}, \ldots, \mathbf{o}\#\mathbf{k}_{K_i}^{(t)}$	j^{th}-vehicle access	$\mathbf{o} \in S_{occ}^4 \wedge j \in \{1, \ldots, K_i\}$	S_{veh}^4
$\mathbf{o}\#\mathbf{k}^{(t)}$	access all vehicles	$\mathbf{o} \in S_{occ}^4$	$\mathcal{P}_{veh}^4(S_{veh}^4)$
$\mathbf{F}_Y(a,b)$	vehicle filter by lane	$S_{occ}^4 \times \mathfrak{Y}_{veh}^4$	$\mathcal{P}_{veh}^4(S_{veh}^4)$, Eq. 3.26
$\mathbf{F}_{Y\times X}(a,b,c)$	vehicle lane/position	$S_{occ}^4 \times \mathfrak{Y}_{veh}^4 \times \mathfrak{X}_{veh}^4$	S_{veh}^4, Eq. 3.27
$\langle \mathbf{k}_1^{(t)}, \ldots, \mathbf{k}_9^{(t)}\rangle^{nbrs}$	neighbours structure	$(S_{veh}^4)^9$	S_{nbrs}^4, Eq. 3.32
$\mathbf{b}\#\mathbf{k}_{l=3i+j+4}^{(t)}, \mathbf{b}\#\mathbf{k}_{i,j}^{(t)}$	neighbouring vehicle	$i, j \in \mathbb{B}_{cell}^{\pm} \wedge \mathbf{b} \in S_{nbrs}^4$	S_{veh}^4
$\mathbf{F}_{Y\times X}^{>}(a,b), \mathbf{F}_{Y\times X}^{<}(a,b)$	leaders/followers	$\mathcal{P}_{veh}^4(S_{veh}^4) \times \mathfrak{Y}_{veh}^4 \times \mathfrak{X}_{veh}^4$	$\mathcal{P}_{veh}^4(S_{veh}^4)$, Eq. 3.30
$\mathbf{F}_{Y\times X}^{<,>}(a,b,c)$	vehicles between and	$\mathcal{P}_{veh}^4(S_{veh}^4) \times \mathfrak{Y}_{veh}^4 \times (\mathfrak{X}_{veh}^4)^2$	$\mathcal{P}_{veh}^4(S_{veh}^4)$
$\mathbf{F}_{area}^{<,>}(a,b)$	vehicles between and	$\mathcal{P}_{veh}^4(S_{veh}^4) \times S_{area}^4$	$\mathcal{P}_{veh}^4(S_{veh}^4)$
$\mathbf{F}_{nbrs}(a,b)$	neighbouring vehicles	$S_{occ}^4 \times S_{veh}^4$	S_{nbrs}^4, Eq. 3.31
$\mathbf{F}_{nbrs*}(a,b)$	neighbours and successor	$S_{occ}^4 \times S_{veh}^4$	S_{nbrs}^4, Eq. 3.31

3.26 presents the definition of the former while Equation 3.27 presents the latter. The definition of the latter uses the minimum function for the vehicles domain which is defined in the usual way except for the comparison that in this case relies on comparing the vehicle positions, i.e. $(\mathbf{k}_j^{(t)}\#\mathbf{q}^{(t)})\#x \leqslant (\mathbf{k}_i^{(t)}\#\mathbf{q}^{(t)})\#x$ with $\mathbf{k}_j^{(t)}, \mathbf{k}_i^{(t)} \in S_{veh}^4$.

$$\mathbf{F}_Y(\mathbf{o}, y) := \{\mathbf{k}_j^{(t)} \in \mathbf{o}\#\mathbf{k}^{(t)} \mid (\mathbf{k}_j^{(t)}\#\mathbf{q}^{(t)})\#y = y\} \,. \tag{3.26}$$

$$\mathbf{F}_{Y\times X}(\mathbf{o}, y, x) := \min \{\mathbf{k}_j^{(t)} \in \mathbf{F}_Y(\mathbf{o}, y) \mid (\mathbf{k}_j^{(t)}\#\mathbf{q}^{(t)})\#x = x\} \,. \tag{3.27}$$

In addition to the access of vehicles, the occupation tables provide two functions to insert or remove vehicles into or from it, respectively. The insert function $\mathbf{F}_{insert} : \mathfrak{R}_{occ}^4 \times S_{roadway}^4 \times (S_{veh}^4)^{K_i+1} \longrightarrow S_{occ}^4$ takes an occupation table $\mathbf{o}$ and a vehicle $\mathbf{k}_{K_i+1}^{(t)}$ as its arguments and returns a cloned occupation table with the inserted vehicle. The remove function is defined analogously. Equations 3.28 and 3.29 introduce their definitions.

$$\mathbf{F}_{insert}(\mathbf{o}, \mathbf{k}_r^{(t)}) := \mathbf{F}_{clone}(\mathbf{o}, \mathbf{k}^{(t)}, \mathbf{o}\#\mathbf{k}^{(t)} \cup \mathbf{k}_r^{(t)}) \,. \tag{3.28}$$

$$\mathbf{F}_{remove}(\mathbf{o}, \mathbf{k}_r^{(t)}) := \mathbf{F}_{clone}(\mathbf{o}, \mathbf{k}^{(t)}, \mathbf{o}\#\mathbf{k}^{(t)}\backslash\mathbf{k}_r^{(t)}) \,. \tag{3.29}$$

The occupation tables represent atomic units for the overall update process. Section 3.7.3 will discuss this in more detail. Consequently, the main task of any occupation table $\mathbf{o} \in S_{occ}^4$ is the updating of each vehicle that drives on the contained roadway $\mathbf{w}_i$. The occupation tables therefore make use of a very similar defined and equally named function $\mathbf{F}_{Y\times X}^{>} : \mathcal{P}_{veh}^4(S_{veh}^4) \times \mathfrak{X}_{veh}^4 \longrightarrow \mathcal{P}_{veh}^4(S_{veh}^4)$ to the one from above that operates on vehicle sets in contrast to the occupation tables. It filters a set of vehicles $\mathbf{k}^{(t)}$ for specific positions that are greater than the supplied position $x \in S_{veh}^4$. Whenever the update procedure requires the leading vehicles of the neighbour lanes $l_i = ((\mathbf{o}\#\mathbf{k}_n^{(t)})\#\mathbf{q}^{(t)})\#y \pm 1$ to a vehicle $\mathbf{o}\#\mathbf{k}_n^{(t)}$ with index n in the occupation table $\mathbf{o} \in S_{\mathrm{occ}}^4$, the procedure can determine the next leading vehicle $\mathbf{k}_{m,i}^{(t)}$ with the help of this function. It does this by simply choosing the minimum of the function evaluation as described by the following term $\mathbf{k}_{m,i}^{(t)} = \min \mathbf{F}_{Y \times X}^{>}(\mathbf{F}_Y(\mathbf{o}, l_i), x)$ with the position $x = ((\mathbf{o}\#\mathbf{k}_n^{(t)})\#\mathbf{q}^{(t)})\#x$. Equation 3.30 introduces function $\mathbf{F}_{Y \times X}^{>}$. The functions $\mathbf{F}_{Y \times X}^{<}$, $\mathbf{F}_{Y \times X}^{<,>}$, and $\mathbf{F}_{\mathrm{area}}^{<,>}$ can be defined analogously. The first one filters a set of vehicles to have a position smaller than the supplied value and the second one filters for a position between a lower and an upper position.

$$\mathbf{F}_{Y \times X}^{>}(\mathbf{k}^{(t)}, x) := \{\mathbf{k}_j^{(t)} \in \mathbf{k}^{(t)} \mid (\mathbf{k}_j^{(t)}\#\mathbf{q}^{(t)})\#x > x\} \, . \tag{3.30}$$

As a part of the occupation table update procedure, the vehicle update procedures depend on a vehicle's preceding and neighbouring vehicles. The so far introduced functions contribute in determining the preceding and the neighbouring vehicles of a vehicle $\mathbf{k}_n^{(t)}$. In particular, the next two preceding vehicles on the same lane are $\mathbf{k}_{n+1}^{(t)}$ and $\mathbf{k}_{n+2}^{(t)}$. The leading vehicles $\mathbf{k}_{m,i}^{(t)} = \min \mathbf{F}_{Y \times X}^{>}(\mathbf{F}_Y(\mathbf{o}, l_i), x)$ with $i \in \mathbb{B}_{\mathrm{cell}}^{\pm} \backslash \{-1, 1\}$ on the neighbour lanes $l_i = i + ((\mathbf{o}\#\mathbf{k}_n^{(t)})\#\mathbf{q}^{(t)})\#y$ are the first ones on the neighbour lane l_i. They have the vehicle position $x = ((\mathbf{o}\#\mathbf{k}_n^{(t)})\#\mathbf{q}^{(t)})\#x$ and preceding vehicles $\mathbf{k}_{m+1,i}^{(t)}$ as well as succeeding vehicles $\mathbf{k}_{m-1,i}^{(t)}$, too. The function $\mathbf{F}_{\mathrm{nbrs}} : S_{\mathrm{occ}}^4 \times S_{\mathrm{veh}}^4 \longrightarrow S_{\mathrm{nbrs}}^4$ with $S_{\mathrm{nbrs}}^4 \subseteq (S_{\mathrm{veh}}^4)^9$ returns an element of the neighbouring structure S_{nbrs}^4. Equation 3.32 presents the definition of the neighbour structure, while Equation 3.31 presents the definition of function $\mathbf{F}_{\mathrm{nbrs}}$. Figure 3.6 shows such a scenario with vehicles on several lanes. A special variant exists also with $\mathbf{F}_{\mathrm{neighbours}*}(\mathbf{o}, \mathbf{k}_n^{(t)}) := \mathbf{F}_{\mathrm{clone}}(\mathbf{F}_{\mathrm{nbrs}}(\mathbf{o}, \mathbf{k}_n^{(t)}), \mathbf{k}_n^{(t)}, \mathbf{k}_{n-1}^{(t)})$ that is particularly useful for the tuning elements.

$$\mathbf{F}_{\mathrm{nbrs}}(\mathbf{o}, \mathbf{k}_n^{(t)}) := \langle \mathbf{k}_{m-1,-1}^{(t)}, \mathbf{k}_{m,-1}^{(t)}, \mathbf{k}_{m+1,-1}^{(t)}, \mathbf{k}_n^{(t)}, \mathbf{k}_{n+1}^{(t)}, \mathbf{k}_{n+2}^{(t)}, \mathbf{k}_{m-1,1}^{(t)}, \mathbf{k}_{m,1}^{(t)}, \mathbf{k}_{m+1,1}^{(t)} \rangle^{\mathrm{nbrs}} \, . \tag{3.31}$$

The neighbour structure elements contains all of the altogether 9 preceding, neighbouring, and succeeding vehicles. With the relative lane index specification $i \in \mathbb{B}_{\mathrm{cell}}^{\pm}$ for vehicles on the neighbour lanes and the relative position index $j \in \mathbb{B}_{\mathrm{cell}}^{\pm}$ for vehicles on the same lane, the vehicles $\mathbf{k}_{1,i}^{(t)}, \mathbf{k}_{2,i}^{(t)}, \mathbf{k}_{3,i}^{(t)}$ of any neighbour structure element $\mathbf{b} \in S_{\mathrm{nbrs}}^4$ can be accessed also by the following access rule $\mathbf{b}\#\mathbf{k}_{i,j}^{(t)} := \mathbf{b}\#\mathbf{k}_{3*(i+1)+j+1}^{(t)}$. Equation 3.32 displays the definition for the neighbour structure S_{nbrs}^4. Finally, Table 3.3.5 summarises all relevant functions, structure definitions, variables, and constants for this section.

$$S_{\mathrm{nbrs}}^4 := \langle \mathbf{k}_1^{(t)}, \dots, \mathbf{k}_9^{(t)} \rangle^{\mathrm{nbrs}} \, . \tag{3.32}$$

Section 3.4 will introduce the vehicle update procedures $\mathbf{F}_{\mathrm{drive}} : \mathfrak{R}_{\mathrm{occ}}^4 \times S_{\mathrm{veh}}^4 \times S_{\mathrm{nbrs}}^4 \times S_{\mathrm{topo}}^4 \longrightarrow \mathfrak{R}_{\mathrm{occ}}^4 \times S_{\mathrm{veh}}^4$ and $\mathbf{F}_{\mathrm{chlane}} : \mathfrak{R}_{\mathrm{occ}}^4 \times S_{\mathrm{veh}}^4 \times S_{\mathrm{nbrs}}^4 \times S_{\mathrm{topo}}^4 \longrightarrow \mathfrak{R}_{\mathrm{occ}}^4 \times S_{\mathrm{veh}}^4$ for the drive update and the lane change update. They differ in their result type but rely both only on the random number generator state, the neighbour structure elements, and the topology structure. As the neighbour structure and the topology structure have been defined already, it is possible (and done in the following) to define the vehicle update for an occupation table. Even though the vehicle update procedure updates all vehicles synchronously and, theoretically, could be done concurrently or in parallel, the definition of the vehicle update for the occupation tables depends on a sequential update as the random number generator requires the resulting state of the previous update. Consequently, the update results in a sequence of pairs $(s_{\mathrm{rand},N}, \mathbf{k}_N^{(t+\Delta t)}), \dots, (s_{\mathrm{rand},1}, \mathbf{k}_1^{(t+\Delta t)})$ with the updated random number generator state and vehicle pair $\mathbf{u}_i^{(t+\Delta t)} := (s_{\mathrm{rand},i}, \mathbf{k}_i^{(t+\Delta t)}) := \mathbf{F}_{\mathrm{drive}}(s_{\mathrm{rand},i-1}, \mathbf{k}_i^{(t)}, \mathbf{b}_i, \mathbf{p}_i)$, the initial random number state $s_{\mathrm{rand},0} := \mathbf{o}_r^{(t)}\#s_{\mathrm{rand}}$, the neighbouring vehicles $\mathbf{b}_i := \mathbf{F}_{\mathrm{nbrs}}(\mathbf{o}_r^{(t)}, \mathbf{k}_i^{(t)})$, the topological information $\mathbf{p}_i := \mathbf{F}_{\mathrm{topo}}(\mathbf{o}_r^{(t)}\#\mathbf{w}^{(t)}, \mathbf{k}_i^{(t)})$, the number of vehicles $N := |\mathbf{o}_r^{(t)}\#\mathbf{k}^{(t)}|$, and the vehicles to update $\mathbf{k}_i^{(t)} \in \mathbf{o}_r^{(t)}\#\mathbf{k}^{(t)}$. The occupation table update procedure uses a recursion formula to generate the sequence that contains the updated vehicles $\mathbf{u}_i^{(t+\Delta t)} := (s_{\mathrm{rand},i}, \mathbf{k}_i^{(t+\Delta t)}) := \mathbf{F}_{\mathrm{drive}}(s_{\mathrm{rand},i-1}, \mathbf{k}_i^{(t)}, \mathbf{b}_i, \mathbf{p}_i)$.

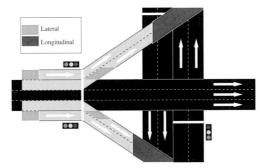

Figure 3.4: Switching areas connect roadways dynamically laterally and longitudinally (source: [30]).

The recursion formula manifests the sequential character of the vehicle update using a pseudo-random number generator. Section 4.2.5 discusses some approaches to transform the sequential operation in a more parallel version.

With the sequence that contain the updated vehicles $\mathbf{u}_i^{(t+\Delta t)}$, the definition of the Function $\mathbf{F}_{\text{drive}}$: $\mathfrak{T}_{\text{secs}} \times S_{\text{occ}}^4 \longrightarrow S_{\text{occ}}^4$ becomes available. The function performs the drive update of time step t for an occupation table $\mathbf{o}_r^{(t)}$. Equation 3.33 presents the definition of the drive update.

$$\mathbf{F}_{\text{drive}}(t, \mathbf{o}_r^{(t)}) := \mathbf{F}_{\text{clone}}(\mathbf{F}_{\text{clone}}(\mathbf{o}_r^{(t)}, \mathbf{k}^{(t)}, \bigcup_{i=1}^{N} \mathbf{u}_i^{(t+\Delta t)} \# \mathbf{k}_i^{(t+\Delta t)}), s_{\text{rand}}, \mathbf{u}_N^{(t+\Delta t)} \# s_{\text{rand}, N-1}) . \tag{3.33}$$

The definition of the Function $\mathbf{F}_{\text{chlane}}$: $\mathfrak{T}_{\text{secs}} \times S_{\text{occ}}^4 \longrightarrow S_{\text{occ}}^4$ is analogous to the drive update. In simulation scenarios with switching areas, the update procedures require additional variants the occupation table update procedures. The variants provide the ability to filter the set of vehicles prior to updating them. The filtering enables to exclude vehicles on certain areas. The occupation table update functions then take the signature as for example $\mathbf{F}_{\text{drive}}$: $\mathfrak{T}_{\text{secs}} \times S_{\text{occ}}^4 \times \mathcal{P}_{\text{area}}(S_{\text{area}}^4) \longrightarrow S_{\text{occ}}^4$.

3.3.6 Switching areas

OLSIMv3 connected tracks statically for both potential connection types, i.e. lateral and longitudinal connections. The shortcomings of the track depending approach have already been mentioned. Section 3.7 will discuss them in more detail. Even though OLSIMv4 does not instantiate the switching areas explicitly, it overcomes the already mentioned shortcomings as well as those presented in the following with the concept of the switching areas. As the tracks in OLSIMv3 could model only static lateral and longitudinal connections between each other, the OLSIMv3 network model could not model complex intersection scenarios with dynamic connections between several incoming tracks that led onto one final destination track. As these kinds of connections are not found on German highways, this may be of minor relevance only. However, as OLSIMv4 aims at network platform portability, this is of importance for OLSIMv4. Another scenario that was impossible to model with the OLSIMv3 network model is a roundabout with several entrance road segments each having multiple lanes. For these kind of network elements, VISSIM uses connectors and conflict areas [61, Sec. 2.2.2.1, 2.3.3.1, Fig. 2.9]. However, the VISSIM connectors can connect lanes only statically and the conflict areas resolve conflicts by additional behavioural logic. The OLSIMv4 network model overcomes all these kind of shortcomings with the switching area network elements. Careful use of the switching areas enables the multicomponent simulation model to be used over the otherwise obligatory discrete time system simulation model. In addition to a more detailed system specification, this model has the advantage that a more parallel interpretation is available for it. Section 3.7 discusses this in more detail. Another advantage aspect of switching areas over statically connected tracks lies in the flexibility of this approach. Whenever it is unclear, whether lateral connections between roadways will be required or not, dynamic connections provide a really flexible approach.

Switching areas interconnect regions of roadways pair-wise for interchanging and lateral motion or in cases where multiple roadways lead onto the same roadway in the longitudinal direction. Figure 3.4 shows such a scenario. The longitudinal connections therein lead onto road segments that have already predecessors. In other words, these road segments have multiple entries. OLSIMv4 models them by the switching areas that provide dynamic connections. As OLSIMv4 uses the switching area information to construct occupation tables dynamically and the occupation tables already have all necessary vehicle update logic, this approach does not require any additional behavioural logic. As an advantage, it does not introduce or change any simulation algorithm logic.

Equation 3.34 defines the switching area structure S^4_{swarea}. It connects the area $s^{\text{from}}_{\text{area}} \in S^4_{\text{area}}$ on roadway $f^{\text{roadway}}_{\text{name}} \in \mathcal{L}^4_{\text{roadway}}$ with the area $s^{\text{to}}_{\text{area}} \in S^4_{\text{area}}$ on roadway $t^{\text{roadway}}_{\text{name}} \in \mathcal{L}^4_{\text{roadway}}$. Additionally, the switching area elements provide an initial state for a random number generator $s_{\text{rand}} \in \mathfrak{R}^4_{\text{occ}}$.

$$S^4_{\text{swarea}} := \langle s_{\text{rand}}, \Delta y_{\text{lc}}, f^{\text{roadway}}_{\text{name}}, s^{\text{from}}_{\text{area}}, t^{\text{roadway}}_{\text{name}}, s^{\text{to}}_{\text{area}} \rangle^{\text{swarea}} . \tag{3.34}$$

The instantiations of the temporary occupation tables require to supply a random number state. As the initial state—when chosen carefully—enables the use of a more parallel interpretation, it is of great importance for the simulation. Section 3.7.2 discusses this in more detail. As its central part, the $\Delta y_{\text{lc}} \in \mathbb{B}^{\pm}_{\text{cell}}$ element defines the connection direction between the two roadways. The switching areas hold the constraint $((\mathbf{F}_{|X|}(s^{\text{from}}_{\text{area}}) = \mathbf{F}_{|X|}(s^{\text{to}}_{\text{area}})) \wedge (|\Delta y_{\text{lc}}| = 1)) \vee ((\mathbf{F}_{|Y|}(s^{\text{from}}_{\text{area}}) = \mathbf{F}_{|Y|}(s^{\text{to}}_{\text{area}})) \wedge (|\Delta y_{\text{lc}}| = 0))$. Except for the component for the RNG state s_{rand} switching areas remain constant.

As part of the switching area's vehicle update for the longitudinal and the lateral motion which is sketched in the following, the switching areas also provide a function $\mathbf{F}_{\text{join-occs}} : S^4_{\text{occ}} \times S^4_{\text{swarea}} \times S^4_{\text{occ}} \longrightarrow S^4_{\text{occ}}$ to instantiate a fresh occupation table with the dimensions of the switching area. It relies on the function $\mathbf{F}_{\text{join-areas}} : S^4_{\text{occ}} \times S^4_{\text{swarea}} \times S^4_{\text{occ}} \longrightarrow S^4_{\text{occ}}$ from Equation 3.35 to construct the resulting switching area. It depends on the switching areas $s^{\text{to}}_{\text{area}} := \mathbf{a} \# s^{\text{to}}_{\text{area}}$ and $s^{\text{from}}_{\text{area}} := \mathbf{a} \# s^{\text{from}}_{\text{area}}$.

$$\mathbf{F}_{\text{join-areas}}(\mathbf{a}) := \begin{cases} \langle 1, \mathbf{F}_{|Y|}(s^{\text{to}}_{\text{area}}), 1, \mathbf{F}_{|X|}(s^{\text{from}}_{\text{area}}) + \mathbf{F}_{|X|}(s^{\text{to}}_{\text{area}}) \rangle^{\text{area}} & , \text{ if } |\Delta y_{\text{lc}}| = 0 \text{ , or} \\ \langle 1, \mathbf{F}_{|Y|}(s^{\text{from}}_{\text{area}}) + \mathbf{F}_{|Y|}(s^{\text{to}}_{\text{area}}), 1, \mathbf{F}_{|X|}(s^{\text{from}}_{\text{area}}) \rangle^{\text{area}} & , \text{ otherwise.} \end{cases} \tag{3.35}$$

To construct the joined occupation table, the switching area update functions $\mathbf{F}_{\text{swa-drive}} : \mathfrak{T}_{\text{secs}} \times S^4_{\text{occ}} \times S^4_{\text{swarea}} \times S^4_{\text{occ}} \longrightarrow S^4_{\text{swarea}} \times \mathcal{P}^4_{\text{veh}}(S^4_{\text{veh}})$ and $\mathbf{F}_{\text{swa-chlane}} : \mathfrak{T}_{\text{secs}} \times S^4_{\text{occ}} \times S^4_{\text{swarea}} \times S^4_{\text{occ}} \longrightarrow S^4_{\text{swarea}} \times \mathcal{P}^4_{\text{veh}}(S^4_{\text{veh}})$ use the joined area as constructed by $\mathbf{F}_{\text{join-areas}}(\mathbf{a})$, the sections and the vehicles from each occupation table. They transform the coordinates of the sections and the vehicles to fit into the joined area. And, finally, they use the random number generator state as provided by the switching area argument $\mathbf{a} \# s_{\text{rand}}$ to construct the joined occupation table $\mathbf{o}^{(t)}_j$. The switching area vehicle update functions then wrap the corresponding vehicle update function for the freshly instantiated occupation table. After the update, they filter the vehicles on the corresponding part of the occupation table and transform their positions back into the originating occupation table. The vehicle update functions update the switching areas random number generator state with the one provided by the updated joined occupation table $\mathbf{F}_{\text{clone}}(\mathbf{a}, s_{\text{rand}}, \mathbf{o}^{(t)}_j \# s_{\text{rand}})$. The switching area update results, finally, in the pair consisting of the updated state of the random number generator and the set of updated vehicles.

Table 3.3.6 summarises all relevant functions, structure definitions, variables, and constants for this section.

3.4 Traffic model

OLSIMv3 successfully used the brakelight model [118, 115] for years. As common to many other microscopic traffic models, too, the model distinguishes between longitudinal and lateral motion. Additionally, the OLSIMv3 implementation of the model employed a two step algorithm of the otherwise atomic state transition function for the longitudinal motion. In the first step, the update process determined and updated the velocities of all vehicles. In the second step, the update process moves the

Table 3.3.6: Summary of relevant functions, structures, variables, and constants for Section 3.3.6.

definition	description	domain	range
$\langle a,b,c,d,e,f \rangle^{\text{swarea}}$	switching area	$\mathfrak{R}_{\text{occ}}^4 \times \mathbb{B}_{\text{cell}}^{\pm} \times (\mathfrak{L}_{\text{rway}}^4 \times S_{\text{area}}^4)^2$	S_{swarea}^4, see Eq. 3.34
$\mathbf{a}_{\text{switch}} \# s_{\text{rand}}$	state of RNG	$\mathbf{a}_{\text{switch}} \in S_{\text{swarea}}^4$	$\mathfrak{R}_{\text{occ}}^4$
$\mathbf{a}_{\text{switch}} \# \Delta y_{\text{lc}}$	connection direction	$\mathbf{a}_{\text{switch}} \in S_{\text{swarea}}^4$	$\mathcal{B}_{\text{cell}}^{\pm}$
$\mathbf{a}_{\text{switch}} \# f_{\text{name}}^{\text{roadway}}$	from roadway name	$\mathbf{a}_{\text{switch}} \in S_{\text{swarea}}^4$	$\mathfrak{L}_{\text{roadway}}^4$
$\mathbf{a}_{\text{switch}} \# t_{\text{name}}^{\text{roadway}}$	to roadway name	$\mathbf{a}_{\text{switch}} \in S_{\text{swarea}}^4$	$\mathfrak{L}_{\text{roadway}}^4$
$\mathbf{a}_{\text{switch}} \# s_{\text{area}}^{\text{from}}$	from area	$\mathbf{a}_{\text{switch}} \in S_{\text{swarea}}^4$	S_{area}^4
$\mathbf{a}_{\text{switch}} \# s_{\text{area}}^{\text{to}}$	to area	$\mathbf{a}_{\text{switch}} \in S_{\text{swarea}}^4$	S_{area}^4
$\mathbf{a}_{\text{switch}} \# \mathbf{k}^{(t)}$	access all vehicles	$\mathbf{a}_{\text{switch}} \in S_{\text{swarea}}^4$	$\mathcal{P}_{\text{veh}}^4(S_{\text{veh}}^4)$
$\mathbf{F}_{\text{swa-drive}}(t,a,b,c)$	drive update	$\mathfrak{T}_{\text{secs}} \times S_{\text{occ}}^4 \times S_{\text{swarea}}^4 \times S_{\text{occ}}^4$	$S_{\text{swarea}}^4 \times \mathcal{P}_{\text{veh}}^4(S_{\text{veh}}^4)$
$\mathbf{F}_{\text{swa-chlane}}(t,a,b,c)$	lane change update	$\mathfrak{T}_{\text{secs}} \times S_{\text{occ}}^4 \times S_{\text{swarea}}^4 \times S_{\text{occ}}^4$	$S_{\text{swarea}}^4 \times \mathcal{P}_{\text{veh}}^4(S_{\text{veh}}^4)$

vehicles to their next position. This kind of implementation is also common to SUMO [19]. The two step algorithms differ from the herein presented one step algorithms in the following characteristics. Whereas the one step algorithm treats the longitudinal update as a single atomic step, the two step algorithm breaks the state transition into two separate steps. As the vehicles hold the state between the two separate steps, they hold an inconsistent state after the first of the two steps. The updated state becomes only valid when the vehicles can in fact move to their aimed positions based on the already updated velocities. Otherwise a rollback would be required. In case of roadways modelled as partitioned into segments, the vehicles may not be able to move onto their aimed positions as those might be blocked. Additionally, the two step algorithm requires synchronisation after the velocity update in a parallel interpretation and due to the reason explained below.

The idea behind the two step algorithms is to separate the model-dependent calculation of the velocity for the next time step from the widely model-independent moving of the vehicles onto the next position. However, moving the vehicles onto their next positions turns out to depend on the model, too. In cases where vehicles enter the region in front of a yellow traffic light and detect a potential collision or are forced to brake by other reasons, they slow down which involves determining the model-dependent maximum braking capability. OLSIMv4 therefore introduces a different vehicle update model. The vehicle update model generalises the longitudinal and the lateral motion by an appropriate interface. This interface is suitable for any traffic model that considers a vehicle and its two predecessors. OLSIMv4 prevents the traffic model from being mixed with topological concepts. As a consequence, implementing a yellow traffic light for OLSIMv4 requires either to provide the information with the topological information or to tune ghost vehicles in front of the affected car that is forced to initiate a full braking manoeuvre. Similarly, the traffic model does not support explicit routing. Where necessary, the topological analysis has to provide this information encoded in an element of the topological structure information S_{topo}^4 which is part of the traffic model interface.

The brakelight model [118] is in good accordance with empirical findings. As Knorr's publication [109] provides an excellent formal description of the brakelight model, and as the latter can also use the interface of this section, it is not described again here. The shortcomings in the model development have been discussed already in Section 2.2.3. As a consequence, OLSIMv4 intended and prepared the use of the microscopic traffic model by Lee, Barlovic, Schreckenberg and Kim [124] that Pottmeier extended [164] for multi-lane traffic. Because of the asymmetric behaviour that is induced by German traffic regulations and that is found on German highways, Habel [70, 71] [1] extended the multi-lane model by the adaption of the asymmetric lane changing rules. As the original publication of the Lee *et al.* model describes the computation of the velocity for the next time step with the help of an inequality, the computation involved an iterative search for the largest value that still satisfied the inequality. With

---

[1] Habel published the asymmetric lane change rules in [71] after the work on this section was already complete.

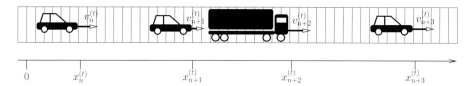

Figure 3.5: Single lane cell model for longitudinal motion of cars and trucks.

respect to the real-time requirements and the millions of vehicle updates for each simulation minute, such an iterative search is too inefficient. Section 3.4.1, thus, presents an optimised algorithm to compute the velocity directly and does not require any incremental search.

The model distinguishes between longitudinal and lateral motion, too. The distinction results in a two step update. Both updates require random number generators that are not discussed as part of the conceptual design. As the microscopic traffic models do not pose any specific characteristics on the random number generators, Subsection 4.2.2.1 discusses them as part of the implementation issues. Section 3.4.1 details the vehicles' longitudinal motion and Section 3.4.2 unveils all details concerning the lateral motion. More precisely, the following sections describe the three functions $\mathbf{F}_{\text{drive}}$, $\mathbf{F}_{\text{safe-ins}}$, and $\mathbf{F}_{\text{chlane}}$, all having signature $\mathbf{F}_{\text{chlane}} : \mathfrak{R}_{\text{occ}}^4 \times S_{\text{veh}}^4 \times S_{\text{nbrs}}^4 \times S_{\text{topo}}^4 \longrightarrow \mathfrak{R}_{\text{occ}}^4 \times S_{\text{veh}}^4$, for the drive update and the lane change update. They can be implemented by the brakelight model or any other microscopic traffic model that follows the two step update process. By providing a network-independent microscopic traffic model, OLSIMv4 boosts the network platform portability .

3.4.1 Longitudinal motion

This section describes the functions for the longitudinal vehicle update $\mathbf{F}_{\text{drive}} : \mathfrak{R}_{\text{occ}}^4 \times S_{\text{veh}}^4 \times S_{\text{nbrs}}^4 \times S_{\text{topo}}^4 \longrightarrow \mathfrak{R}_{\text{occ}}^4 \times S_{\text{veh}}^4$ and for the safe insertion of vehicles into a bulk of neighbouring vehicles $\mathbf{F}_{\text{safe-ins}} : \mathfrak{R}_{\text{occ}}^4 \times S_{\text{veh}}^4 \times S_{\text{nbrs}}^4 \times S_{\text{topo}}^4 \longrightarrow \mathfrak{R}_{\text{occ}}^4 \times S_{\text{veh}}^4$. The functions rely on the microscopic traffic model by Lee, Barlovic, Schreckenberg and Kim which is introduced in Subsection 3.4.1.1. As OLSIMv4 has required an implementation of this model and still uses it, the description focuses not only on the model rules but explains and details also an algorithm for an efficient calculation of the vehicle update result in Subsection 3.4.1.2. Subsection 3.4.1.3 completes this section with a discussion of the model's shortcomings, possible solutions, and further optimisations.

3.4.1.1 Classic model

Function $\mathbf{F}_{\text{drive}}$ takes four arguments which are described in the following. The first one, $s_{\text{rand}} \in \mathfrak{R}_{\text{occ}}^4$, represents the state of a random number generator. It consists of a generator function $\mathbf{F}_{\text{random}} : \mathfrak{R}_{\text{occ}}^4 \longrightarrow S_{\text{random}}^4$ and the random structure $S_{\text{random}}^4 := \langle r_{\text{num}}, s_{\text{rand}} \rangle^{\text{random}}$ with $r_{\text{num}}, s_{\text{rand}} \in \mathfrak{R}_{\text{occ}}^4$. A single evaluation of Function results in an element $\mathbf{F}_{\text{random}}(s_{\text{rand}}) = g_{\text{random}} \in S_{\text{random}}^4$ of the random structure. The definition of the set of random numbers $\mathfrak{R}_{\text{occ}}^4$ made use of the upper bound R_{max} for the random numbers. Table 3.4.2 presents the value chosen for OLSIMv4. The occupation tables store their states for the random number generators, and the simulation update function supplies them to both vehicle update functions, $\mathbf{F}_{\text{drive}}$ and $\mathbf{F}_{\text{chlane}}$. The second argument of Function $\mathbf{F}_{\text{drive}}$ specifies the vehicle $\mathbf{k}_n^{(t)} \in S_{\text{veh}}^4$ which will become the subject of the update. Its index n refers to the order of the vehicles on its current lane $(\mathbf{k}_n^{(t)} \# \mathbf{q}^{(t)}) \# y$ abbreviated as $y_n^{(t)}$ from now on. The third argument to the vehicle update function contains the preceding and the neighbouring vehicles in form of a neighbour structure element $\mathbf{b} \in S_{\text{nbrs}}^4$ to vehicle $\mathbf{k}_n^{(t)}$. The longitudinal update in the presented form requires only the preceding vehicles $\mathbf{k}_{n+1}^{(t)} := \mathbf{b} \# \mathbf{k}_5^{(t)}$ and $\mathbf{k}_{n+2}^{(t)} := \mathbf{b} \# \mathbf{k}_6^{(t)}$ of the neighbour structure element. The definitions $x_{n+i}^{(t)} := (\mathbf{k}_{n+i}^{(t)} \# \mathbf{q}^{(t)}) \# x$ for the positions, for the lanes $y_{n+i}^{(t)} := (\mathbf{k}_{n+i}^{(t)} \# \mathbf{q}^{(t)}) \# y$, and for the velocities $y_{n+i}^{(t)} := (\mathbf{k}_{n+i}^{(t)}\#\mathbf{q}^{(t)})\#y$ of all vehicles in the neighbour structure element follow analogously to the abbreviation scheme of the vehicle's current lane $y_n^{(t)}$ from above. The abbreviation scheme applies also to the other vehicle structure components such as the length $L_{n+i} := \mathbf{k}_{n+i}^{(t)}\#L$ and the maximum braking capability $D_{n+i} := \mathbf{k}_{n+i}^{(t)}\#D$. The fourth argument of the update function contains the information that the topological analysis encodes as a topology structure element $\mathbf{p} \in S_{\mathrm{topo}}^4$.

The original publication by Lee *et al.* [124] used the inequality in the form of 3.36 to determine the "safe" maximum velocity for a vehicle $\mathbf{k}_n^{(t)}$ with index n and its next two predecessors $\mathbf{k}_{n+1}^{(t)}, \mathbf{k}_{n+2}^{(t)}$ with indices $n+1$, $n+2$ for the next time step $t+\Delta t$ with $\Delta t = 1\,\mathrm{s}$ and the current time $t \in \mathfrak{T}_{\mathrm{secs}} = \{i\,\mathrm{s} \mid i \in \mathbb{N}_0\}$ in seconds. The largest value for the velocity type value $c_n^{(t+\Delta t)} \in \mathfrak{V}_{\mathrm{veh}}^4$ in cells per second that still satisfies Inequality 3.36 represents the "safe" maximum velocity for the next time step $t + \Delta t$. The original publication used $c_n^{(t+1)} := c_n^{(t+\Delta t)}$ and $v_n^{(t+1)} := v_n^{(t+\Delta t)}$ as well as $x_n^{(t+1)} := x_n^{(t+\Delta t)}$ to denote the velocities and the positions for the next time step. The vehicles in the original publication have the length $L_n = L_{n+1} := 5$ cells. They stand on their front positions $x_n^{(t)} \leqslant x_n^{(t)} - L_{n+1} \in \mathbb{Z}$ that represent cell indices of size $\Delta x := 1$ cells $:= 1.5\,\mathrm{m}$ of a single lane. Figure 3.5 depicts such a scenario. The value $v_{n+1}^{(t)} \in \mathfrak{V}_{\mathrm{veh}}^4$ denotes the current velocity of the predecessor to the vehicle with index n. The symbols D_n, $D_{n+1} \in \mathbb{N}_0$ represent the maximum braking capabilities for the vehicles with indices n and $n + 1$. Respectively the original model assigned identical maximum braking capabilities to all vehicles, i.e. $\forall n : D = D_n = D_{n+1} := 2$ cells/s². The form of Inequality 3.36 might look a little bit confusing, as the terms in the upper bounds of the sums might suggest that they range over the natural numbers.

$$x_n^{(t)} + \Delta_n + \sum_{i=0}^{\tau_f(c_n^{(t+1)})} \left(c_n^{(t+1)} - D_n i\right) \leqslant x_{n+1}^{(t)} + \sum_{i=1}^{\tau_l(v_{n+1}^{(t)})} \left(v_{n+1}^{(t)} - D_{n+1} i\right) \tag{3.36}$$

However, the functions denoting the upper boundaries of the summations, $\tau_f : \mathfrak{V}_{\mathrm{veh}}^4 \longrightarrow \mathfrak{T}_{\mathrm{xupd}}^4 := \{t \in \mathfrak{T}_{\mathrm{secs}} \mid t \leqslant \lfloor v_{\mathrm{max}}/D \rfloor\}$ and $\tau_l : \mathfrak{V}_{\mathrm{veh}}^4 \longrightarrow \mathfrak{T}_{\mathrm{xupd}}^4$ with $v_{\mathrm{max}} = 20$ cells/s, result in the range of the time in seconds $\mathfrak{T}_{\mathrm{secs}}$. Therefore, OLSIMv4 substitutes the terms in the upper boundaries of the sums by the two variables $N_f := \tau_f(c_n^{(t+\Delta t)})/\Delta t$ and $N_t := \tau_l(v_{n+1}^{(t)})/\Delta t$. Both functions will be defined later in the context of Equation 3.41. The calculation of the upper boundaries value N_f for the left sum requires to evaluate function τ_f with the function argument $c_n^{(t+\Delta t)}$. As this parameter is also the subject of the optimisation problem posed by the inequality, a closed term that solves the inequality is more difficult to find. The solution to this problem will be unveiled with the optimisation. For the moment, finding the largest value that solves the inequality requires iterating over all possible values of $c_n^{(t+\Delta t)} \in \mathfrak{V}_{\mathrm{veh}}^4$ and choosing the largest one that solves the inequality. Additionally, in order to define the range for the set $\mathfrak{T}_{\mathrm{xupd}}^4$ with respect to the needs of OLSIMv4, the maximum velocity constant requires adjustment to $v_{\mathrm{max}} := \mathbf{s}_{\mathrm{vehtype}}^{\mathrm{any}}\#V_{\mathrm{max}}$. When denoted properly, the inequality takes the form of 3.37.

$$x_n^{(t)} + \Delta_n + \sum_{i=0}^{N_f} \Delta t \left(c_n^{(t+\Delta t)} - i\Delta t D_n\right) \leqslant x_{n+1}^{(t)} + \sum_{i=1}^{N_t} \Delta t \left(v_{n+1}^{(t)} - i\Delta t D_{n+1}\right). \tag{3.37}$$

The left hand side of inequality 3.37 uses the minimal relative safety headway Δ_n between two subsequent vehicles that the model determines with the help of Equation 3.38. The value for the headway depends on the constant parameter $g_{\mathrm{add}} = 4$ cells and on the value of $\gamma_n^{(t)} \in \mathbb{B}_{\mathrm{model}}^+ := \{0, 1\}$ that models optimistic or pessimistic (or "defensive state" [124]) driving behaviour for a vehicle with index n and the length L_{n+1} of its predecessor.

$$\Delta_n := L_{n+1} + \gamma_n^{(t)} \cdot \max\left\{0\,\mathrm{cells}, \min\left\{g_{\mathrm{add}}, v_n^{(t)}\Delta t - g_{\mathrm{add}}\right\}\right\}. \tag{3.38}$$

Equation 3.40 defines the cases that distinguish between optimistic and pessimistic driving behaviour, i.e. $\gamma_n^{(t)} = 0$ and $\gamma_n^{(t)} = 1$. The distinction depends on the velocity values $v_n^{(t)}$, $v_{n+1}^{(t)}$, $v_{n+2}^{(t)}$ of any three vehicles in a sequence. The original publication used only a part of the second case from Equation 3.40

Table 3.4.1: Summary of relevant functions, structures, variables, and constants for Section 3.4.1.

definition	description	domain	range
$\mathfrak{T}_{\text{secs}}$	time in seconds		$\{i\,\mathrm{s} \mid i \in \mathbb{N}_0\}$
$v_{\max}$	maximum velocity		$\mathbf{s}_{\text{vehtype}}^{\text{any}} \# V_{\max}$
$v_{\max,n}$	$v_{\max}$ for vehicle $\mathbf{k}_n^{(t)}$	$\mathbf{k}_n^{(t)} \in S_{\text{veh}}^4$	see Equation 3.39
$v_{\text{fast},n}$	velocity still optimistic	$\mathbf{k}_n^{(t)} \in S_{\text{veh}}^4$	$v_{\max} - 1$ cells/s
$\mathfrak{T}_{\text{xupd}}^4$	braking time horizons		$\{t \mid t \in \mathfrak{T}_{\text{secs}} \wedge t \leqslant \lfloor \frac{v_{\max}}{D} \rfloor\}$
$R_{\max}$	maximum for RNG		$\mathfrak{R}_{\text{occ}}^4$
$\mathbb{B}_{\text{model}}^+$	binary digit		$\{0,1\}$
$\mathfrak{D}_{\text{dawdle}}^4$	stochastic parameters		$\{d \in \mathbb{R} \mid 0 \leqslant d < 1\}$
Δ_n	relative safety headway		$\mathfrak{X}_{\text{rway}}^4$, see Equation 3.38
N_f	braking steps follower		$\tau_f(c_n^{(t+\Delta t)})/\Delta t$
N_l	braking steps leader		$\tau_l(v_{n+1}^{(t)})/\Delta t$
$\mathbf{F}_{\text{drive}}(a,b,c,d)$	drive update	$\mathfrak{R}_{\text{occ}}^4 \times S_{\text{veh}}^4 \times S_{\text{nbrs}}^4 \times S_{\text{topo}}^4$	$\mathfrak{R}_{\text{occ}}^4 \times S_{\text{veh}}^4$
$\mathbf{F}_{\text{safe--ins}}(a,b,c,d)$	safe insert	$\mathfrak{R}_{\text{occ}}^4 \times S_{\text{veh}}^4 \times S_{\text{nbrs}}^4 \times S_{\text{topo}}^4$	$\mathfrak{R}_{\text{occ}}^4 \times S_{\text{veh}}^4$
$\langle a,b \rangle^{\text{random}}$	the random structure	$\mathfrak{R}_{\text{occ}}^4 \times \mathfrak{R}_{\text{occ}}^4$	S_{random}^4
$\mathbf{g} \# r_{\text{num}}$	random number	$\mathbf{g} \in S_{\text{random}}^4$	$\mathfrak{R}_{\text{occ}}^4$
$\mathbf{g} \# s_{\text{rand}}$	next state	$\mathbf{g} \in S_{\text{random}}^4$	$\mathfrak{R}_{\text{occ}}^4$
$\gamma_n^{(t)}$	optimistic/pessimistic	$\mathbf{k}_n^{(t)}, \mathbf{k}_{n+1}^{(t)}, \mathbf{k}_{n+2}^{(t)} \in S_{\text{veh}}^4$	$\mathbb{B}_{\text{model}}^+$, see Equation 3.40
$\mathbf{F}_\tau(a,b)$	braking time horizon	$\mathfrak{V}_{\text{veh}}^4 \times \mathbb{B}_{\text{model}}^+$	$\mathfrak{T}_{\text{xupd}}$, see Equation 3.41
$c_n^{(t+\Delta t)}$	"safe" maximum velocity		$\mathfrak{V}_{\text{veh}}^4$ see Algorithm 1
$\mathbf{F}_{\text{dawdle}}(v)$	dawdle parameter	$\mathfrak{V}_{\text{veh}}^4$	$\mathfrak{D}_{\text{dawdle}}^4$, see Equation 3.43
$\tilde{v}_n^{(t+\Delta t)}$	velocity w.r.t. acc./dec.		see Eqs. 3.42 and 3.48

as additional rule for optimistic driving behaviour. That rule employed "a constant v_{fast} slightly below $v_{\max}$" [124] with $v_{\text{fast}} = 19$ cells/s. The constant describes a lower threshold for a velocity difference at which optimistic driving behaviour is still adequate. The publication [164, Sec. 5.6, Eq. 5.29] and, consequently, OLSIMv4 also use this constant. However, OLSIMv4 employs the two different vehicle types, cars and trucks, and increased the value of the overall maximum velocity $v_{\max}$. Additionally, OLSIMv4 respects the topological constraints. Thus, the value for v_{fast} requires adjustment and individualisation. To distinguish it from the original publication and to denote the individualisation, it is named $v_{\text{fast},n}$ in the context of OLSIMv4. The constant still holds the original relation $v_{\text{fast},n} = v_{\max,n} - 1$ cells/s but now uses a different value for the per vehicle individual maximum velocity $v_{\max,n}$. It is obtained from Equation 3.39 as described in the following.

$$v_{\max,n} := \min\left\{ v_{\max,\text{TP},n}, \lceil v_{\max,\text{CA}} \cdot v_{\text{mul},n} \rceil \right\}. \tag{3.39}$$

The calculation of a potential, resulting maximum velocity $v_{\max,n}$ respects the topological constraint of the maximum velocity $v_{\max,\text{CA}} := \mathbf{p} \# v_{\pm 0,\max}$ as provided by the topological analysis. The other variable in Equation 3.39 $v_{\text{mul},n} := \mathbf{k}_n^{(t)} \# V_{\text{mul}}$ specifies the strictness factor by which the vehicle $\mathbf{k}_n^{(t)}$ or the driver, respectively, obeys to the topological constraints. The product rounded to the nearest natural number $\lceil v_{\max,\text{CA}} \cdot v_{\text{mul},n} \rceil$ represents a vehicle's maximum desired velocity. It is only limited by the velocity capability of its vehicle type, namely $v_{\max,\text{TP},n} := (\mathbf{k}_n^{(t)} \# \mathbf{s}_{\text{vehtype}}) \# V_{\max}$. Equation 3.39 displays the calculation of the resulting maximum velocity $v_{\max,n}$. Even though the strictness factor $v_{\text{mul},n}$ is a vehicle parameter, OLSIMv4 adjusts it only per vehicle type due to the fact that the optimistic to pessimistic ratio would no longer match the original design [70, Sec. 3.3.2]. As a consequence, it would remove the high flow states in the model (compare [70, Fig. 3.12(b)] and [164, Fig. 5.14(a)]).

The other conditions in the second case in Equation 3.40 originate from the multi-lane traffic adaptions by Pottmeier [164, Sec. 5.6, Eq. 5.29] and, thus, differ from the original publication [124].

$$\gamma_n^{(t)} := \begin{cases} 0 & , \text{ if } v_n^{(t)} \leqslant v_{n+1}^{(t)} \leqslant v_{n+2}^{(t)}, \\ 0 & , \text{ if } v_{n+2}^{(t)} \geqslant v_{\text{fast},n} \;\wedge\; v_n^{(t)} - v_{n+1}^{(t)} < D_n \Delta t \;\wedge\; x_{n+1}^{(t)} - x_n^{(t)} - L_{n+1} > 8 \text{ cells}, \\ 1 & , \text{ otherwise.} \end{cases} \qquad (3.40)$$

The description of inequality 3.37 is still incomplete as it lacks the definitions of the two functions $\tau_f : \mathfrak{V}_{\text{veh}}^4 \longrightarrow \mathfrak{T}_{\text{xupd}}^4$ and $\tau_l : \mathfrak{V}_{\text{veh}}^4 \longrightarrow \mathfrak{T}_{\text{xupd}}^4$ from the original publication. They determine the time horizons in an assumed braking manoeuvre. Their definitions can be summarised in a more general function $\mathbf{F}_\tau : \mathfrak{V}_{\text{veh}}^4 \times \mathbb{B}_{\text{model}}^+ \longrightarrow \mathfrak{T}_{\text{xupd}}$. This function calculates the braking manoeuvre time horizon for a given vehicle's velocity $v \in \mathfrak{V}_{\text{veh}}^4$. Equation 3.41 introduces Function $\mathbf{F}_\tau$. With its help, the two calculations of the time horizons write as $N_f \Delta t = \tau_f(c_n^{(t+\Delta t)}) = \mathbf{F}_\tau(c_n^{(t+\Delta t)}, 1)$ and $N_l \Delta t = \tau_l(v_{n+1}^{(t)}) = \mathbf{F}_\tau(v_{n+1}^{(t)}, 0)$. Function $\mathbf{F}_\tau$ makes use of another parameter, namely t_{safe}. It represents the minimum time horizon in an assumed braking manoeuvre for the optimistic driver behaviour. In the original model the value for that constant t_{safe} was chosen as $t_{\text{safe}} := 3\Delta t$ whereas in OLSIMv4 the value $t_{\text{safe}} := 4\Delta t$ turned out to better match the requirements for German highway traffic [70, Sec. 3.3.1]. Due to the generally higher maximum velocities that are found on German highways when compared with other European countries, OLSIMv4 uses $v_{\text{max}} = 22 \text{ cells/s} = 118.8 \text{ km/h}$ for the maximum velocity whereas the original publication by Lee $et\ al.$ used only $v_{\text{max}} = 20 \text{ cells/s} = 108 \text{ km/h}$. The vehicle type diversification gives another reason for this choice.

$$\mathbf{F}_\tau(v, \delta_{\text{fl}}) := \gamma_n^{(t)} \cdot \frac{v}{D} + (1 - \gamma_n^{(t)}) \cdot \max\left\{ 0\,\text{s}, \min\left\{ \left\lfloor \frac{v}{D} \right\rfloor, t_{\text{safe}} - \delta_{\text{fl}}\Delta t \right\} \right\}. \qquad (3.41)$$

The definitions for τ_f and τ_l complete the prerequisites for the computation of $c_n^{(t+\Delta t)}$ which represents the "safe" maximum velocity. The computation requires checking several values to find those that solve the inequality. The largest value that solves the inequality results in the final value for the "safe" maximum velocity in the next time step. A solution in the form of a closed term was neither supplied in the original publication nor in the later model extensions by Pottmeier and Habel. Algorithm 1 however can be represented as a closed term.

Once the value for the "safe" maximum velocity $c_n^{(t+\Delta t)}$ is available, the remaining steps are limited to the following ones. First, the computation of the velocity considers the limited acceleration and deceleration capabilities. Second, it takes dawdling of the vehicles into account. And, finally, it moves the vehicle to its next position. In the following, these steps are explained in more detail. Respecting the limited acceleration and deceleration capabilities requires to calculate a resulting maximum velocity value $v_{\text{max},n}$ first. The potential maximum velocity $c_n^{(t+\Delta t)}$ for the next time step has an upper limit that is denoted by the vehicle's current velocity plus its maximum acceleration capability $v_n^{(t)} + \Delta t A_n$. Analogously, does a vehicle's current velocity decreased by its maximum deceleration capability $v_n^{(t)} - \Delta t D_n$ truncate the velocity value at the lower bound. Finally, the resulting velocity still has to be smaller than $v_{\text{max},n}$. The definition of the velocity $\tilde{v}_n^{(t+1)}$ in Equation 3.42 uses the maximum acceleration capability A_n that is defined analogously to the maximum deceleration capability as $A := A_n := A_{n+1} = 1 \text{ cells/s}^2$.

$$\tilde{v}_n^{(t+1)} = \min\left\{ v_{\text{max},n}, v_n^{(t)} + \Delta t A_n, \max\left\{ 0\,\text{cells/s}, v_n^{(t)} - \Delta t D_n, c_n^{(t+\Delta t)} \right\} \right\} \qquad (3.42)$$

After consideration of the limited acceleration and deceleration capabilities, the vehicle dawdling step takes place. This step is also common to other probabilistic traffic cellular automaton models such as Nagel-Schreckenberg-Model [147] and the brakelight model [118]. The dawdling step models velocity fluctuations of the vehicles that occur due to non-constant driving behaviour. Possible reasons for this driving behaviour range from the drivers' lack of full concentration to ascendant slopes in the road topology. The velocity fluctuations are necessary to reproduce spontaneous jam formation and metastable traffic dynamics.

The computation of the dawdling requires the random number generator environment that Section 3.3.5 and the beginning of this section have already introduced. The random number generator function

$\mathbf{F}_{\text{random}}$ helps to determine the next random number value $z_{\text{rand}} := (\mathbf{F}_{\text{random}}(s_{\text{rand}}))\#r_{\text{num}}$ as well as the next random number state $s'_{\text{rand}} := (\mathbf{F}_{\text{random}}(s_{\text{rand}}))\#s_{\text{state}}$. The next state will become a part of the drive update $\mathbf{F}_{\text{drive}}$ function result. The stochastic parameter p is another prerequisite of the dawdling step. Function $\mathbf{F}_{\text{dawdle}} : \mathfrak{V}_{\text{veh}}^4 \longrightarrow \mathfrak{D}_{\text{dawdle}}^4$ from Equation 3.43 with $\mathfrak{D}_{\text{dawdle}}^4 := \{d \in \mathbb{R} \mid 0 \leqslant d < 1\}$ linearly interpolates the stochastic parameter $p := \mathbf{F}_{\text{dawdle}}(v_n^{(t)})$ as a generalisation of the slow-to-start rule [124, 16] in the velocity interval $[0\,\text{cells/s}; v_{\text{slow}}]$ between the parameters $p_0 = 0.32$ and $p_d = 0.11$. The original publication as well as OLSIMv4 use the slow-to-start rule generalisation up to $v_{\text{slow}} := 5\,\text{cells/s}$. The interpolation further depends on a velocity v. To evaluate Function $\mathbf{F}_{\text{dawdle}}$, the Lee et al. model uses the current time step's velocity $v_n^{(t)}$ of the nth vehicle $\mathbf{k}_n^{(t)}$ rather than the "safe" maximum velocity $\tilde{v}_n^{(t+1)}$ for the next time step.

$$\mathbf{F}_{\text{dawdle}}(v) := \max\left\{p_d, p_0 - v \cdot \frac{p_0 - p_d}{v_{\text{slow}}}\right\}. \tag{3.43}$$

The value of the stochastic parameter p decides whether the vehicle will dawdle or not, i.e. whether the value of δ_{dawdle} will be greater than zero or not. Equation 3.44 distinguishes the two cases. The random number value z_{rand} in combination with the stochastic parameter p allows to define δ_{dawdle}.

$$\delta_{\text{dawdle}} := \begin{cases} 1 & , \text{if } z_{\text{rand}} \leqslant pR_{\max}, \\ 0 & , \text{otherwise.} \end{cases} \tag{3.44}$$

With δ_{dawdle} the prerequisites for the calculation of a vehicle's velocity in the next time step $t + \Delta t$ are complete. Equation 3.45 presents the calculation of a vehicle's velocity in the next time step. The calculation prevents the velocity values from falling below the value of the current velocity reduced by the maximum braking capability. Thus, the vehicle dawdles only when the velocity update to $\tilde{v}_n^{(t+\Delta t)}$ did not consider the maximum braking capability.

$$v_n^{(t+\Delta t)} := \max\left\{0\,\text{cells/s}, v_n^{(t)} - D\Delta t, \tilde{v}_n^{(t+\Delta t)} - \delta_{\text{dawdle}}\Delta t\right\} \tag{3.45}$$

Finally, Equation 3.46 determines a vehicle's next position $x_n^{(t+\Delta t)}$ for time step t simply by adding the covered distance to its current position. Table 3.4.2 summarises all parameters of the single lane model. Most of them should remain unchanged otherwise intrinsic model characteristics are likely to break.

$$x_n^{(t+\Delta t)} = x_n^{(t)} + v_n^{(t+\Delta t)}\Delta t. \tag{3.46}$$

The velocity and the position for the next time step $t + \Delta t$ result in a new coordinate structure element $\mathbf{q}_n^{(t+\Delta t)} := \langle y_n^{(t)}, x_n^{(t+\Delta t)}, v_n^{(t+\Delta t)}, b_n^{(t)}\rangle^{\text{coord}}$ for the use by the updated vehicle $\mathbf{k}_n^{(t)}$. As the vehicles contain only a fixed number of past coordinates, the updated vehicle $\mathbf{k}_n^{(t+\Delta t)}$ re-uses every coordinate element from before the update except for the oldest coordinate element $\mathbf{k}_n^{(t)}\#\mathbf{q}_j$ with $j = |\mathbf{k}_n^{(t)}\#\mathbf{q}| - 1$. The updated vehicle $\mathbf{k}_n^{(t+\Delta t)}$ takes the form of Equation 3.47.

$$\mathbf{k}_n^{(t+\Delta t)} := \langle s_{\text{vehtype}}, L, V_{\max}, A_{\max}, D_{\max}, V_{\text{mul}}, \mathbf{q}^{(t+\Delta t)}, \mathbf{q}^{(t)}, \mathbf{q}_1, \ldots, \mathbf{q}_{j-1}\rangle^{\text{veh}}. \tag{3.47}$$

Finally, the evaluation of Function $\mathbf{F}_{\text{drive}} : \mathfrak{R}_{\text{occ}}^4 \times S_{\text{veh}}^4 \times S_{\text{nbrs}}^4 \times S_{\text{topo}}^4 \longrightarrow \mathfrak{R}_{\text{occ}}^4 \times S_{\text{veh}}^4$ for the arguments state of the random number generator $s_{\text{rand}} \in \mathfrak{R}_{\text{occ}}^4$, vehicle to update $\mathbf{k}_n^{(t)} \in S_{\text{veh}}^4$, neighbouring vehicles $\mathbf{b} \in S_{\text{nbrs}}^4$, and topological constraints $\mathbf{p} \in S_{\text{topo}}^4$ results in the pair of the updated state of the random number generator s'_{rand} and the updated vehicle $\mathbf{k}_n^{(t+\Delta t)}$, i.e. $\mathbf{F}_{\text{drive}}(s_{\text{rand}}, \mathbf{k}_n^{(t)}, \mathbf{b}, \mathbf{p}) := (s'_{\text{rand}}, \mathbf{k}_n^{(t+\Delta t)})$ with $s'_{\text{rand}} := (\mathbf{F}_{\text{random}}(s_{\text{rand}}))\#s_{\text{state}}$ and $\mathbf{k}_n^{(t+\Delta t)}$ from Equation 3.47.

The safe insert function $\mathbf{F}_{\text{safe}-\text{ins}}$ calculates the vehicle for the next time step $\mathbf{k}_n^{(t+\Delta t)}$ very similar to the $\mathbf{F}_{\text{drive}}$ function. But instead of limiting the velocity for the next time step $\tilde{v}_n^{(t+\Delta t)}$ by the maximum acceleration and deceleration capability as in Equation 3.42 for the drive update, it defines the velocity according to Equation 3.48. The vehicle's final velocity for the next time step

Table 3.4.2: Parameters of the single lane model for OLSIMv4.

name	description	scope	value
Δt	time discretisation	global	$1\,\mathrm{s}$
Δx	spatial discretisation	global	$1.5\,\mathrm{m}$
L_{car}	vehicle length of cars	global	$5\,\mathrm{cells}$
L_{truck}	vehicle length of trucks	global	$10\,\mathrm{cells}$
A	maximum acceleration capability	global	$1\,\mathrm{cells/s^2}$
D	maximum deceleration capability	global	$2\,\mathrm{cells/s^2}$
v_{mul}^{n}	strictness to obey topological constraints	vehicle	1.15
p_0	slow-to-start dawdle parameter	global	0.32
p_{d}	driving dawdle parameter	global	0.11
v_{slow}	upper bound for interpolation of dawdle parameter	global	$5\,\mathrm{cells/s}$
v_{fast}	velocity for optimistic driving	global	$v_{\mathrm{max},n} - 1\,\mathrm{cells/s}$
t_{safe}	time horizont for collision-free dynamics in optimistic state	global	$4\,\mathrm{s}$
g_{add}	additional gap in defensive state	global	$4\,\mathrm{cells}$
R_{max}	maximum value for random numbers	global	10^6

$v_n^{(t+\Delta t)}$ results then also from the dawdling as described by Equation 3.45. However, the vehicle's next position coordinate does not necessarily need to reflect the updated velocity and use instead $\mathbf{q}_n^{(t+\Delta t)} := \langle y_n^{(t)}, x_n^{(t)}, v_n^{(t+\Delta t)}, b_n^{(t+\Delta t)} \rangle^{\mathrm{coord}}$ instead. The resulting pair of the safe insertion function that consists of the updated random number state and the updated vehicle is then obtained as $\mathbf{F}_{\mathrm{safe-ins}}(s_{\mathrm{rand}}, \mathbf{k}_n^{(t)}, \mathbf{b}, \mathbf{p}) := (s'_{\mathrm{rand}}, \mathbf{k}_n^{(t+\Delta t)})$ with $s'_{\mathrm{rand}} := (\mathbf{F}_{\mathrm{random}}(s_{\mathrm{rand}})) \# s_{\mathrm{state}}$ and $\mathbf{k}_n^{(t+\Delta t)}$ from Equation 3.47. Other variants of the safe insert function may want to insert more aggressively or may want to limit the impact of disturbing the following vehicles as described in Equation 3.65.

$$\tilde{v}_n^{(t+\Delta t)} = \min\left\{ v_{\mathrm{max},n}, \max\left\{ 0\,\mathrm{cells/s}, c_n^{(t+\Delta t)} \right\} \right\} \tag{3.48}$$

To ease the reference to this section's definitions, Table 3.4.1 summarises the relevant functions, structure definitions, variables, and constants of this section.

3.4.1.2 Optimisation

Inequality 3.37 describes the position trajectories of the two successive vehicles in an extreme braking manoeuvre. The inequality guarantees that no collisions occur for the next $N_f = \tau_f(c_n^{(t+\Delta t)})/\Delta t$ time steps with the "safe" maximum velocity $c_n^{(t+\Delta t)}$.

In the presented form, Inequality 3.37 involves an iterative search for the largest values of $c_n^{(t+\Delta t)}$ in the worst case for each possible value of $c_n^{(t+\Delta t)}$ in the interval $\left[v_n^{(t)} + A; v_n^{(t)} - D \right]$. Calculating all values of $c_n^{(t+\Delta t)}$ for all possible vehicle constellations in advance and applying the dawdling to the precalculated values might seem to be an optimisation at hand. As the value $c_n^{(t+\Delta t)}$ depends on the velocities of $v_n^{(t)}$, $v_{n+1}^{(t)}$, and $v_{n+2}^{(t)}$ as well as the headway $x_{n+1}^{(t)} - x_n^{(t)} - L_{n+1} > 8\,\mathrm{cells}$, the precalculation would result in a lookup table of roughly 5×10^5 entries which is not a very large table but also not a large optimisation as it requires frequent key lookups with a four component key. In a parallel interpretation it is questionable where exactly such a lookup table should be made available— as a common data structure or instantiated separately for each thread. Additionally, in a simulation network such as the NRW highway network, roughly 12×10^4 vehicles move simultaneously in the rush-hour. Summarising the 60 simulation runs for one real-time minute requires, thus, approximately 16×10^6 computations or lookups of the value for $c_n^{(t+\Delta t)}$ solely for the longitudinal motion. Taking the lateral motion also into account would consume another two to four times of the 16×10^6 computations according to Equation 3.65. Considering the real-time requirements of OLSIMv4 and the number of computations or lookups, the computation methods are insufficient.

The calculation of the "safe" maximum velocity $c_n^{(t+\Delta t)}$ hides the optimisation potential as described hereafter. The optimised calculation yields Algorithm 1 which consists of 7 to 12 steps. The optimisation first requires to transform Inequality 3.37 into the form of Inequality 3.49. After the transformation, the sums therein are represented by simpler terms through elimination of the $\mathbf{F}_\tau(v, \delta_{\mathrm{fl}})$ functions from Equation 3.41. The elimination removes the minimum and maximum functions in $\mathbf{F}_\tau(v_j, \delta_{\mathrm{fl}})$ by splitting the domain for v into distinct sub-intervals. The simplified terms allow solving the inequality by simply solving a quadratic equation for each of the distinct sub-intervals.

$$\mathbf{X}_{\mathrm{n}}^{\mathrm{B}}(c_n^{(t+\Delta t)}) \leqslant g_n^{(t)} + \mathbf{X}_{\mathrm{n+1}}^{\mathrm{B}}(v_{n+1}^{(t)}) \text{, where } g_n^{(t)} := x_{n+1}^{(t)} - x_n^{(t)} - \Delta_n \ . \tag{3.49}$$

The terms in Inequality 3.49 for the covered braking distances of the follower, $\mathbf{X}_{\mathrm{n}}^{\mathrm{B}}(c_n^{(t+\Delta t)})$, and of the leader, $\mathbf{X}_{\mathrm{n+1}}^{\mathrm{B}}(v_{n+1}^{(t)})$ are defined by Equation 3.50 and Equation 3.51, respectively. The superscript letter B in the braking distances $\mathbf{X}_{\mathrm{n}}^{\mathrm{B}}(c_n^{(t+\Delta t)})$ and $\mathbf{X}_{\mathrm{n+1}}^{\mathrm{B}}(v_{n+1}^{(t)})$ indicates the braking distance. The latter depends on the maximum "safe" velocity of the follower $c_n^{(t+\Delta t)}$ for the next time step and of the leader $v_{n+1}^{(t)}$ for the current time stamp.

$$\mathbf{X}_{\mathrm{n}}^{\mathrm{B}}(c_n^{(t+\Delta t)}) := \Delta t \sum_{i=0}^{\tau_f(c_n^{(t+\Delta t)})} \left(c_n^{(t+\Delta t)} - \Delta t D_n i \right) \text{ and} \tag{3.50}$$

$$\mathbf{X}_{\mathrm{n+1}}^{\mathrm{B}}(v_{n+1}^{(t)}) := \Delta t \sum_{i=1}^{\tau_l(v_{n+1}^{(t)})} \left(v_{n+1}^{(t)} - \Delta t D_{\mathrm{n+1}} i \right) \ . \tag{3.51}$$

The value for the effective gap $g_n^{(t)}$ depends on the the headway $h_n^{(t)} := x_{n+1}^{(t)} - x_n^{(t)} - L_{\mathrm{n+1}}$, on the driving behaviour $\gamma_n^{(t)}$, and on the current velocity $v_n^{(t)}$ of vehicle n. Moreover, the value for the effective gap $g_n^{(t)}$ depends monotonously on the current velocity $v_n^{(t)}$. Thus, the velocity part of the domain for the effective gap ranges in the interval $V_g = 1/\Delta t \, [0 \text{ cells}; v_{\max}\Delta t + \epsilon g_{\mathrm{add}}[$ with $0 \text{ cells} < \epsilon g_{\mathrm{add}} < 1 \text{ cells}$. Splitting the interval at the locations $\varsigma_i \in V_g$ with $\varsigma_i \Delta t := 0g_{\mathrm{add}}, 1g_{\mathrm{add}}, 2g_{\mathrm{add}}, v_{\max}\Delta t + \epsilon g_{\mathrm{add}}$ and $i = 0, \ldots, N_\varsigma$ into $N_\varsigma := 3$ sub-intervals $[\varsigma_i; \varsigma_{i+1}[$ yields the definition of the velocity-dependent, effective gap Function $\mathbf{F}_{\mathrm{gap}} : \mathfrak{V}_{\mathrm{veh}}^4 \longrightarrow \mathfrak{X}_{\mathrm{rway}}^4$ of Equation 3.53. The modified Heaviside-Function Θ is required in the following. It is defined in the usual way except for the case $x = 0$ where instead of the established $\Theta(0) = 1/2$ herein the value $\Theta(0) := 0$ is used. Additionally, it is herein used in an extended form of a generic function over the real numbers, the velocities $\mathfrak{V}_{\mathrm{veh}}^4$, and the time in seconds $\mathfrak{T}_{\mathrm{secs}}^4$ domains. With the help of the modified Heaviside-Function Θ and the interval factor Functions $\sigma^i : \mathfrak{V}_{\mathrm{veh}}^4 \longrightarrow \mathbb{B}_{\mathrm{bit}}^+$ from Equation 3.52, determining the effective gap becomes equivalent to evaluating $g_n^{(t)} := \mathbf{F}_{\mathrm{gap}}(v_n^{(t)})$ with Function $\mathbf{F}_{\mathrm{gap}}$ from Equation 3.53. Furthermore, evaluating the Functions σ^i for each $i \in I$ from Equation 3.52 for the current velocity $v_n^{(t)}$ defines the interval factors $\sigma_0(v_n^{(t)})$, $\sigma_1(v_n^{(t)})$, $\sigma_2(v_n^{(t)})$, and $\sigma_3(v_n^{(t)})$ uniquely.

$$\sigma_i(v) := \Theta(\varsigma_i - v) \prod_{j \in J(i)} [1 - \sigma_j(v)] \text{, with } i \in I = \{0, \ldots, N_\varsigma\} \wedge J(i) = \{j \in I \mid j < i\} \ . \tag{3.52}$$

Eliminating the maximum and minimum functions in the calculation of the effective gap is possible now and it is done for the purpose of defining Equation 3.53. As an advantage, Equation 3.53 allows to distinguish the conditions and the intervals for the piecewise defined function $\mathbf{F}_{\mathrm{gap}}$ into the five disjunct cases that hold either $\gamma_n^{(t)} = 1$, or $(1 - \gamma_n^{(t)})\sigma_0(v_n^{(t)}) = 1$, or $(1 - \gamma_n^{(t)})\sigma_1(v_n^{(t)}) = 1$, or $(1 - \gamma_n^{(t)})\sigma_2(v_n^{(t)}) = 1$, or $(1 - \gamma_n^{(t)})\sigma_3(v_n^{(t)}) = 1$.

$$\mathbf{F}_{\mathrm{gap}}(v) := h_n^{(t)} + \gamma_n^{(t)} \left\{ \sigma_2(v) \left(g_{\mathrm{add}} - v\Delta t \right) - \sigma_3(v)g_{\mathrm{add}} \right\} \ . \tag{3.53}$$

Algorithm 1 (Computation of the "safe" maximum velocity $c_n^{(t+\Delta t)}$ for the next time step.).

1. Determine the headway $h_n^{(t)} := x_{n+1}^{(t)} - x_n^{(t)} - L_{n+1}$ (steps 2 and 3 depend on it).

2. Determine the value for optimistic or pessimistic driving behaviour $\gamma_n^{(t)}$ by Equation 3.40.

3. Determine the value for the gap $g_n^{(t)} = \mathbf{F}_{\text{gap}}(v_n^{(t)})$ by Equation 3.53.

4. Determine the covered braking distance $\mathbf{X}_{n+1}^{\text{B}}(v_{n+1}^{(t)})$ according to Equation 3.62.

5. Determine the total covered braking distance $s_n^{(t)} = g_n^{(t)} + \mathbf{X}_{n+1}^{\text{B}}(v_{n+1}^{(t)})$ for vehicle n.

6. Determine the value of the constant $Q = \frac{1}{2}Dt_{\text{safe}}(t_{\text{safe}} + \Delta t)$.

7. The definitions in Equation 3.54 for the "safe" maximum velocity for the next time step use the unique quotient $\alpha(c) \in \mathfrak{T}_{\text{xupd}}^4$ and remainder $\beta(c) \in \mathfrak{V}_{\text{veh}}^4$ representation of the velocity value $c = c_n^{(t+\Delta t)} \in \mathfrak{V}_{\text{veh}}^4$ modulo the braking capability D as it has been demonstrated in similar context for Equation 3.60. The superscript letters ^{o} and ^{p} denote optimistic and pessimistic driver behaviour. Equations 3.55, 3.56, 3.57, 3.58, and 3.59 provide the definitions.

$$
c_n^{(t+\Delta t)} := \begin{cases} \alpha_0^p D + \beta_0^p & \text{, if } \gamma_n^{(t)} = 1 \wedge 0 \leqslant s_n^{(t)} \text{ (see Sec. 3.4.1.3), or} \\ 0 \text{ cells/s} & \text{, if } \gamma_n^{(t)} = 1 \wedge s_n^{(t)} \leqslant 0 \text{, or} \\ c_2^o & \text{, if } \gamma_n^{(t)} = 0 \wedge Q \leqslant s_n^{(t)} \text{, or} \\ \alpha_1^o D + \beta_1^o & \text{, if } \gamma_n^{(t)} = 0 \wedge D\Delta t^2 \leqslant s_n^{(t)} < Q \text{, or} \\ \beta_0^o & \text{, if } \gamma_n^{(t)} = 0 \wedge 0 \leqslant s_n^{(t)} < D\Delta t^2. \end{cases}
\tag{3.54}
$$

8. Equation 3.55 defines the "safe" maximum velocity in the optimistic case $\gamma_n^{(t)} = 0 \wedge Q \leqslant s_n^{(t)}$:

$$
c_2^o := \left\lfloor \frac{2s_n^{(t)} + Dt_{\text{safe}}(t_{\text{safe}} - \Delta t)}{2t_{\text{safe}}} \right\rfloor.
\tag{3.55}
$$

9. Equation 3.56 defines the quotient of the value for the "safe" maximum velocity divided by D in the optimistic case $\gamma_n^{(t)} = 0 \wedge D\Delta t^2 \leqslant s_n^{(t)} < Q$ and in the pessimistic case $\gamma_n^{(t)} = 1$:

$$
\alpha_1^o = \alpha_0^p := \left\lfloor \frac{1}{2} \left(\sqrt{\Delta t^2 + 8s_n^{(t)}\frac{1}{D}} - \Delta t \right) \right\rfloor.
\tag{3.56}
$$

10. Equation 3.57 defines the remainder of the value for the potential maximum velocity divided by D in the optimistic case $\gamma_n^{(t)} = 0 \wedge D\Delta t^2 \leqslant s_n^{(t)} < Q$:

$$
\beta_1^o := \left\lfloor \frac{2s_n^{(t)} - D\alpha_1^o(\alpha_1^o + \Delta t)}{2\alpha_1^o} \right\rfloor.
\tag{3.57}
$$

11. Equation 3.58 defines the remainder of the value for the potential maximum velocity divided by D in the pessimistic case $\gamma_n^{(t)} = 1$:

$$
\beta_0^p := \left\lfloor \frac{2s_n^{(t)} - D\alpha_0^p(\alpha_0^p + \Delta t)}{2(\alpha_0^p + \Delta t)} \right\rfloor.
\tag{3.58}
$$

12. Equation 3.59 defines the remainder of the value for the potential maximum velocity divided by D in the optimistic case $\gamma_n^{(t)} = 0 \wedge 0 \leqslant s_n^{(t)} < D\Delta t^2$:

$$
\beta_0^o := \max\left\{0 \text{ cells/s}, s_n^{(t)}/\Delta t\right\}.
\tag{3.59}
$$

As the three cases $\gamma_n^{(t)} = 0$, $(1 - \gamma_n^{(t)})\sigma_0(v_n^{(t)}) = 1$, and $(1 - \gamma_n^{(t)})\sigma_1(v_n^{(t)}) = 1$ result in the same value for the effective gap, namely $g_n^{(t)} = \mathbf{F}_{\text{gap}}(v_n^{(t)}) = h_n^{(t)}$, the five cases result in only three possible, distinct values for $g_n^{(t)} = \mathbf{F}_{\text{gap}}(v_n^{(t)})$. Additionally, except for the case $(1 - \gamma_n^{(t)})\sigma_2(v_n^{(t)}) = 1$, the result values of the effective gap do not depend on the current velocity $v_n^{(t)}$.

Applying the same technique to Equation 3.41 with a different discretisation eliminates the maximum and minimum functions there, too. The discretisation relies on the Euclidean division theorem that states that "for every pair of integers a, b where $b \neq 0$ there exist unique natural integers q, r with $r < b$ such that $a = qb + r$ and $0 \leqslant r < |b|$" [168]. The Euclidean division is widely known as an algorithm to determine the remainder r and the quotient q. Both exist and are unique which follows from the well-ordering property of the integer numbers [84]. The uniqueness forms the important property in this context. The proof of the unique property is usually done using a contradiction proof. Applied to the domain of the velocities in Equation 3.41, the two operations, remainder and quotient, can be rewritten as functions $\lambda^{(v)} = v \text{ div } D$ and $\mu^{(v)} = v \bmod D$ where $v = \lambda^{(v)}D + \mu^{(v)}$ and o cells/s $\leqslant \mu^{(v)} < D\Delta t$. The interval splitting of the time horizon $\mathbf{F}_\tau(v_j, \delta_{\text{fl}})$ is similar to the one of Equation 3.52. It uses the time slots $\tau_i := 0\Delta t, 1\Delta t, t_{\text{safe}}, \lfloor v_{\text{max}}/D \rfloor + \epsilon\Delta t$ with $0 < \epsilon < 1$, $\tau_i \in \mathfrak{T}_{\text{xupd}}^4$, and $i = 0, \ldots, 3$ for $N_\tau := 3$ disjunct sub-intervals $[\tau_i; \tau_{i+1}[$. Equation 3.60 uses then the sub-intervals to define the interval factorising functions $\vartheta_i : \mathfrak{V}_{\text{veh}}^4 \longrightarrow \mathbb{B}_{\text{bit}}^+$.

$$\vartheta_i(v) := \Theta(\lambda^{(v)} - \tau_i) \prod_{j \in J(i)} (1 - \vartheta_j(v)) \text{, with } i \in I = \{0, \ldots, N_\vartheta\} \land J(i) = \{j \in I \mid j < i\} . \quad (3.60)$$

With the interval factors $\vartheta_i(v)$ for a given velocity v, the definition of the time horizon from Equation 3.41 is equivalent to the form of Equation 3.61 where the maximum and minimum functions have been eliminated successfully. Similar to the case with Equation 3.53, it distinguishes only four resulting cases, namely $\gamma_n^{(t)} = 1$, or $(1 - \gamma_n^{(t)})\vartheta_0(v) = 1$ or $(1 - \gamma_n^{(t)})\vartheta_1(v) = 1$, or $(1 - \gamma_n^{(t)})\vartheta_2(v) = 1$.

$$\mathbf{F}_\tau(v, d_{\text{fl}}) = \gamma_n^{(t)}\lambda_n^{(v)} + (1 - \gamma_n^{(t)})\left(\vartheta_2(v)[t_{\text{safe}} - d_{\text{fl}}\Delta t] + \vartheta_1(v)[\lambda^{(v)} - d_{\text{fl}}\Delta t]\right) \quad (3.61)$$

Using the definitions from above, the transformation of the calculation of the covered braking distance $\mathbf{X}_{n+1}^B(v_{n+1}^{(t)})$ for the leader vehicle according to Equation 3.51 results in a function evaluation of Function $\mathbf{X}_{n+1}^B : \mathfrak{V}_{\text{veh}}^4 \longrightarrow \mathfrak{X}_{\text{rway}}^4$ that uses the definition as presented by Equation 3.62. Therein the two cases $\gamma_n^{(t)} = 1$ and $(1 - \gamma_n^{(t)})\vartheta_1(v) = 1$ in Equation 3.51 lead to the same formula when determining the resulting covered braking distance $\mathbf{X}_{n+1}^B(v_{n+1}^{(t)})$ for the leading vehicle with index $n + 1$.

$$
\begin{aligned}
\mathbf{X}_{n+1}^B(v) = \quad & \left(\gamma_n^{(t)} + (1 - \gamma_n^{(t)})\vartheta_1(v)\right)\lambda^{(v)}\left(\mu^{(v)} + \frac{1}{2}D(\lambda^{(v)} - \Delta t)\right) \\
+ \quad & \left(1 - \gamma_n^{(t)}\right)\vartheta_2(v)\left(vt_{\text{safe}} + \frac{1}{2}Dt_{\text{safe}}(t_{\text{safe}} - \Delta t)\right) .
\end{aligned}
\quad (3.62)
$$

Similar considerations apply to the calculation of the covered braking distance for the follower and for any arbitrary value for the "safe" maximum velocity $c_n^{(t+\Delta t)}$ in the next time step $t + \Delta t$. The calculation uses the same interval factor functions ϑ_i for the calculation of the sub-interval factors as Equation 3.62. However, instead of applying these functions to the leading vehicle's velocity, this time, it applies them to the "safe" maximum velocity $c_n^{(t+\Delta t)}$ for the next time step.

Additionally, it evaluates Function $\mathbf{X}_n^B : \mathfrak{V}_{\text{veh}}^4 \longrightarrow \mathfrak{X}_{\text{rway}}^4$ instead of the one from above. Equation 3.63 presents the transformed calculation for Function $\mathbf{X}_n^B$.

$$
\begin{aligned}
\mathbf{X}_n^B(c) = \quad & \gamma_n^{(t)}\left((\lambda^{(c)} + \Delta t)(\mu^{(c)} + \frac{1}{2}D\lambda^{(c)})\right) + (1 - \gamma_n^{(t)})\vartheta_0(c)\mu^{(c)} \\
+ \quad & (1 - \gamma_n^{(t)})\vartheta_1(c)\lambda^{(c)}\left(\mu^{(c)} + \frac{1}{2}D(\lambda^{(c)} + \Delta t)\right) \\
+ \quad & (1 - \gamma_n^{(t)})\vartheta_2(c)\left(ct_{\text{safe}} - \frac{1}{2}Dt_{\text{safe}}(t_{\text{safe}} - \Delta t)\right) .
\end{aligned}
\quad (3.63)
$$

Using the Equations 3.63 and 3.62 in combination with Functions ϑ_i of Equation 3.60, the value of $c_n^{(t+\Delta t)} := \lambda^{(c)} D + \mu^{(c)}$ with the quotient $\lambda^{(c)}$ and the remainder $\mu^{(c)}$ of $c = c_n^{(t+\Delta t)} \mod D$ can be derived by solving the quadratic equations for the distinct cases which then results in Algorithm 1.

Algorithm 1 provides a straightforward way to calculate the "safe" maximum velocity $c_n^{(t+\Delta t)}$ of the vehicle with index n for the next time step. The conditions for the cases with the appropriate formula for $c_n^{(t+\Delta t)}$ in step 7 depend only on the values of the constant Q and the effective covered braking distance $s_n^{(t)}$. As a first test, the value for $\gamma_n^{(t)}$ determines the choice of the pessimistic case and the choice of the optimistic path. The most frequent optimistic case depends only on the comparison of the constant Q from step 6 with the value for the effective covered braking distance $s_n^{(t)}$ from step 5. In case that it is greater than or equal to the constant, step 8 is chosen. Otherwise, if $s_n^{(t)}$ is greater than or equal to D, steps 9 and 10 are chosen. Otherwise, step 12 is chosen. Assuming almost equal computation costs for a step in the iterative search based on Inequality 3.37 and for a path in Algorithm 1, the latter is two to four times faster.

Additionally and compared to the description with Inequality 3.37, Algorithm 1 provides deeper insights into the relation between the "safe" maximum velocity in the next time step and the velocity of the leading vehicle. The computation of the "safe" maximum velocity in the optimistic case of Equation 3.55 where $c_n^{(t+\Delta t)} = c_2^o$ holds, depends only linearly on the velocity of the leading vehicle $v_{n+1}^{(t)}$. In contrary, the pessimistic case and the optimistic case with $D\Delta t^2 \leqslant s_n^{(t)} < Q$ have a square root dependence on the velocity of the leading vehicle. Both formulas differ only in the calculation of the remainder β_0^p and β_1^o, respectively.

3.4.1.3 Further considerations

Algorithm 1 supports a straightforward calculation of the value for the next safe velocity $c_n^{(t+\Delta t)}$. However, there are certain configurations where Inequality 3.37 is undefined and does not provide a valid solution for $c_n^{(t+\Delta t)}$. This follows from the definition of the capital sigma notation $\sum_i^n a_i = 0$ for the special case $n < i$ [159]. When applied to Inequality 3.37, it results in the constraint that the upper bound of the summation for the term involving $c_n^{(t+\Delta t)}$ has to be equal or greater than 0. Otherwise the term for the next velocity $c_n^{(t+\Delta t)}$ erases which means that, depending on the validity of the inequality, the inequality for the chosen upper bound always holds or never holds. In the former case the inequality holds and does not depend on the concrete value for $c_n^{(t+\Delta t)}$ as long as it is in the scope of the upper bound. In the latter case, the value that still satisfies the inequality does not exist $c_n^{(t+\Delta t)}$. When the right hand side of the inequality in form of 3.49 is less than o cells, the inequality can become invalid.

As a consequence of the invalid inequality and the non-existent value for $c_n^{(t+\Delta t)}$, the optimised algorithm will not provide any valid solution either and instead provide an imaginary value that is obviously invalid with respect to the range of $c_n^{(t+\Delta t)}$. The affected Equations are 3.38 and 3.53. The problematic configurations occur when the headway plus safety distance $g_n^{(t)}$ becomes negative for the pessimistic case when also $\sigma_3(v_n^{(t)}) = 1 \Rightarrow v_n^{(t)} \geqslant 2g_{\text{add}}$ or $\sigma_2(v_n^{(t)}) = 1 \Rightarrow g_{\text{add}} \leqslant v_n^{(t)} < 2g_{\text{add}}$ from the context of Equation 3.52 holds. As stated above, this is the case when $x_{n+1}^{(t)} - x_n^{(t)} - L_{n+1} - g_{\text{add}} + \mathbf{X}_{n+1}^B(v_{n+1}^{(t)}) < 0$. Example configuration exist with $v_n^{(t)} = 9$ cells/s, $v_{n+1}^{(t)} = o$ cells/s, $x_{n+1}^{(t)} - x_n^{(t)} - L_{n+1} = 3$, $v_n^{(t)} = 9$ cells/s, $v_{n+1}^{(t)} = 4$ cells/s, $x_{n+1}^{(t)} - x_n^{(t)} - L_{n+1} = 1$, and $v_n^{(t)} = 8$ cells/s, $v_{n+1}^{(t)} = 2$ cells/s, $x_{n+1}^{(t)} - x_n^{(t)} - L_{n+1} = 3$. These sort of configurations will most likely lead to a collision in one of the next time steps. To make the calculation robust, the pessimistic case in Equation 3.54 requires the additional restriction $0 \leqslant s_n^{(t)}$ as well as an additional clause that defines the velocity to o cells/s for $\gamma_n^{(t)} = 1 \wedge s_n^{(t)} \leqslant 0$. Another approach might be to look ahead after an update and forbid certain configurations as invalid states.

Several circumstances may lead to the negative headways among which the probably most interesting question is whether the traffic model might produce such configurations. In scenarios with periodic boundary conditions, the values for the system length and the number of vehicles may force a system

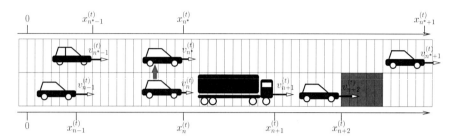

Figure 3.6: Multi-lane cell model for lateral motion of cars and trucks.

initialisation where some vehicles do not get sufficiently large headways. This might be visible at system initialisation or occur only thereafter. Tuning elements and lane changes may also provoke these situations. Additionally, the latter calculates the value for the next safe velocity as part of Equation 3.65 for any possible configuration including those that do not reflect the result of successive drive updates. The critical configurations can, however, also arise as the result of the toggling between optimistic and pessimistic driver behaviour. As the first optimistic case of Equation 3.40 allows short distance headways with equal velocities for the three and even four vehicles in a series, dawdling can provoke these kind of situation within a few timestamps. The toggling is not limited to the case when the leader dawdles but also to the case when the leader of a four vehicles series dawdles or brakes. The second last vehicle in the series toggles then from optimistic to pessimistic driving behaviour which will result in a braking manoeuvre. Even though the optimistic and the pessimistic case use the same formulae Equation 3.56 to determine $\alpha_1^o = \alpha_0^p$, the former never applies it to negative values for $s_n^{(t)}$. The development of an inverse function to $\mathbf{F}_{\mathrm{drive}}$ can help to investigate the collisions.

Figure 5.4 depicts the acceleration (on a ten times finer scale and shifted by ± 1 for optimistic and pessimistic, respectively) in relation to the velocity difference and the distance headway in the deterministic case where the vehicles do not dawdle. Concluding from the figure, there may be further optimisations to calculate the next "safe" velocity. The figure shows that for each acceleration value $a_n^{(t+\Delta t)} \in [-D; A]$, the set of input parameter configurations takes the form of a convex hull. Consequently, it is possible to derive a set of intervals for the velocity differences and the headway values that define the acceleration value adequately. Such a definition would then further enhance the computational cost for determining the next value for $c_n^{(t+\Delta t)}$. Such a definition would also provide more insights when the next velocity before the dawdling results in an acceleration or in a deceleration.

3.4.2 Lateral motion

In the context of OLSIMv4, lateral motion of vehicles is limited to changing between discrete lanes. This is in contrast to traffic flow without lane discipline that is found in countries such as India. The publication [136] may serve as an example for a traffic model that covers traffic flow streams without lane discipline. Vehicles in traffic flow with lane discipline can move in accordance with either symmetric or asymmetric lane changing rules. However, OLSIMv4 models asymmetric lane change behaviour by an extension of the Pottmeier multi-lane model [164] as introduced by Habel [71, 70]. Many of the thoughts that this section presents are influenced by the work that introduced the asymmetric lane change extensions [70, Sec. 4.2]. He recently published his extensions in [71] after the work on this section was already complete.

German traffic regulations induce a lane changing behaviour that motivates the asymmetric lane changing rules. On German highways and apart from some special situations, vehicles are not allowed to overtake on a right lane. As a consequence, vehicles first have to change onto a left lane, then overtake the previously leading vehicle, and thereafter change back again onto the right lane whenever

Table 3.4.3: Parameters of the asymmetric lane changing model for OLSIMv4.

name	description	scope	value
D_{tol}	maximum deceleration tolerance of a follower	global	1 cells/s²
Δt_{lc}	minimum time headway for lane changes	global	1.8 s
v_{sym}	velocity threshold for symmetric lane change rules	global	6 cells/s
p_{sym}	probability for lane changes with the symmetric rules	global	1
v_{ovt}	velocity threshold for enabling overtaking on the right	global	16 cells/s

they find a gap that is large enough. The exceptions to the overtaking on the left rule permit overtaking on the right whenever vehicles queue on both lanes. In this case, overtaking is allowed and in most cases does not involve lane changing. This sort of lane changing behaviour due to overtaking on the left results in a characteristic traffic flow per lane distribution. Usually, it increases from right to left significantly. It is thus called "lane usage inversion" [115].

Figure 3.6 portraits a typical lane change scenario as it may occur on German highways. The vehicles in German highway traffic obey lane discipline and, thus, modelling the lanes as discrete values is adequate. At the beginning of this section, the signature for the vehicle update procedures has already been introduced. It is the same as for the drive update, namely $\mathbf{F}_{\text{chlane}} : \mathfrak{R}_{\text{occ}}^4 \times S_{\text{veh}}^4 \times S_{\text{nbrs}}^4 \times S_{\text{topo}}^4 \longrightarrow \mathfrak{R}_{\text{occ}}^4 \times S_{\text{veh}}^4$. Section 3.4.1 has introduced the domains for the arguments to an evaluation already. This section uses the same names for the function arguments and, thus, introduces here again only briefly. The first one, $s_{\text{rand}} \in \mathfrak{R}_{\text{occ}}^4$, represents the state of a random number generator. The second argument to Function $\mathbf{F}_{\text{chlane}}$ specifies the vehicle $\mathbf{k}_n^{(t)} \in S_{\text{veh}}^4$ which will become the subject of the lane change update. Its index n refers to the order of the vehicles on its current lane $y_n^{(t)} := (\mathbf{k}_n^{(t)} \# \mathbf{q}^{(t)}) \# y$. Its current position is $x_n^{(t)} := (\mathbf{k}_n^{(t)} \# \mathbf{q}^{(t)}) \# x$ and its current velocity $v_n^{(t)} := (\mathbf{k}_n^{(t)} \# \mathbf{q}^{(t)}) \# v$. As it has been the case for the vehicle variables in Section 3.4.1, this abbreviation scheme applies also for this section and also to the other vehicle structure components such as the length $L_{\text{n}} := \mathbf{k}_n^{(t)} \# L$ and the maximum braking capability $D_n := \mathbf{k}_n^{(t)} \# D$. The third argument to the vehicle update function contains the preceding and the neighbouring vehicles in the form of a neighbourhood structure element $\mathbf{b} \in S_{\text{nbrs}}^4$. The fourth argument to the update function contains the information that the topological analysis encodes as a topology structure element $\mathbf{p} \in S_{\text{topo}}^4$. However, the drive update function of Section 3.4.1 results in a pair of a random number state and an updated vehicle $\mathbf{k}_n^{(t+\Delta t)}$ for the next time step. Due to the fact, that the exact evaluation order of the two vehicle update procedures has not yet been discussed, this section ignores that a drive update has already taken place and the vehicles may already have an update coordinate structure element for the time step $t + \Delta t$. When the time step $t + \Delta t$ is interpreted as an absolute value, this may be confusing. However, t is the time variable and it may refer also to an absolute time that matches a time step $t + \Delta t$ of another function evaluation.

This section examines first the lane changes to the left and thereafter the lane changes to the right. Whenever a vehicle $\mathbf{k}_n^{(t)}$ with index n, position $x_n^{(t)}$, and lane $y_n^{(t)}$ changes onto a neighbouring lane $y_{i,n} := y_n^{(t)} + i$ with $i \in \mathbb{B}_{\text{cell}}^{\pm} \backslash \{0\}$, its index may change as it refers to the order of the vehicles in a single-lane. Therefore, the index n^* refers to the vehicle's index on the neighbouring lane in the following. Figure 3.6 illustrates such a scenario for two lanes. From the context will become clear, whether the vehicle changes onto the left single-lane $y_{+1,n}$ or onto the right single-lane $y_{-1,n}$. According to the definition from Equation 3.32, the vehicles in the neighbourhood structure element $\mathbf{b} \in S_{\text{nbrs}}^4$ have the special properties that the vehicles $\mathbf{b} \# \mathbf{k}_2^{(t)}$ and $\mathbf{b} \# \mathbf{k}_8^{(t)}$ each have a position that is greater than the one of vehicle $\mathbf{k}_n^{(t)}$ whereas the positions of the vehicles $\mathbf{b} \# \mathbf{k}_1^{(t)}$ and $\mathbf{b} \# \mathbf{k}_7^{(t)}$ are smaller. This follows from the construction rules of the neighbourhood structure elements and the choice of the lengths of the vehicles $L > 1$ cells. Thus, when vehicle $\mathbf{k}_n^{(t)}$ considers a change onto the neighbouring lane, the vehicle becomes a part of the neighbouring vehicles on that lane. As a consequence, the neighbouring vehicles together with the lane changing vehicle form a sequence of 4 vehicles $\mathbf{k}^{(t)}_{n*-1}$, $\mathbf{k}^{(t)}_{n*}$, $\mathbf{k}^{(t)}_{n*+1}$, and $\mathbf{k}^{(t)}_{n*+2}$. As an example, in case of a change onto the left lane the follower and leader vehicles of vehicle $\mathbf{k}^{(t)}_{n*} := \mathbf{F}_{\text{clone}}(\mathbf{k}^{(t)}_{n}, \mathbf{q}^{(t)}, \mathbf{F}_{\text{clone}}(\mathbf{q}^{(t)}, y, y_{+1,n}))$ in the sequence are $\mathbf{k}^{(t)}_{n*-1} := \mathbf{b} \# \mathbf{k}^{(t)}_{7}$, $\mathbf{k}^{(t)}_{n*+1} := \mathbf{b} \# \mathbf{k}^{(t)}_{8}$, and $\mathbf{k}^{(t)}_{n*+2} := \mathbf{b} \# \mathbf{k}^{(t)}_{9}$. For these vehicles, the same abbreviation scheme as the one from above applies.

The following example explains how a lane change takes place on German highways. Suppose a vehicle with index n on a right lane on position $x^{(t)}_{n}$ reaches a truck in front of it. It speculates to change to the left lane for overtaking the truck. Before in fact changing to the left lane, the model assesses the benefit and the safety of such a manoeuvre. While assessing the benefit of the lane change, it compares its current velocity to the one achievable when driving as vehicle n^* on position $x^{(t)}_{n*}$ with velocity $v^{(t)}_{n*}$ on the left lane. In order to assess the safety, the model computes the gaps and the resulting velocities between the three vehicles on the destination lane. The gap has to match at least the vehicle length.

However, prior to checking for the benefit of a potential lane change, the model examines the topological constraints that contain the information whether lane changing is permitted or not. The topologic constraints consist of the four values encoded in the topology structure element $\mathbf{p} \in S^4_{\text{topo}}$, namely the three maximum velocities $v_{-1,\text{max}}, := \mathbf{p} \# v_{-1,\text{max}}$, $v_{+1,\text{max}} := \mathbf{p} \# v_{\pm 0,\text{max}}$, and $v_{+1,\text{max}} := \mathbf{p} \# v_{+1,\text{max}}$ as well as the lane change priority $u_{\text{prio}} := \mathbf{p} \# u_{\text{prio}}$. A velocity value $v_{i,\text{max}} > 0$ cells/s with $i \in \mathbb{B}^{\pm}_{\text{cell}}$ indicates that lane changing from lane $y^{(t)}_{n}$ onto lane y_i is permitted and that the potential maximum velocity on the destination lane has the value $v_{i,\text{max}}$. In contrast, a maximum velocity value $v_{i,\text{max}} = 0$ cells/s indicates that lane y_i is closed. The other topologic variable, $u_{\text{prio}} \in \mathfrak{P}^4_{\text{cell}}$, prioritises whether a lane change is preferred over staying on the current lane or not. Equation 3.64 displays the conditions for whom the topological constraints allow overtaking.

$$v_{i,\text{max}} > 0 \text{ cells/s }, \text{ where } i \in \mathbb{B}^{\pm}_{\text{cell}} \backslash \{0\} . \tag{3.64}$$

Whenever the topology permits lane changing, the two criteria that a lane change has to meet are whether a potential lane change is safe and, in case it is, whether it results in a velocity gain. Thus, only two kinds of conditions are necessary to assess a potential lane change—namely, safety rules and benefit tests. Assuming that a potential lane change is safe, its benefit depends only on the resulting velocity gain for the time horizon of the next single time step. The criteria that assess whether a lane change results in a velocity gain or not, will be described with Equation 3.66. At first, Equation 3.65 displays the safety rules for a lane change. These safety rules apply also for scenarios when tuning elements intend to insert a vehicle into a lane. Thus, a special Function $\mathbf{F}_{\text{lc-safe?}} : S^4_{\text{veh}} \times S^4_{\text{veh}} \times S^4_{\text{veh}} \times S^4_{\text{veh}} \times \mathfrak{V}^4_{\text{veh}} \longrightarrow B^4_{\text{occ}}$ provides these information. The first four arguments are the vehicles in the sequence. Its last argument is the maximum velocity of the topological constraint $v_{i,\text{max}}$ with the appropriate value for i. Its purpose is to limit the vehicles' velocities to the one provided by the topological constraint in cases where the vehicles drive with higher velocities. The function result matches an evaluation of the rules from Equation 3.65 that are explained in the following.

$$
\begin{aligned}
x^{(t)}_{n*} - x^{(t)}_{n*-1} &> L_{n*} & \wedge \quad x^{(t)}_{n*+1} - x^{(t)}_{n*} &> L_{n*+1} \quad \wedge \\
v^{(t)}_{n} - D_n \Delta t &\leqslant \min \left\{ v_{\text{max},n*}, c^{(t+\Delta t)}_{n*} \right\} & \wedge \quad v^{(t)}_{n*-1} - D_{\text{tol}} \Delta t &\leqslant \min \left\{ v_{\text{max},n*-1}, c^{(t+\Delta t)}_{n*-1} \right\} .
\end{aligned}
\tag{3.65}
$$

The first two safety rules of Equation 3.65 assert that the pair-wise headways on the destination lane between vehicle with index n^*+1 and its follower n^*-1 offer a large enough gap for the vehicle initiating the lane change $\mathbf{k}^{(t)}_{n*}$. The last two rules use the "safe" maximum velocity determined by Algorithm 1. The third rule from Equation 3.65 ensures that the vehicle's "safe" maximum velocity $c^{(t+\Delta t)}_{n*}$ on the destination lane during the next time step respects the vehicle's limited braking capability as well as the topological constraints represented by $v_{\text{max},n*}$. The calculation of this value uses the formula from Equation 3.39 but with the values $v_{\text{max},\text{TP},n*}$ and $v_{\text{max},\text{CA}} := v_{i,\text{max}}$ and i chosen appropriately according to Equation 3.64. The last rule in Equation 3.65 verifies that the lane change will not disturb the follower vehicle n^*-1 too much. The parameter D_{tol} limits the maximum tolerable disturbance by a lane change. It also uses the definition of Equation 3.39 but this time with the values $v_{\text{max},\text{TP},n*-1}$

Table 3.4.4: Summary of relevant functions, structures, variables, and constants for Section 3.4.2.

definition	description	domain	range
$\mathbf{F}_{\text{chlane}}(a, b, c, d)$	lateral motion	$\mathfrak{R}^4_{\text{occ}} \times S^4_{\text{occ}} \times S^4_{\text{veh}} \times S^4_{\text{nbrs}} \times S^4_{\text{topo}}$	$\mathfrak{R}^4_{\text{occ}} \times S^4_{\text{veh}}$
$\mathbf{F}_{\text{lc-safe?}}(k_1, k_2, k_3, k_4, v_{\max})$	lane change safe?	$S^4_{\text{veh}} \times S^4_{\text{veh}} \times S^4_{\text{veh}} \times S^4_{\text{veh}} \times \mathfrak{V}^4_{\text{veh}}$	$\mathbb{B}^\pm_{\text{occ}}$
$\Delta y^{(t+\Delta t)}_{lc,n}$	relative lane number		$\mathbb{B}^\pm_{\text{cell}}$
v_{ov}	right overtaking vel.		Equation 3.71

and $v_{\max,\text{CA}} := v_{\text{i},\max}$ for the calculation. Table 3.4.3 lists all the parameters that influence the lane change behaviour.

Whenever the safety rules are fulfilled, the velocity benefit test of a lane change follows. Its assessment depends only on the comparison of the safe maximum velocity on the current lane $\tilde{v}^{(t+\Delta t)}_n$ and the velocity value that is achievable on the destination lane $\tilde{v}^{(t+\Delta t)}_{n*}$. Equation 3.42 serves to determine both velocities values. Equation 3.66 formulates the velocity benefit test.

$$\tilde{v}^{(t+\Delta t)}_{n*} > \tilde{v}^{(t+\Delta t)}_n \ . \tag{3.66}$$

On German highways, overtaking of vehicles is permitted only on the left lanes. Thus, the velocities on the leftmost lane is generally higher compared to the one of the right lanes. Equation 3.66 provides only a test for lane changes to the left. As soon as any chosen lane change does no longer provide a velocity gain for the initiating vehicle on the left lane, the vehicle is required to change back onto the right lane. The lane change to the right lane, however, must not result in another velocity gain as this would be a violation of the rule not to overtake on the right lane. Thus, the lane change to the right is only favourable when the velocities on both lanes are equal.

$$\tilde{v}^{(t+\Delta t)}_{n*} = \tilde{v}^{(t+\Delta t)}_n \ . \tag{3.67}$$

A vehicle that changes to the right lane in accordance with the above condition may encounter another slow driving leader vehicle in the next time step that follows the lane change to the right. Therefore the new leading vehicle might force the following vehicle to another lane change for overtaking again. The conditions in Equation 3.68 introduce and require an additional time headway to be met when changing from the left to the right. The purpose of the conditions is to avoid a so-called ping-pong effect. This effect can occur whenever vehicles change to the right as demanded by the German highway traffic regulations and back again for overtaking the next vehicle. Whenever the conditions hold, the behaviour results, however, in the following vehicle motion. Vehicles that change lanes from the left to the right choose the lane change only, when they have enough free distance headway in front of them. The model by Habel uses the time headway parameter Δt_{lc} to estimate the distance headway based on the current velocity $v^{(t)}_n$. Equation 3.68 specifies the conditions that a lane change to the right requires in addition to Equation 3.67.

$$x^{(t)}_n - x^{(t)}_{n-1} - L_n > v^{(t)}_n \Delta t_{\text{lc}} \quad \wedge \quad x^{(t)}_{n*+1} - x^{(t)}_{n*} - L_{n*+1} > v^{(t)}_n \Delta t_{\text{lc}} \quad . \tag{3.68}$$

The situation in which the vehicles queue but constantly move with lower velocities on both lanes substantiates an exception to the rule of no overtaking on the right lane. The threshold velocity parameter v_{sym} specifies the velocity boundary where the traffic model uses symmetric lane changing behaviour instead of the so far presented asymmetric lane changing behaviour. With the symmetric lane change behaviour a potential lane change is chosen whenever it results in a velocity gain. Equation 3.69 specifies the test for this behaviour. However, the model limits the number of chosen lane changes by introducing a lane change probability p_{sym} for these kinds of lane changes. The lane change condition from Equation 3.69 uses this probability parameter as well as a random number $z_{\text{rand}} := (\mathbf{F}_{\text{random}}(s_{\text{rand}})) \# r_{\text{num}}$.

To support synchronicity between several random number generator states, the lane change update function generates a random number in each evaluation step—no matter whether it is required or not. This is, however, only required when choosing the multicomponent discrete time system specification as the simulation model. Section 3.7.2 discusses this in more detail. The updated state of the random number generator is then part of the $\mathbf{F}_{\text{chlane}}$ function result $s'_{\text{rand}} := (\mathbf{F}_{\text{random}}(s_{\text{rand}})) \# s_{\text{state}}$.

$$\tilde{v}_{n*}^{(t+\Delta t)} > \tilde{v}_n^{(t+\Delta t)} \ \wedge \ z_{\text{rand}} \leqslant p_{\text{sym}} R_{\text{max}} \ . \tag{3.69}$$

As a consequence of the "lane usage inversion," the vehicles on a right lane have–with increasing density–significantly larger distance headways in front of them compared to the average headway on a left lane. In such a scenario, the right lane vehicles could thus drive at higher velocities and could be willing to overtake on the right the vehicles on the left lane. As this is forbidden by the German traffic rules, the model has to prevent the vehicles from overtaking on the right. Because the vehicles in the presented model have only limited braking capability, further braking of the vehicles in the drive update step could potentially result in decelerations greater than the maximum deceleration capability. Thus, the present model restricts the possible maximum velocity for vehicles on the right lane whenever they enter a critical region. The critical region is defined as the spatial extent on the right lane up to the back position of the neighbouring vehicle on the left. Figure 3.6 has coloured the critical region that starts at position $x_{n+2}^{(t)}$ and ends at position $x_{n*+1}^{(t)} - L_{n*+1}$. Equation 3.70 specifies the test whether a vehicle on the right lane at position $x_n^{(t)}$ has entered the critical region.

$$x_{n*+1}^{(t)} - x_n^{(t)} - L_{n*+1} < v_n^{(t)} \Delta t \ . \tag{3.70}$$

Whenever a vehicle enters the critical region, the lane change update procedure adjusts its velocity with the help of Equation 3.71. Equation 3.71 facilitates overtaking on the right lane with only slightly higher velocities, i.e. 2 cells/s , compared to the velocity of the vehicle on the left lane. The velocity parameter v_{ovt} limits overtaking on the right lane to velocities that result in a behaviour similar to that on German highways where overtaking on the right is enabled up to a range of 70 km/h to 80 km/h .

$$v_{\text{ov}} = \begin{cases} \min\left\{v_{\text{ovt}}, v_{n*+1}^{(t)} + 2\,\text{cells/s}\right\} & \text{, if } v_{n*+1}^{(t)} < v_{\text{ovt}} \ , \ \wedge \text{ Eq. } 3.70 \text{ holds, or} \\ v_{n*+1}^{(t)} + 2\,\text{cells/s} & \text{, if } v_{n+1}^{(t)} \geqslant v_{\text{ovt}} \ , \ \wedge \text{ Eq. } 3.70 \text{ holds, or} \\ v_{\text{max},n} & \text{, if Eq. } 3.70 \text{ does not hold.} \end{cases} \tag{3.71}$$

The resulting velocity $v_{\text{eff},n}$ of the current time step follows from Equation 3.71. Equation 3.72 presents this velocity. It is used by all further lane change considerations in the current time step and requires recalculation in each time step.

$$v_{\text{eff},n} = \min\left\{v_{\text{max},n}, v_{\text{ov}}\right\} \ . \tag{3.72}$$

The lane change priority variable $u_{\text{prio}} \in \mathfrak{P}_{\text{cell}}^4$ influences the lane changing scenario in the way as described in the following. When $u_{\text{prio}} = 1$ a lane change is always chosen. If $u_{\text{prio}} = 2$ the lane change is only chosen when the safety rules from Equation 3.65 hold. When the lane change priority has the value $u_{\text{prio}} = 3$, it results in assessing the lane change for safety and velocity gain. Otherwise $u_{\text{prio}} = 4$ holds and the lane change is not chosen. Finally, this results in the relative lane number $\Delta y_{lc,n}^{(t+\Delta t)} \in \mathbb{B}_{\text{cell}}^{\pm}$ from Equation 3.73.

$$\Delta y_{lc,n}^{(t+\Delta t)} := \begin{cases} +1 & \text{, if Eqs. } (\ 3.64 \ \wedge \ u_{\text{prio}} = 1\ ) \vee (\ 3.64 \ \wedge \ 3.65 \ \wedge \ u_{\text{prio}} = 2\ ) \ , \\ +1 & \text{, if Eqs. } 3.64 \ \wedge \ 3.65 \ \wedge \ 3.66 \ \wedge \ u_{\text{prio}} = 3 \ , \\ +1 & \text{, if Eqs. } 3.64 \ \wedge \ y_{n*}^{(t)} = y_n^{(t)} + 1 \ \wedge \ 3.65 \ \wedge \ 3.69 \ \wedge \ v_n^{(t)} \leqslant v_{\text{sym}} \ \wedge \ u_{\text{prio}} \leqslant 3 \ , \\ -1 & \text{, if Eqs. } 3.64 \ \wedge \ y_{n*}^{(t)} = y_n^{(t)} - 1 \ \wedge \ 3.65 \ \wedge \ 3.69 \ \wedge \ v_n^{(t)} \leqslant v_{\text{sym}} \ \wedge \ u_{\text{prio}} \leqslant 3 \ , \\ -1 & \text{, if Eqs. } 3.64 \ \wedge \ 3.65 \ \wedge \ 3.67 \ \wedge \ 3.68 \ \wedge \ u_{\text{prio}} = 3 \ , \\ -1 & \text{, if Eqs. } (\ 3.64 \ \wedge \ u_{\text{prio}} = 1\ ) \vee (\ 3.64 \ \wedge \ 3.65 \ \wedge \ u_{\text{prio}} = 2\ ) \ , \\ 0 & \text{, otherwise.} \end{cases}$$

$$\tag{3.73}$$

The relative lane number $\Delta y_{lc,n}^{(t+\Delta t)} \in \mathbb{B}_{\text{cell}}^{\pm}$ from Equation 3.73 assists now in determining the velocity $v_n^{(t+\Delta t)}$ for the next time step. Equation 3.74 presents the two cases.

$$v_n^{(t+\Delta t)} := \begin{cases} (\mathbf{k}_n^{(t)} \# \mathbf{q}_0) \# v & \text{, if } \Delta y_{lc,n}^{(t+\Delta t)} = 1 \text{ ,} \\ v_{\text{eff},n} & \text{, otherwise.} \end{cases} \tag{3.74}$$

The relative lane number $\Delta y_{lc,n}^{(t+\Delta t)} \in \mathbb{B}_{\text{cell}}^{\pm}$ from Equation 3.73 and the velocity $v_n^{(t+\Delta t)}$ for the next time step from Equation 3.74 enable then to update the vehicle's coordinates for the next time step. Equation 3.75 displays the manipulation procedure of vehicle $\mathbf{k}_n^{(t)}$. The two vehicle's coordinate structure elements v and y wear the updated values. The evaluation of the lane change update function results then, finally, in the pair of the updated state of the random number generator s_{rand} and the updated vehicle $\mathbf{k}_n^{(t+\Delta t)}$, i.e. $\mathbf{F}_{\text{chlane}}(s_{\text{rand}}, \mathbf{k}_n^{(t)}, \mathbf{b}, \mathbf{p}) := (s_{\text{rand}}', \mathbf{k}_n^{(t+\Delta t)})$ with $s_{\text{rand}}' := (\mathbf{F}_{\text{random}}(s_{\text{rand}})) \# s_{\text{state}}$ and $\mathbf{k}_n^{(t+\Delta t)}$ from Equation 3.75. Table 3.4.4 summarises the relevant functions, structure definitions, variables, and constants of this section.

$$\mathbf{k}_n^{(t+\Delta t)} := \mathbf{F}_{\text{clone}}(\mathbf{k}_n^{(t)}, \mathbf{q}_0, \mathbf{F}_{\text{clone}}(\mathbf{F}_{\text{clone}}(\mathbf{q}_0, v, v_n^{(t+\Delta t)}), y, y_n^{(t)} + \Delta y_{lc,n}^{(t+\Delta t)})) \text{ .} \tag{3.75}$$

3.5   Detector model

Detectors handle input and output from or to the simulation. However, most traffic simulations do not follow this definition very strictly and provide several input methods. OLSIMv3 used the same data structure for the detectors and the tuning elements as both were positioned at the same location. SUMO [19] uses detectors to output traffic observables from the simulation and handles input as part of a "demand modelling" [119] that covers origin/destination matrices, random trips, and activity based demand generation. Most market solutions handle vehicle generation similarly and do not use tuning elements [31] that rely on detectors to handle traffic data input. That results in the concrete disadvantage of having no comparable input/output pairs and, thus, being most often hardly or sometimes even not at all able to verify simulation results numerically and formally.

Even though OLSIMv3 used the measuring method that is similar to the one from the global detectors it did not use functionally encapsulated detectors to collect the traffic information from the simulation. A combination of a tuning method, a loop detector, and several calculation methods collected the traffic information. Additionally, the OLSIMv3 loop detectors used the positioning scheme that has been listed in Listing 2.1.2. It consists of the components of the detector identification, the relative detector position, and the relative lane number of the track. However, comparable input/output traffic data pairs from flexible and advanced simulation setups such as the one described in Section 2.1.4 are not to get with the OLSIMv3 approach. Only a clear distinction between the various measuring and calculation methods in combination with each method having encapsulated in a detector model is able to support that. One common denominator to support the advanced setups is a common positioning scheme. The OLSIMv3 positioning scheme used a combination of lane and position. It is applicable to single lane detectors only. As OLSIMv4 relies at least on the detector types loop detector, local detector, and global detector this combination is insufficient. OLSIMv4, thus, employs an area structure to describe a detector's position inside a roadway element. And it distinguishes the various measuring and calculation methods. It combines them with having each method encapsulated in a detector model as introduced in the following.

This section presents at first the general detector operations with Section 3.5.1. This includes the specification of the measurements structure, the measuring method, and the detector structure. As a first concrete detector type with the specified detector operations, Section 3.5.2 introduces the widely known loop detectors. The traffic data range that they present to the simulation is also subject of Section 3.5.2. As a special variant of them, Section 3.5.3 presents the prognosis detectors that range over the same sorts of traffic data. While the two previous detector model types rely more or less on real traffic data and real traffic data may be hard to get in testing and verification scenarios, traffic data generators

Table 3.5.1: Summary of relevant prerequisites definitions for Section 3.5.1.

definition	description	domain	range
$\mathfrak{U}^4_{\text{obsv}}$	observable units		$\{\text{1 s}, \text{1 km/h}, \dots\}$
$\mathfrak{Q}^4_{\text{obsv}}$	quantity sizes		$\mathbb{N}^{255}_0\}$
$\mathfrak{L}^4_{\text{dettype}}$	detector type names		$\{\texttt{loop},\texttt{local},\texttt{global},\dots\}$
$\mathfrak{L}^4_{\text{det}}, \mathfrak{L}^4_{\text{detector}}$	detector names		$\{\texttt{d00001},\dots\}$
$\mathfrak{T}_{\text{mins}}$	time in minutes	$\mathfrak{T}_{\text{secs}}$	$\{t \bmod 60 = \text{0 s}\}$
$\langle a, b\rangle^{\text{obsv}}$	observable specification	$S^4_{\text{vehtype}} \times \mathfrak{U}^4_{\text{obsv}}$	S^4_{obsv}, see Equation 3.76
$\mathbf{p} \# \mathbf{s}_{\text{vehtype}}$	vehicle type	$\mathbf{p} \in S^4_{\text{obsv}}$	S^4_{vehtype}
$\mathbf{p} \# \Delta u$	observable unit	$\mathbf{p} \in S^4_{\text{obsv}}$	$\mathfrak{U}^4_{\text{obsv}}$
$\langle a, b, \dots\rangle^{\text{dettype}}$	detector type structure	$\mathfrak{L}^4_{\text{dettype}} \times (S^4_{\text{obsv}})^{M_i}$	S^4_{dettype}, see Eq. 3.77
$\mathbf{s}_{\text{dettype}} \# \mathbf{t}^{\text{dettype}}_{\text{name}}$	detector type name	$\mathbf{s}_{\text{dettype}} \in S^4_{\text{dettype}}$	$\mathfrak{L}^4_{\text{dettype}}$
$\mathbf{s}_{\text{dettype}} \# \mathbf{p}_j$	observable specification	$\mathbf{s}_{\text{dettype}} \in S^4_{\text{dettype}}$	$S^4_{\text{obsv}}, j \in 1, \dots, M_i$
$\mathbf{s}_{\text{dettype}} \# \mathbf{p}$	observable specifications	$\mathbf{s}_{\text{dettype}} \in S^4_{\text{dettype}}$	$(S^4_{\text{obsv}})^{M_i}$
$\mathbf{s}_{\text{dettype}} \# M_i$	number of observables	$\mathbf{s}_{\text{dettype}} \in S^4_{\text{dettype}}$	$\mathbb{N}, \mathbf{s}_{\text{dettype}} \# M_i := \lvert \mathbf{s}_{\text{dettype}} \# \mathbf{p}\rvert$
$\langle a, b, \dots\rangle^{\text{measure}}$	measurements structure	$\mathfrak{L}^4_{\text{dettype}} \times \mathfrak{T}^4_{\text{secs}} \times (\mathfrak{Q}^4_{\text{obsv}})^{M_i}$	S^4_{measure}, see Eq. 3.78
$\mathbf{m}^{(t)} \# \mathbf{t}^{\text{dettype}}_{\text{name}}$	detector type	$\mathbf{m}^{(t)} \in S^4_{\text{measure}}$	$\mathfrak{L}^4_{\text{dettype}}$
$\mathbf{m}^{(t)} \# t$	measurement time	$\mathbf{m}^{(t)} \in S^4_{\text{measure}}$	$\mathfrak{T}^4_{\text{secs}}$
$\mathbf{m}^{(t)} \# b^{(t)}_j$	observable quantity	$\mathbf{m}^{(t)} \in S^4_{\text{measure}}$	$\mathbb{N}, j \in 1, \dots, M_i$
$\mathbf{m}^{(t)} \# \mathbf{b}^{(t)}$	observable quantities	$\mathbf{m}^{(t)} \in S^4_{\text{measure}}$	$(\mathbb{N})^{M_i}$
$\mathbf{M}_{\text{zero}}(t)$	empty measurement	$\mathfrak{T}_{\text{secs}}$	S^4_{measure}

may come in handy. With the invariant detectors, Section 3.5.4 introduces the representant of the traffic data generators for the context of this thesis. The invariant detectors also present traffic data to the simulation. This property changes with the introduction of the local detectors in Section 3.5.5. They collect traffic data from the simulation that is comparable to the data of the loop detectors. Moreover, they try to assimilate the behaviour of the loop detectors. Finally and as the last detector type in this enumeration, Section 3.5.6 explains the specification of the global detectors. They provide spatio-temporal traffic information from the simulation to the user while all other aforementioned detector types provided only stationary traffic data.

3.5.1 Detector operations

All OLSIMv4 detectors provide measurements periodically as input to or as output from the simulation. For each input and output value a specification exists. It combines a vehicle type name $\mathbf{s}_{\text{vehtype}} \in S^4_{\text{vehtype}}$ with a generic unit $\Delta u \in \mathfrak{U}^4_{\text{obsv}} := \{ \text{ 1 s}, \text{ 1 vehs/min}, \text{ 1 km/h}, \text{ 1 SL}, \text{ 1 \%}, \dots \}$ into an element of the observables structure. Equation 3.76 introduces the latter.

$$S^4_{\text{obsv}} := \langle \mathbf{s}_{\text{vehtype}}, \Delta u\rangle^{\text{obsv}} . \tag{3.76}$$

Each traffic observable value contained in a detector measurement refers, thus, to a particular observable specification. Furthermore, each traffic observable value can be written as a product $q\Delta u$ of the quantity $q \in \mathfrak{Q}^4_{\text{obsv}}$ and the unit Δu to which the traffic observable refers to. In the context of OLSIMv4, the quantity set is restricted to $\mathfrak{Q}^4_{\text{obsv}} := \{i \in \mathbb{N}_0 \mid i \leqslant 255\}$. As the measurement domains do only vary between detector types and do not vary over time, the unit specification is contained in the detector type definition as explained next.

The detector type structure models detector types. Equation 3.77 introduces it. It consists of a detector type name $\mathbf{t}^{\text{dettype}}_{\text{name}} \in \mathfrak{L}^4_{\text{dettype}}$ with $\mathbf{t}^{\text{dettype}}_{\text{name}} \in \mathfrak{L}^4_{\text{dettype}} := \{ \texttt{loop}, \texttt{local}, \texttt{global}, \texttt{const}, \texttt{func}, \dots \}$ and a fixed number M_i of observable domain specifications $\mathbf{p}_1, \dots, \mathbf{p}_{M_i} \in S^4_{\text{obsv}}$. The index i in the number of observable domain specifications M_i refers to the index in the sequence of

Table 3.5.2: Summary of relevant detector definitions and operations for Section 3.5.1.

definition	description	domain	range
$\mathbb{N}^{59}$	nat. numbers $\leqslant 59$		$\{i \in \mathbb{N} \mid i \leqslant 59\}$
$\mathfrak{M}^4(n,d)$	avail. measurements	$\mathfrak{L}^4_{\text{dettype}} \times \mathfrak{L}^4_{\text{det}}$	$\mathcal{P}(S^4_{\text{measure}})$
N_{aggvals}	num. aggregate values		60
$\langle a, b, c, \ldots \rangle^{\text{det}}$	detector structure	$\mathfrak{L}^4_{\text{dtpe}} \times S^4_{\text{dtpe}} \times S^4_{\text{area}} \times \ldots$	S^4_{det}, Eq. 3.79
$\mathbf{d}_i^{(t)} \# \mathsf{t}_{\text{name}}^{\text{detector}}$	detector name	$\mathbf{d}_i^{(t)} \in S^4_{\text{det}}$	$\mathfrak{L}^4_{\text{det}}$
$\mathbf{d}_i^{(t)} \# \mathsf{s}_{\text{dettype}}$	detector type	$\mathbf{d}_i^{(t)} \in S^4_{\text{det}}$	S^4_{dettype}
$\mathbf{d}_i^{(t)} \# \mathsf{t}_{\text{name}}^{\text{roadway}}$	roadway name	$\mathbf{d}_i^{(t)} \in S^4_{\text{det}}$	$\mathfrak{L}^4_{\text{rway}}$
$\mathbf{d}_i^{(t)} \# \mathbf{r}_{\text{area}}$	spatial extent	$\mathbf{d}_i^{(t)} \in S^4_{\text{det}}$	S^4_{area}
$\mathbf{d}_i^{(t)} \# \mathbf{m}^{(t-j\Delta t)}, \mathbf{d}_i^{(t)} \# \mathbf{m}_j$	measurement value	$\mathbf{d}_i^{(t)} \in S^4_{\text{det}}, j \in \mathbb{N}_0^{59}$	S^4_{measure}
$\mathbf{d}_i^{(t)} \# \mathbf{M}^{(\tau)}$	agg. measurement	$\mathbf{d}_i^{(t)} \in S^4_{\text{det}}$	S^4_{measure}
$\mathbf{d}_i^{(t)} \# \mathbf{m}$	all measurements	$\mathbf{d}_i^{(t)} \in S^4_{\text{det}}$	$(S^4_{\text{measure}})^{\mathbb{N}_0^{59}}$
$\mathbf{F}_{\text{measure}}$	measure operation	$\mathfrak{T}_{\text{secs}} \times S^4_{\text{occ}} \times S^4_{\text{det}}$	S^4_{det}, Eq. 3.85
$\mathbf{F}_{\text{measurement}}$	measure operation	$\mathfrak{T}_{\text{secs}} \times S^4_{\text{occ}} \times S^4_{\text{det}}$	S^4_{measure}, Eq. 3.84
$\mathbf{F}_{\text{out-measure}}$	measure operation	$\mathfrak{T}_{\text{secs}} \times S^4_{\text{occ}} \times S^4_{\text{det}}$	S^4_{measure}, Eq. 3.81
$\mathbf{F}_{\text{det-update}}$	update detector	$\mathfrak{T}_{\text{secs}} \times S^4_{\text{occ}} \times S^4_{\text{det}} \times S^4_{\text{measure}}$	S^4_{det}, Eq. 3.82
$\mathbf{D}_{\text{zero}}(a,b,c,d)$	empty detector	$\mathfrak{T}_{\text{secs}} \times \mathfrak{L}^4_{\text{det}} \times S^4_{\text{dettype}} \times S^4_{\text{area}}$	S^4_{det}
$\mathbf{F}_{\text{insert}}(t,d,m)$	add measurement for t	$\mathfrak{T}_{\text{secs}} \times S^4_{\text{det}} \times S^4_{\text{measure}}$	S^4_{det}
$\mathbf{F}_{\text{insert}}(d,m)$	add measurement	$S^4_{\text{det}} \times S^4_{\text{measure}}$	S^4_{det}
$\mathbf{F}_{\text{insert}}(d,M,m)$	add aggregate+measrm.	$S^4_{\text{det}} \times S^4_{\text{measure}} \times S^4_{\text{measure}}$	S^4_{det}
$\mathbf{F}_{\text{output}}(t,d)$	detector output	$\mathfrak{T}_{\text{secs}} \times S^4_{\text{det}}$	S^4_{measure}, Eq. 3.80
$\mathbf{F}_{\text{input}}(t,d)$	measurement input	$\mathfrak{T}_{\text{secs}} \times S^4_{\text{det}}$	S^4_{measure}, Eq. 3.83
$\mathbf{F}_{\text{in-det?}}$	input/output pred.	$\mathfrak{L}^4_{\text{dettype}}$	$\mathfrak{B}^4_{\text{occ}}$

all detector types—1 for the first detector type will have M_1 observable domain specifications and so on up to $N = |\mathfrak{L}^4_{\text{dettype}}|$ observable domain specifications. As usual, each observable domain specification of a detector type $\mathsf{s}_{\text{dettype}} \in S^4_{\text{dettype}}$ is accessible as $\mathsf{s}_{\text{dettype}} \# \mathbf{p}_j$ with $j \in 1, \ldots, M_i$. The accessor $\mathsf{s}_{\text{dettype}} \# \mathbf{p}$ provides the sequence of all specifications. Consequently, the number of traffic observables in a measurement provided by detector type $\mathsf{s}_{\text{dettype}}$ is retrieved by counting them in the detector type specification $M_i := |\mathsf{s}_{\text{dettype}} \# \mathbf{p}|$. Using the detector type names $\mathsf{t}_{\text{name}}^{\text{dettype}} \in \mathfrak{L}^4_{\text{dettype}}$, the detector types $\mathsf{s}_{\text{dettype}}$ can be uniquely distinguished into input detector types and output detector types using the predicate function $\mathbf{F}_{\text{input-det?}} : \mathfrak{L}^4_{\text{dettype}} \longrightarrow \mathfrak{B}^4_{\text{occ}}$. Each of the various detector types implements the latter.

$$S^4_{\text{dettype}} := \langle \mathsf{t}_{\text{name}}^{\text{dettype}}, \mathbf{p}_1, \ldots, \mathbf{p}_{M_i} \rangle^{\text{dettype}} . \tag{3.77}$$

For each detector type $\mathsf{s}_{\text{dettype}} \in S^4_{\text{dettype}}$, the measurement value structure consists of the detector type name $\mathsf{t}_{\text{name}}^{\text{dettype}} \in \mathfrak{L}^4_{\text{dettype}}$, the time $t \in \mathfrak{T}_{\text{secs}}$, and exactly M_i traffic observables $b_j^{(t)} \in \mathfrak{Q}^4_{\text{obsv}}$ with $j \in 1, \ldots, M_i$ in the corresponding order of the traffic observable specifications. All traffic observables in one measurement value refer to the same timestamp $t \in \mathfrak{T}_{\text{secs}}$. Equation 3.78 defines the measurement structure. Similar to above, the accessor $\mathbf{m}^{(t)} \# \mathbf{b}_t$ provides access to the sequence of all traffic observables $b_1^{(t)}, \ldots, b_{M_i}^{(t)}$ of the measurement $\mathbf{m}^{(t)}$. Each of them originates from a detector of type $\mathsf{s}_{\text{dettype}}$ and has the value $b_j^{(t)} \cdot (\mathsf{s}_{\text{dettype}} \# \mathbf{p}_j) \# \Delta u$. Each measurement structure definition also provides a generic function $\mathbf{M}_{\text{zero}} : \mathfrak{T}_{\text{secs}} \longrightarrow S^4_{\text{measure}}$ that returns an empty or default zero value $\mathbf{M}_{\text{zero}}(t) := \langle t, 0, \ldots, 0 \rangle^{\text{measure}}$.

$$S^4_{\text{measure}} := \langle \mathsf{t}_{\text{name}}^{\text{dettype}}, t, b_1^{(t)}, \ldots, b_{M_i}^{(t)} \rangle^{\text{measure}} . \tag{3.78}$$

The detector type structure from Equation 3.77 and the measurement structure from Equation 3.78 serve as prerequisites for the definition of the detectors. Further prerequisites are the area structure S_{area}^4 from Equation 3.13 and the set of the detector names $\mathfrak{L}_{\text{det}}^4$. A definition for the set of detector names $\mathfrak{L}_{\text{det}}^4$ that suits the needs for OLSIMv4 consists, for example, of a concatenation of the letter d and a five digit number with leading zeros. For the definition of the detectors by Equation 3.79, the only limitation is that each detector needs a unique name. With the help of the prerequisites, Equation 3.79 defines a detector as a combination of the detector name $t_{\text{name}}^{\text{detector}} \in \mathfrak{L}_{\text{det}}^4$, the detector type $s_{\text{dettype}} \in S_{\text{dettype}}^4$, the roadway name $t_{\text{name}}^{\text{roadway}} \in \mathfrak{L}_{\text{roadway}}^4$ to which the detector refers to, the spatial extent of the detector $r_{\text{area}} \in S_{\text{area}}^4$, an aggregated measurement $\mathbf{M}^{(\tau)}$ of the previous meaurement cycle, and a variable number of measurements $\mathbf{m}^{(t)}, \ldots, \mathbf{m}^{(t-j\Delta t)}$ with $j \in \mathbb{N}^{59} := \{i \in \mathbb{N} \mid i \leqslant 59\}$. In each second, the number of measurements in a detector structure element increases by 1. A detector can contain at most 60 measurements. The time τ of the aggregated measurement is older than each of the times of the measurements, i.e. $\tau < t - j\Delta t$, $\forall j$. Similar to the traffic observables in a measurement, a detector structure element $\mathbf{d}_i^{(t)} \in S_{\text{det}}^4$ provides an accessor $\mathbf{d}_i^{(t)} \# \mathbf{m}$ for the sequence of all measurements $\mathbf{m}^{(t)}, \ldots, \mathbf{m}^{(t-j\Delta t)}$. Additionally, the accessors $\mathbf{d}_i^{(t)} \# \mathbf{m}_j := \mathbf{d}_i^{(t)} \# \mathbf{m}^{(t-j\Delta t)}$ with $j \in \mathbb{N}_0^{59}$ return any measurement of a detector structure element $\mathbf{d}_i^{(t)}$. The aggregated measurement $\mathbf{M}^{(\tau)}$ of the previous measurement cycle contains a timestamp τ with $\tau \in \mathfrak{T}_{\text{mins}} := \{t \in \mathfrak{T}_{\text{secs}} \mid t \bmod 60 = 0\,\text{s}\}$ that holds the constraint $(\mathbf{d}_i^{(t)} \# \mathbf{M}^{(\tau)}) \# \tau := (\mathbf{d}_i^{(t)} \# \mathbf{m}_0) \# t - (|\mathbf{d}_i^{(t)} \# \mathbf{m}| - 1)\Delta t$. The constraint ensures that in each minute only one and exactly one measurement value is aggregated. The definition of the set of minutes $\mathfrak{T}_{\text{mins}}$ that is used herein, is a simplification as it does not consider leap seconds. The detector structure definition provides also a function $\mathbf{D}_{\text{zero}} : \mathfrak{T}_{\text{secs}} \times \mathfrak{L}_{\text{det}}^4 \times S_{\text{dettype}}^4 \times S_{\text{area}}^4 \longrightarrow S_{\text{det}}^4$ that constructs an empty detector as $\mathbf{D}_{\text{zero}}(t, t_{\text{name}}^{\text{detector}}, s_{\text{dettype}}, r_{\text{area}}) := \langle t_{\text{name}}^{\text{detector}}, s_{\text{dettype}}, r_{\text{area}}, \mathbf{M}_{\text{zero}}(t), \mathbf{M}_{\text{zero}}(t) \rangle^{\text{det}}$. Additionally, the detector structure definitions also support adding a single measurement value by functions $\mathbf{F}_{\text{insert}} : \mathfrak{T}_{\text{secs}} \times S_{\text{det}}^4 \times S_{\text{measure}}^4 \longrightarrow S_{\text{det}}^4$ and $\mathbf{F}_{\text{insert}} : S_{\text{det}}^4 \times S_{\text{measure}}^4 \longrightarrow S_{\text{det}}^4$. The function evaluates as follows $\mathbf{F}_{\text{insert}}(\mathbf{d}_i^{(t-\Delta t)}, \mathbf{m}^{(t)}) := \langle t_{\text{name}}^{\text{detector}}, s_{\text{dettype}}, r_{\text{area}}, \mathbf{d}^{(t-\Delta t)} \# \mathbf{M}^{(\tau)}, \mathbf{m}^{(t)}, \mathbf{d}^{(t-\Delta t)} \# \mathbf{m} \rangle^{\text{det}}$. As the last variant of the input functions, function $\mathbf{F}_{\text{insert}} : S_{\text{det}}^4 \times S_{\text{measure}}^4 \times S_{\text{measure}}^4 \longrightarrow S_{\text{det}}^4$ allows to construct a detector with an aggregated value and a fresh measurement. The function evaluates into the following term $\mathbf{F}_{\text{insert}}(\mathbf{d}_i^{(t-\Delta t)}, \mathbf{M}^{(\tau)}, \mathbf{m}^{(t)}) := \langle t_{\text{name}}^{\text{detector}}, s_{\text{dettype}}, r_{\text{area}}, \mathbf{M}^{(\tau)}, \mathbf{m}^{(t)} \rangle^{\text{det}}$.

$$S_{\text{det}}^4 := \langle t_{\text{name}}^{\text{detector}}, s_{\text{dettype}}, t_{\text{name}}^{\text{roadway}}, r_{\text{area}}, \mathbf{M}^{(\tau)}, \mathbf{m}^{(t)}, \ldots, \mathbf{m}^{(t-j\Delta t)} \rangle^{\text{det}} . \tag{3.79}$$

Equation 3.79 describes a detector under the aspect of its name, its detector type, its spatial extent, and the measurements it contains. Additional aspects are the operations that a detector supports. In addition to the construction and access methods that have already been introduced, all detectors in this context support the measure and the aggregate method as well as the output method. The methods are explained in the following for the general case. The exact behaviour of the methods depends on the detector type. Whenever it differs from the general case, the definition of the specific behaviour is given in the sections following to this one. Whithout limiting the general case, all OLSIMv4 detectors have a cycle of 60 s and rely on the time discretisation of 1 s . The detectors that provide output from the simulation measure several traffic observables in each time step by means of function $\mathbf{F}_{\text{output}} : S_{\text{det}}^4 \longrightarrow S_{\text{measure}}^4$ in form of Equation 3.80.

$$\mathbf{F}_{\text{output}}(t, \mathbf{d}_i^{(t)}) := \mathbf{d}_i^{(t)} \# \mathbf{M}^{(\tau)} . \tag{3.80}$$

The detectors collect these measurements $\mathbf{m}^{(t)}, \mathbf{m}^{(t-\Delta t)}, \ldots \in S_{\text{measure}}^4$ during the whole measurement cycle, and aggregate them at the end of the cycle. The current aggregated measurement $\mathbf{M}^{(\tau)}$ with $\tau \in \mathfrak{T}_{\text{mins}}$ is available in each time step but is updated only once in a cycle. The detectors that provide input from external sources to the simulation during the aggregate step use a generalisation of the measure operation. The detectors that feed data from the simulation back into the simulation have a delay of a 60 s full measurement cycle. The delay is important for the classification as "closed under coupling" of discrete time specified networks [216, Sec. 7.3]. Section 3.7 discusses this in more detail.

The detectors perform the measuring in each second by Function $\mathbf{F}_{\text{measure}} : \mathfrak{T}_{\text{secs}} \times S^4_{\text{occ}} \times S^4_{\text{det}} \longrightarrow S^4_{\text{det}}$. In general, the measuring operation consists of two steps. In the first step, the detector determines a new measurement value either by fetching it from an external resource or by reading the vehicle properties such as velocity, vehicle type, and length of those vehicles that drive on the part of the roadway that is associated with the spatial extent of the detector. After reading the vehicle properties, the detector averages, summarises, and aggregates the properties into several traffic observables per vehicle type and, finally, into a new measurement value $\mathbf{m}^{(t)}$. In the second step, the measure operation updates the detector by adding the measurement value to the detector's list of measurements $\mathbf{m}^{(t)}, \mathbf{m}^{(t-\Delta t)}, \dots, \mathbf{m}^{(t-j\Delta t)} \in S^4_{\text{measure}}$ with $j \in \mathbb{N}^{59}$. The functions for the two steps of the measure operation are $\mathbf{F}_{\text{out}-\text{measure}}$ for determining the measurement value and $\mathbf{F}_{\text{det}-\text{update}}$ for updating the detectors measurements list. Function $\mathbf{F}_{\text{insert}}$ assists in adding the new measurement to the list of already existing measurements. The measure operation depends on the parameters current time $t \in \mathfrak{T}_{\text{secs}}$, the occupation table $\mathbf{o}_r^{(t)} \in S^4_{\text{occ}}$, and the detector itself $\mathbf{d}_i^{(t)} \in S^4_{\text{det}}$. Section 3.3.5 describes the occupation table structure S^4_{occ}. Equation 3.81 presents the signature of the first one. Its implementation depends on the detector type and is, thus, part of the following subsections. The second one, serves only for adding the new measurement value into the detector's list of measurements.

$$\mathbf{F}_{\text{out}-\text{measure}} : \mathfrak{T}_{\text{secs}} \times S^4_{\text{occ}} \times S^4_{\text{det}} \longrightarrow S^4_{\text{measure}} . \tag{3.81}$$

The operation $\mathbf{F}_{\text{aggregate}} : \mathfrak{T}_{\text{mins}} \times S^4_{\text{det}} \longrightarrow S^4_{\text{measure}}$ aggregates the detector's list of measurements $\mathbf{m}^{(t)}, \mathbf{m}^{(t-\Delta t)}, \dots, \mathbf{m}^{(t-j\Delta t)} \in S^4_{\text{measure}}$ with $j \in \mathbb{N}^{59}$ into a single measurement value $\mathbf{M}^{(\tau)}$ and it is evaluated as the last step in a cycle. The OLSIMv4 detectors apply the aggregate operation every $60\,\text{s}$ in order to replace the $\mathbf{M}^{(\tau)}$ by $\mathbf{M}^{(\tau+\Delta\tau)}$ with $\Delta\tau := 60\,\text{s}$. Depending on the detector type, the number of aggregate values $N_{\text{aggvals}} = 60$ that triggers the aggregate function may vary. The detector's concrete behaviour depends on the detector type and in particular on the kinds of traffic observables that the detector's measurements include. In general, a barycentric formula or another averaging method is implemented. As the aggregate operation does not depend on resources other than the detector, the aggregate operation is used implicitly by the measure operation. The definition of the measure operation from above requires, thus, adjustment. Equation 3.82 presents the adjusted definition. Any evaluation of function $\mathbf{F}_{\text{det}-\text{update}}$ depends on the measurement value for the next time step $\mathbf{m}^{(t+\Delta t)}$. Function $\mathbf{F}_{\text{det}-\text{update}}$ consumes the measurement value as an argument. Equation 3.84 presents later the calculation of the measurement value.

$$\mathbf{F}_{\text{det}-\text{update}}(t, \mathbf{o}_r^{(t)}, \mathbf{d}_i^{(t)}, \mathbf{m}^{(t)}) := \begin{cases} \mathbf{F}_{\text{insert}}(\mathbf{d}_i^{(t)}, \mathbf{F}_{\text{aggregate}}(t, \mathbf{d}_i^{(t)}), \mathbf{m}^{(t)}) & \text{, if } |\mathbf{d}_i^{(t)} \# \mathbf{m}| = N_{\text{aggvals}}, \\ \mathbf{F}_{\text{insert}}(\mathbf{d}_i^{(t)}, \mathbf{m}^{(t)}) & \text{, otherwise.} \end{cases}$$
$$\tag{3.82}$$

The measurement value $\mathbf{m}^{(t)}$ for the current time step relies on the detector type. In case of an input detector $\mathbf{d}_i^{(t)} \in \mathbf{L}_{\text{det}}^{\text{in}}$ such as the loop and the prognosis detectors, the measurement value comes from an external resource, e.g. $\mathbf{m}_{\text{in}}^{(t)} := \mathbf{F}_{\text{input}}(t, \mathbf{d}_i^{(t)})$ with $\mathbf{F}_{\text{input}} : \mathfrak{T}_{\text{secs}} \times S^4_{\text{det}} \longrightarrow S^4_{\text{measure}}$, and $\mathfrak{M}^4((\mathbf{d}_i^{(t)} \# \mathbf{s}_{\text{dettype}}) \# \mathbf{t}_{\text{name}}^{\text{dettype}}, \mathbf{d}_i^{(t)} \# \mathbf{t}_{\text{name}}^{\text{detector}})$ as the set of available measurements for the loop detector $\mathbf{d}_i^{(t)} \# \mathbf{t}_{\text{name}}^{\text{detector}}$ of type $(\mathbf{d}_i^{(t)} \# \mathbf{s}_{\text{dettype}}) \# \mathbf{t}_{\text{name}}^{\text{dettype}}$, and $\mathbf{F}_{\text{input}}$ from the definition of Equation 3.83.

$$\mathbf{F}_{\text{input}}(t, \mathbf{d}_i^{(t)}) := \min \{\mathbf{m}^{(t)} \in \mathfrak{M}^4((\mathbf{d}_i^{(t)} \# \mathbf{s}_{\text{dettype}}) \# \mathbf{t}_{\text{name}}^{\text{dettype}}, \mathbf{d}_i^{(t)} \# \mathbf{t}_{\text{name}}^{\text{detector}})\} \tag{3.83}$$

An output detector $\{\mathbf{d}_i^{(t)} \in \mathbf{L}_{\text{det}}^{\text{out}}\}$ determines that value by $\mathbf{m}_{\text{out}}^{(t)} := \mathbf{F}_{\text{out}-\text{measure}}(t, \mathbf{o}_r^{(t)}, \mathbf{d}_i^{(t)})$. Equation 3.84 summarises the calculation of the measurement value. It uses the boolean value $x_{\text{in?}} := \mathbf{F}_{\text{in}-\text{det?}}((\mathbf{d}_i^{(t)} \# \mathbf{s}_{\text{dettype}}) \# \mathbf{t}_{\text{name}}^{\text{dettype}})$ to determine whether detector $\mathbf{d}_i^{(t)}$ is an input detector or not.

$$\mathbf{F}_{\text{measurement}}(t, \mathbf{o}_r^{(t)}, \mathbf{d}_i^{(t)}) := \begin{cases} \mathbf{F}_{\text{input}}(t, \mathbf{d}_i^{(t)}) & \text{, if } x_{\text{in?}}, \\ \mathbf{F}_{\text{out}-\text{measure}}(t, \mathbf{o}_r^{(t)}, \mathbf{d}_i^{(t)}) & \text{, otherwise.} \end{cases} \tag{3.84}$$

Finally, Equation 3.85 presents the definition of the resulting measure operation that has the signature $\mathbf{F}_{\text{measure}} : \mathfrak{T}_{\text{secs}} \times S^4_{\text{occ}} \times S^4_{\text{det}} \longrightarrow S^4_{\text{det}}$. The function applies the aggregate measure operation implicitly

Table 3.5.3: Loop detector data type and traffic observables for OLSIMv4.

definition	description	domain	range
$\langle \tau, a, b, c, d, e \rangle^{\text{loop}}$	loop detector data type	$\mathfrak{T}_{\text{mins}} \times (S_{\text{obsv}}^4)^5$	S_{loop}^4, see Equation 3.87
$\langle \tau, a, b, c, d, e, f, g \rangle^{\text{loop}*}$	loop* detector data type	$\mathfrak{T}_{\text{mins}} \times (S_{\text{obsv}}^4)^7$	$S_{\text{loop}*}^4$, see Equation 3.89
$\mathbf{d}_i^{(t)} \# j_{\text{any}}^{(t)}$	traffic flow, any vehtype	$\mathbf{d}_i^{(t)} \in S_{\text{loop}}^4$	$\mathbb{N}_0^{80} \subset S_{\text{obsv}}^4$
$\mathbf{d}_i^{(t)} \# j_{\text{car}}^{(t)}$	traffic flow, car vehtype	$\mathbf{d}_i^{(t)} \in S_{\text{loop}}^4$	$\mathbb{N}_0^{80} \subset S_{\text{obsv}}^4$
$\mathbf{d}_i^{(t)} \# j_{\text{truck}}^{(t)}$	traffic flow, truck vehtype	$\mathbf{d}_i^{(t)} \in S_{\text{loop}}^4$	$\mathbb{N}_0^{60} \subset S_{\text{obsv}}^4$
$\mathbf{d}_i^{(t)} \# v_{\text{any}}^{(t)}$	average velocity, any vehtype	$\mathbf{d}_i^{(t)} \in S_{\text{loop}}^4$	$\mathbb{N}_0^{240} \subset S_{\text{obsv}}^4$
$\mathbf{d}_i^{(t)} \# v_{\text{car}}^{(t)}$	average velocity, car vehtype	$\mathbf{d}_i^{(t)} \in S_{\text{loop}}^4$	$\mathbb{N}_0^{240} \subset S_{\text{obsv}}^4$
$\mathbf{d}_i^{(t)} \# v_{\text{truck}}^{(t)}$	velocity, truck vehtype	$\mathbf{d}_i^{(t)} \in S_{\text{loop}}^4$	$\mathbb{N}_0^{180} \subset S_{\text{obsv}}^4$
$\mathbf{d}_i^{(t)} \# p_{\text{occ}}^{(t)}$	occupation time, any vehtype	$\mathbf{d}_i^{(t)} \in S_{\text{loop}}^4$	$\mathbb{N}_0^{100} \subset S_{\text{obsv}}^4$
$\mathbf{F}_{\text{ext}-\text{loop}}$	extended measurement	S_{loop}^4	$S_{\text{loop}*}^4$, see Equation 3.90

and distinguishes between input and output detectors by use of Function $\mathbf{F}_{\text{measurement}}$ from above. In case of input detectors other than the loop detectors, the Function $\mathbf{F}_{\text{measurement}}$ requires adjustment as demonstrated in Section 3.5.3.

$$\mathbf{F}_{\text{measure}}(t, \mathbf{o}_r^{(t)}, \mathbf{d}_i^{(t)}) := \mathbf{F}_{\text{det}-\text{update}}(t, \mathbf{o}_r^{(t)}, \mathbf{d}_i^{(t)}, \mathbf{F}_{\text{measurement}}(t, \mathbf{o}_r^{(t)}, \mathbf{d}_i^{(t)})). \tag{3.85}$$

The following sections present the various detector types. At first come the loop detectors that present traffic data from devices to the simulation. The subsection thereafter, portraits the prognosis detectors as variants of the loop detectors. Whenever no real data is available, the detectors with time invariant behaviour might be useful. They follow directly after the prognosis detectors. This section then continues with the description of the cross-section measuring local detectors. They are the pendants in the simulation to the loop detectors in the reality. Finally, this section closes with the presentation of the global detectors. They do not measure cross-section wise but instead measure road-segment wise. The list of detectors presented in this section is by no means complete. Detectors have broad application fields. The latter cover the spectrum from emission estimation, over weather data provision, to scientific simulation evaluation.

3.5.2 Loop detectors

Loop detectors provide data from roadway single lanes. Table 2.1.2 presented the loop detector data for OLSIMv3. As the amount of available data did not change since then, the data remains the same in the context of OLSIMv4. However, the average netto time headway, the variance for velocity, and the exponentially smoothened velocity turned out to lack stability in information quality. As a consequence, OLSIMv4 does not consider them for use and each measurement of the loop detectors has only five traffic observables. Equations 3.76 and 3.77 help in defining the loop detector structure. This is accomplished as described in the following. First, the five traffic observable types are defined.

Traffic observables combine vehicle types with units. OLSIMv4 uses the vehicle types from Section 3.2.1 for the loop detectors. The traffic flow for the `any` vehicle type belongs to the traffic observables specification $\mathbf{j}_{\text{any}} := \langle s_{\text{vehtype}}^{\text{any}}, 1 \text{ vehs/min} \rangle^{\text{obsv}}$. A similar traffic observables specification $\mathbf{j}_{\text{truck}} := \langle s_{\text{vehtype}}^{\text{truck}}, 1 \text{ vehs/min} \rangle^{\text{obsv}}$ results for the vehicle type `truck`. The velocity $\mathbf{v}_{\text{car}} := \langle s_{\text{vehtype}}^{\text{car}}, 1 \text{ km/h} \rangle^{\text{obsv}}$ for the `car` vehicle type forms the third traffic observable for the loop detector measurements. Analogously, the velocity $\mathbf{v}_{\text{truck}} := \langle s_{\text{vehtype}}^{\text{truck}}, 1 \text{ km/h} \rangle^{\text{obsv}}$ for the `truck` vehicle type becomes the fourth traffic observable for the loop detector measurements. As the last traffic observable specification, the loop detector measurements list the occupation time value in percent of the 1 min measurement cycle for the `any` vehicle type which results in $\mathbf{p}_{\text{occ}} := \langle s_{\text{vehtype}}^{\text{any}}, 1 \% \rangle^{\text{obsv}}$. The five specifications $\mathbf{j}_{\text{any}}, \mathbf{j}_{\text{truck}}$, $\mathbf{v}_{\text{car}}, \mathbf{v}_{\text{truck}}, \mathbf{p}_{\text{occ}} \in S_{\text{obsv}}^4$ combined with the time in seconds $t \in \mathfrak{T}_{\text{secs}}^4$ form then the prerequisites for the

Table 3.5.4: Summary of relevant functions, structures, variables, and constants for Section 3.5.2.

definition	description	domain	range
$S^4_{\text{lptype}}(a,b,c,d,e)$	loop detector type	$(S^4_{\text{obsv}})^5$	S^4_{dettype}, see Equation 3.86
$S^4_{\text{lptype}*}(a,b,c,d,e,f,g)$	loop* detector type	$(S^4_{\text{obsv}})^7$	S^4_{dettype}, see Equation 3.88
$\mathbf{d}^{(t)}_i \# \mathbf{j}_{\text{any}}$	traffic flow, any vehtype	$\mathbf{d}^{(t)}_i \in S^4_{\text{dettype}}$	S^4_{obsv}
$\mathbf{d}^{(t)}_i \# \mathbf{j}_{\text{car}}$	traffic flow, car vehtype	$\mathbf{d}^{(t)}_i \in S^4_{\text{dettype}}$	S^4_{obsv}
$\mathbf{d}^{(t)}_i \# \mathbf{j}_{\text{truck}}$	traffic flow, truck vehtype	$\mathbf{d}^{(t)}_i \in S^4_{\text{dettype}}$	S^4_{obsv}
$\mathbf{d}^{(t)}_i \# \mathbf{v}_{\text{any}}$	average velocity, any vehtype	$\mathbf{d}^{(t)}_i \in S^4_{\text{dettype}}$	S^4_{obsv}
$\mathbf{d}^{(t)}_i \# \mathbf{v}_{\text{car}}$	average velocity, car vehtype	$\mathbf{d}^{(t)}_i \in S^4_{\text{dettype}}$	S^4_{obsv}
$\mathbf{d}^{(t)}_i \# \mathbf{v}_{\text{truck}}$	average velocity, truck vehtype	$\mathbf{d}^{(t)}_i \in S^4_{\text{dettype}}$	S^4_{obsv}
$\mathbf{d}^{(t)}_i \# \mathbf{p}_{\text{occ}}$	occupation time, any vehtype	$\mathbf{d}^{(t)}_i \in S^4_{\text{dettype}}$	S^4_{obsv}

definition of the OLSIMv4 loop detector measurements structure S^4_{lptype}. Equation 3.86 presents the definition for the lptype-element of the detector type structure S^4_{dettype}. It consists of a set with only one element that defines the domains for the loop detector measurements.

$$S^4_{\text{lptype}} := \langle \text{loop}, \mathbf{j}_{\text{any}}, \mathbf{j}_{\text{truck}}, \mathbf{v}_{\text{car}}, \mathbf{v}_{\text{truck}}, \mathbf{p}_{\text{occ}} \rangle^{\text{dettype}} . \tag{3.86}$$

In contrast to the detector type structure, the loop detector data structure S^4_{loop} defines the range of the detector data as a set of measurement structure elements with the constant detector type name loop $\in \mathfrak{L}^4_{\text{dettype}}$. Equation 3.87 specifies the loop detector data.

$$S^4_{\text{loop}} := \langle \text{loop}, \tau, j^{(\tau)}_{\text{any}}, j^{(\tau)}_{\text{truck}}, v^{(\tau)}_{\text{car}}, v^{(\tau)}_{\text{truck}}, p^{(\tau)}_{\text{occ}} \rangle^{\text{measure}} . \tag{3.87}$$

It consists further of a time stamp $\tau \in \mathfrak{T}_{\text{mins}}$ and the five traffic observables. The first two of them are the traffic flows, $j^{(\tau)}_{\text{any}}$ and $j^{(\tau)}_{\text{truck}}$, for the any and for the truck vehicle type. They represent the number of vehicles and of trucks that passed the loop detector during the last measurement interval. The next two traffic observables contain the average velocities, $v^{(\tau)}_{\text{car}}$ and $v^{(\tau)}_{\text{truck}}$, for the for the car and for the truck vehicle type. As the last value of the measurement's traffic observables the occupation time $p^{(\tau)}_{\text{occ}}$ for the any vehicle type contains the time in percent within the detector was covered or blocked by a vehicle. The values of the loop detector data range each in the set $\mathfrak{O}^4_{\text{obsv}}$. It has been chosen relatively large for the purpose of the loop detector data. As an example, the occupation time $p^{(\tau)}_{\text{occ}}$ cannot have values above 100. Thus, OLSIMv4 limits the values of the traffic observables by the specifying a subset of the set $\mathfrak{O}^4_{\text{obsv}}$ for each of the traffic observables. Table 3.5.3 lists the ranges for the traffic observables. Additionally, OLSIMv4 examines the incoming loop detector data for consistency with the defined ranges. When the value ranges hold, OLSIMv4 verifies additionally the correctness of some correlations such as $j^{(\tau)}_{\text{any}} = 0 \implies j^{(\tau)}_{\text{truck}} = 0 \wedge v^{(\tau)}_{\text{car}} = v^{(\tau)}_{\text{truck}} = 0$. The loop detectors implement the measure operation from Equation 3.85 without any changes. As they belong to the class of the input detectors, they employ the fetch function as part of the implementation of the measurement function from Equation 3.84. The implementation of the latter for the case of the loop detectors results then in the definition $\mathbf{F}_{\text{measurement}}(t, \mathbf{o}^{(t)}_r, \mathbf{d}^{(t)}_i) := \mathbf{F}_{\text{ext}-\text{loop}}(\mathbf{F}_{\text{input}}(t, \mathbf{d}^{(t)}_i))$. The definition puts a small wrapper $\mathbf{F}_{\text{ext}-\text{loop}} : S^4_{\text{loop}} \longrightarrow S^4_{\text{loop}*}$ around the fetch function to provide the more generalised loop detector type data as introduced below.

By using a functional dependency, the loop detector type and, consequently, also the loop detector data can be extended into more generalised versions . This is especially useful for the prognosis detectors and for the comparison of measurements from other detectors. Equation 3.88 presents the extended detector type while Equation 3.89 lists the extended loop detector data specification.

$$S^4_{\text{lptype}*} := \langle \text{loop}*, \mathbf{j}_{\text{any}}, \mathbf{j}_{\text{car}}, \mathbf{j}_{\text{truck}}, \mathbf{v}_{\text{any}}, \mathbf{v}_{\text{car}}, \mathbf{v}_{\text{truck}}, \mathbf{p}_{\text{occ}} \rangle^{\text{dettype}} . \tag{3.88}$$

$$S_{\text{loop}*}^4 := \langle \texttt{loop}*, \tau, j_{\text{any}}^{(\tau)}, j_{\text{car}}^{(\tau)}, j_{\text{truck}}^{(\tau)}, v_{\text{any}}^{(\tau)}, v_{\text{car}}^{(\tau)}, v_{\text{truck}}^{(\tau)}, p_{\text{occ}}^{(\tau)} \rangle^{\text{measure}} . \tag{3.89}$$

In addition to the traffic observables provided by the measurement of the fetch operation, the wrapper $\mathbf{F}_{\text{ext}-\text{loop}}$ for the extended loop detector data provides the traffic flow observable $\mathbf{j}_{\text{car}}$ for the $\texttt{car}$ vehicle type and the velocity for the $\texttt{any}$ vehicle type $\mathbf{v}_{\text{any}}$. The traffic flow observable for the $\texttt{car}$ vehicle type of the loop detector type has the domain $\mathbf{j}_{\text{car}} := \langle \text{s}_{\text{vehtype}}^{\text{car}}, 1 \text{ vehs/min} \rangle^{\text{obsv}}$ and it ranges over the set $\mathbb{N}_0^{80} \subset S_{\text{obsv}}^4$. The average velocity for the $\texttt{any}$ vehicle type of the loop detector type has the domain $\mathbf{v}_{\text{any}} := \langle \text{s}_{\text{vehtype}}^{\text{any}}, 1 \text{ km/h} \rangle^{\text{obsv}}$. It ranges over the set $\mathbb{N}_0^{240} \subset S_{\text{obsv}}^4$.

The functional relationship of the traffic flow between a measurement $\mathbf{m}^{(\tau)} \in S_{\text{loop}}^4$ and the corresponding measurement $\mathbf{m}*^{(\tau)} \in S_{\text{loop}*}^4$ uses the difference between the flow values of the $\texttt{any}$ and the $\texttt{truck}$ vehicle types. The functional relationship for the $\texttt{any}$ vehicle type velocity uses the barycentric formula. Function $\mathbf{F}_{\text{ext}-\text{loop}} : S_{\text{loop}}^4 \longrightarrow S_{\text{loop}*}^4$ from Equation 3.90 presents the functional dependency between the two measurement values. Its definition uses the following values. The traffic flow values $j_{\text{any}}^{(\tau)} := \mathbf{m}^{(\tau)} \# j_{\text{any}}^{(\tau)}$, $j_{\text{truck}}^{(\tau)} := \mathbf{m}^{(\tau)} \# j_{\text{truck}}^{(\tau)}$, and $j_{\text{car}}^{(\tau)} := j_{\text{any}}^{(\tau)} - j_{\text{truck}}^{(\tau)}$ are the first three values. The second three values are the average velocity values $v_{\text{car}}^{(\tau)} := \mathbf{m}^{(\tau)} \# v_{\text{car}}^{(\tau)}$, $v_{\text{truck}}^{(\tau)} := \mathbf{m}^{(\tau)} \# v_{\text{truck}}^{(\tau)}$, and $v_{\text{any}}^{(\tau)} := (j_{\text{car}}^{(\tau)} \cdot v_{\text{car}}^{(\tau)} + j_{\text{truck}}^{(\tau)} \cdot v_{\text{truck}}^{(\tau)})/j_{\text{any}}^{(\tau)}$. The occupation time $p_{\text{occ}}^{(\tau)} := \mathbf{m}^{(\tau)} \# p_{\text{occ}}^{(\tau)}$ forms the last traffic observable value. The corresponding value for the extended measurement data is then the result of the function evaluation $\mathbf{m}*^{(\tau)} := \mathbf{F}_{\text{ext}-\text{loop}}(\mathbf{m}^{(\tau)})$. Tables 3.5.4 and 3.5.3 summarise the relevant functions, structure definitions, variables, and constants for Section 3.5.2.

$$\mathbf{F}_{\text{ext}-\text{loop}}(\mathbf{m}^{(\tau)}) := \langle \texttt{loop}*, \mathbf{m}^{(\tau)} \# \tau, j_{\text{any}}^{(\tau)}, j_{\text{car}}^{(\tau)}, j_{\text{truck}}^{(\tau)}, v_{\text{any}}^{(\tau)}, v_{\text{car}}^{(\tau)}, v_{\text{truck}}^{(\tau)}, p_{\text{occ}}^{(\tau)} \rangle^{\text{measure}} . \tag{3.90}$$

3.5.3 Prognosis detectors

During the development of OLSIMv3, Chrobok developed the 60-minutes short-term and the seven days long-term prognosis [40, Chap. 6] for the context of OLSIM. The method relies on a classification of days into day classes such as working days, weekends, and vacancy days. Once the day part of a timestamp is classified, the prognosis method combines the corresponding historical loop detector data by an exponential smoothing method. OLSIMv3 did not provide a seven days long term prognosis. The implementation of the 30-minutes and 60-minutes was somehow experimental. It lacked stability and covered only a part of the network. Additionally, OLSIMv3 provided the short term prognoses only for a fraction of the time. As part of an improvement, OLSIMv4 provides the 30- and 60-minutes short term as well as the 7-days long term prognosis data for the whole network in each minute.

The OLSIMv4 prognosis detectors use the loop detector data according to the specification from Section 3.5.2 as their input data. They aggregate the input data and provide the aggregated data in the same form as output data. Their input data has the domain as specified by Equation 3.88 and they range over the loop detector data as specified by Equation 3.89. The pure existence of a prognosis method makes them distinct from the loop detectors.

The traffic data prognosis for a given timestamp τ sorts out first the correct class for the day part of the timestamp. Thereafter, it determines the set of the last $\text{N}_{\text{hist}} = 30$ corresponding historical timestamps by evaluating Function $\mathbf{T}_{\text{hist}} : \mathfrak{T}_{\text{mins}} \longrightarrow \mathcal{P}_{\text{mins}}(\mathfrak{T}_{\text{mins}})$ from Equation 3.91. It uses Function $\mathbf{F}_{\text{class}} : \mathfrak{T}_{\text{mins}} \longrightarrow \mathbb{N}$ to determine the class of the day wherein the timestamp belongs to. Additionally, it uses Function $\mathbf{F}_{\text{time}} : \mathfrak{T}_{\text{mins}} \longrightarrow \mathfrak{T}_{\text{mins}}$ to calculate the time of the day. Its operation is roughly equivalent to $\mathbf{F}_{\text{time}}(\tau) := \tau \bmod 1440$ albeit this approach does not consider leap seconds and daylight saving times. The operation of Function $\mathbf{F}_{\text{class}}$ is similar to the one of $\mathbf{F}_{\text{time}}$. As an example, the day of week class uses the calculation $\mathbf{F}_{\text{class}}(\tau) := (\tau \text{ div } 1440) \bmod 7$. The calculation for the easter sunday is, of course, more complex [23] but in principle a straight forward task. Except for Christmas, the other non-business days align themselves relative to the date of the easter sunday. The NRW administration sets the summer holidays and similar. OLSIMv4 collects the affected days in a set. With all these special classes, the classification by Function $\mathbf{F}_{\text{class}}$ results finally in the class with the highest priority, e.g.

Table 3.5.5: Summary of relevant functions, structures, variables, and constants for Section 3.5.3.

definition	description	domain	range
N_{hist}	number of historic mmnts.		30
α_{exp}	weight for smoothing		0.2
t_{hmax}	time horizon boundary		$37\,min$
$\mathbf{T}_{avg}(\tau)$	3 successive minutes	$\tau \in \mathfrak{T}_{mins}$	$\{\tau - \Delta\tau, \tau, \tau + \Delta\tau\}$
$\mathbf{T}_{hist}(\tau)$	N_{hist} historical timestamps	$\tau \in \mathfrak{T}_{mins}$	$\subseteq \mathfrak{T}_{mins}$, see Equation 3.91
$\mathbf{F}_{avg}(M, i, j)$	barycentric formula	$\mathcal{P}_{measure*}(S_{measure*}) \times \mathbb{N}_0^7 \times \mathbb{N}^7$	S_{obsv}^4, see Equation 3.92
$\mathbf{F}_{smooth}$	exponential smoothing	$\mathfrak{T}_{mins} \times S_{det} \times \mathbb{N}_0^7$	S_{obsv}^4, see Equation 3.94
$\mathbf{F}_{smooth}$	exponential smoothing	$\mathfrak{T}_{mins} \times S_{det}$	$S_{measure}^4$, see Equation 3.95
$\mathbf{F}_{mix}$	30 min prognosis	$S_{loop} \times S_{loop} \times S_{loop} \times \mathbb{N}$	S_{obsv}^4, see Equation 3.96

a normal business day during the summer holidays falls into the summer holiday class. Each of the historical timestamps $\mathbf{T}_{hist}(\tau)$ refers to the same time of day as the given timestamp τ.

$$\mathbf{T}_{class}(\tau) := \{\tau_i \in \mathfrak{T}_{mins} \mid \mathbf{F}_{class}(\tau) = \mathbf{F}_{class}(\tau_i) \wedge \mathbf{F}_{time}(\tau) = \mathbf{F}_{time}(\tau_i) \wedge \forall i : \tau_{i+1} < \tau_i < \tau\},$$
$$\mathbf{T}_{hist}(\tau) := \{\tau_i \in \mathbf{T}_{class}(\tau) \mid i \leqslant N_{hist} \wedge \mathbf{T}_{class}(\tau)\backslash\mathbf{T}_{class}(\tau_1) = \{\tau_1\}\}. \tag{3.91}$$

As part of the calculation $\mathbf{F}_{measurement}(t, \mathbf{o}_r^{(t)}, \mathbf{d}_i^{(t)})$ of the next measurement value from Equation 3.84, Function $\mathbf{F}_{measurement}$ wraps the input function $\mathbf{F}_{input}$ for the detector $\mathbf{d}_i^{(t)}$ and the timestamp t to make the loop detector data available. For OLSIMv4, the set of available measurements contains about 10 years of loop detector data. The prognostication method smoothes the historic loop detector data by averaging the data over the minutes in the range between the minute before and after the given timestamp τ. The set of timestamps for the data to be averaged is then $\mathbf{T}_{avg}(\tau)$ with $\mathbf{T}_{avg}(\tau) := \{\tau - \Delta\tau, \tau, \tau + \Delta\tau\}$ as the set of three successive minutes. The smoothening uses the barycentric averaging method. Equation 3.92 presents the Function $\mathbf{F}_{avg} : \mathcal{P}_{measure}(S_{measure}^4) \times \mathbb{N}_0^7 \times \mathbb{N}^7 \longrightarrow S_{obsv}^4$ that calculates the barycentre for the traffic observable values with indices i and j of any given set of measurement values $\mathfrak{M} \subset \mathcal{P}_{measure}(S_{measure}^4)$. As a special case, the evaluation of the barycentric formula with a value of 0 for the first index, it matches the arithmetic average. Equation 3.92 reflects this by artificially setting $\mathbf{m}^{(\tau)}\#b_0^{(\tau)} := 1$.

$$\mathbf{F}_{avg}(\mathfrak{M}, i, j) := \left\lceil \frac{\sum_{\mathbf{m}^{(\tau)}\in\mathfrak{M}} \mathbf{m}^{(\tau)}\#b_i^{(\tau)} \cdot \mathbf{m}^{(\tau)}\#b_j^{(\tau)}}{\sum_{\mathbf{m}^{(\tau)}\in\mathfrak{M}} \mathbf{m}^{(\tau)}\#b_i^{(\tau)}} \right\rceil, \text{ with } \mathbf{m}^{(\tau)}\#b_0^{(\tau)} := 1. \tag{3.92}$$

With the loop detector measurements $\mathbf{F}_{in-avg}(\tau, \mathbf{d}_i^{(\tau)}) := \{\mathbf{F}_{ext-loop}(\mathbf{F}_{input}(\tau_j, \mathbf{d}_i^{(t)})) \mid \tau_j \in \mathbf{T}_{avg}(\tau)\}$ to a given timestamp τ and a given detector $\mathbf{d}_i^{(\tau)}$ in combination with the barycentric formula $\mathbf{F}_{avg}$ from Equation 3.92, the smoothened traffic observable values are the two traffic flow values, $\bar{\jmath}_{any}^{(\tau)} := \mathbf{F}_{avg}(\mathbf{F}_{in-avg}(\tau, \mathbf{d}_i^{(\tau)}), 0, 1)$ and $\bar{\jmath}_{truck}^{(\tau)} := \mathbf{F}_{avg}(\mathbf{F}_{in-avg}(\tau, \mathbf{d}_i^{(\tau)}), 0, 3)$, $\bar{\jmath}_{car}^{(\tau)} := \mathbf{F}_{avg}(\mathbf{F}_{in-avg}(\tau, \mathbf{d}_i^{(\tau)}), 0, 2)$, the three average velocity values, $\bar{v}_{any}^{(\tau)} := \mathbf{F}_{avg}(\mathbf{F}_{in-avg}(\tau, \mathbf{d}_i^{(\tau)}), 1, 4)$, $\bar{v}_{car}^{(\tau)} := \mathbf{F}_{avg}(\mathbf{F}_{in-avg}(\tau, \mathbf{d}_i^{(\tau)}), 2, 5)$ and $\bar{v}_{truck}^{(\tau)} := \mathbf{F}_{avg}(\mathbf{F}_{in-avg}(\tau, \mathbf{d}_i^{(\tau)}), 3, 6)$, as well as $\bar{p}_{any}^{(\tau)} := \mathbf{F}_{avg}(\mathbf{F}_{in-avg}(\tau, \mathbf{d}_i^{(\tau)}), 0, 7)$ for the average occupation time. Equation 3.93 presents then the smoothened measurement value $\overline{m}^{(\tau)}$ for detector $\mathbf{d}_i^{(\tau)}$ as a function of the timestamp τ.

$$\overline{m}^{(\tau)} := \langle \texttt{loop}, \tau, \bar{\jmath}_{any}^{(\tau)}, \bar{\jmath}_{car}^{(\tau)}, \bar{\jmath}_{truck}^{(\tau)}, \bar{v}_{any}^{(\tau)}, \bar{v}_{car}^{(\tau)}, \bar{v}_{truck}^{(\tau)}, \bar{p}_{occ}^{(\tau)} \rangle^{measure}. \tag{3.93}$$

As the last step, the prognosis method exponentially smoothens the averaged historical loop detector data. The data is available as the ordered set $\mathbf{M}_{hist}(\tau, \mathbf{d}_i^{(\tau)}) := \{\overline{m}^{(t)} \in S_{loop}^4 \mid t \in \mathbf{T}_{hist}(\tau)\}$ where each of its measurements refers to equal time of days within the sequence $t_{N_{\text{hist}}} < t_{N_{\text{hist}}-1} < \ldots < t_2 <$ $t_1 < \tau$ and timestamp τ is the youngest. Each measurement of the ordered set $\mathbf{M}_{\text{hist}}(\tau, \mathbf{d}_i^{(\tau)})$ wherein each is accessible by an index that refers to its position in the ordered set, e.g. $\overline{\mathbf{m}}^{(t_i)}$ refers to the ith-position not including τ. Equation 3.94 presents the exponential smoothing function $\mathbf{F}_{\text{smooth}}$: $\mathfrak{T}_{\text{mins}} \times \mathcal{P}_{\text{measure}*}(S_{\text{measure}*}^4) \times \mathbb{N}_0^7 \longrightarrow S_{\text{obsv}}^4$ which is equivalent to the one from [40, Sec. 6.2] and smoothens the measurements' traffic observables as part of the prognosis method. In addition to the time-based index t_i for the measurements of $\mathbf{M}_{\text{hist}}(\tau, \mathbf{d}_i^{(\tau)})$, the smoothing function relies on another index k to access the traffic observables of each measurement in $\mathbf{M}_{\text{hist}}(\tau, \mathbf{d}_i^{(\tau)})$ by $\mathbf{b}_k(t_i) := \overline{\mathbf{m}}^{(t_i)} \# b_k^{(t_i)}$. For the context of OLSIMv4, the exponential smoothing uses the constant parameter $\alpha_{\text{exp}} := 0.2$. The constantness of the sum is a special property of the exponential smoothing function. It holds for the special case of $\mathbf{F}_{\text{smooth}}(\tau, \mathbf{d}_i^{(\tau)}, 0) = 1$ for any given timestamp τ and $\mathbf{b}_k(t_i) = 1, \forall i$.

$$\mathbf{F}_{\text{smooth}}(\tau, \mathbf{d}_i^{(\tau)}, \mathbf{k}) := (1 - \alpha_{\text{exp}}) \sum_{i=1}^{N_{\text{hist}}} \alpha_{\text{exp}}^{i-1} \mathbf{b}_k(t_i) + \alpha_{\text{exp}}^{N_{\text{hist}}} \mathbf{b}_k(t_{N_{\text{hist}}}) . \qquad (3.94)$$

As has been the case for the average method from Equation 3.93, the resulting value $\mathbf{m}_{\text{smooth}}^{(\tau)} \in S_{\text{loop}}^4$ for the prognosis method consists then of the two traffic flow values $\bar{\jmath}_{\text{any}}^{(\tau)} := \mathbf{F}_{\text{smooth}}(\tau, \mathbf{d}_i^{(\tau)}, 1)$ and $\bar{\jmath}_{\text{truck}}^{(\tau)} := \mathbf{F}_{\text{smooth}}(\tau, \mathbf{d}_i^{(\tau)}, 3)$, of the two average velocity values that have the form $\bar{\jmath}_{\text{car}}^{(\tau)} :=$ $\mathbf{F}_{\text{smooth}}(\tau, \mathbf{d}_i^{(\tau)}, 5)$ and $\bar{\jmath}_{\text{truck}}^{(\tau)} := \mathbf{F}_{\text{smooth}}(\tau, \mathbf{d}_i^{(\tau)}, 6)$, and the average occupation time that has the form of $\bar{\mathbf{p}}_{\text{any}}^{(\tau)} := \mathbf{F}_{\text{smooth}}(\tau, \mathbf{d}_i^{(\tau)}, 7)$. The measurement value for the 60-minutes short-term prognosis and for the 7-days long-term prognosis results then from extending the measurement that consists of the previous 5 traffic observables $\mathbf{m}_{\text{smooth}*}^{(\tau)} := \mathbf{F}_{\text{ext-loop}}(\mathbf{m}_{\text{smooth}}^{(\tau)})$ with $\mathbf{m}_{\text{smooth}}^{(\tau)} := \mathbf{F}_{\text{smooth}}(\tau, \mathbf{d}_i^{(\tau)})$ and function $\mathbf{F}_{\text{smooth}} : \mathfrak{T}_{\text{mins}} \times S_{\text{det}}^4 \longrightarrow S_{\text{measure}}^4$.

$$\mathbf{F}_{\text{smooth}}(\tau, \mathbf{d}_i^{(\tau)}) := \langle \text{loop}, \tau, \bar{\jmath}_{\text{any}}^{(\tau)}, \bar{\jmath}_{\text{truck}}^{(\tau)}, \overline{\mathbf{v}}_{\text{car}}^{(\tau)}, \overline{\mathbf{v}}_{\text{truck}}^{(\tau)}, \overline{\mathbf{p}}_{\text{occ}}^{(\tau)} \rangle^{\text{measure}} \qquad (3.95)$$

The extension of the 5 prognosticated traffic observables instead of the direct calculation by means of function $\mathbf{F}_{\text{smooth}}$ ensures consistency across the measurements components. The presented prognosis method applies for all prognoses with a time horizon greater than $t_{\text{hmax}} := 37\,\text{min}$ [40, Sec. 7.3]. In contrast to those, the 30-minutes short-term prognosis of loop detector data for the time $\tau_{\text{p}} := \tau + 30\Delta\tau$ which is $30\,\text{min}$ in the future from τ mixes the current loop detector measurement $\mathbf{m}_{\text{curr}} := \mathbf{F}_{\text{input}}(\tau, \mathbf{d}_i^{(\tau)})$ with the prognosticated measurement for the current timestamp $\mathbf{m}_{\text{hist}} := \mathbf{F}_{\text{smooth}}(\mathbf{F}_{\text{input}}(\tau, \mathbf{d}_i^{(\tau)}))$ and for the timestamp $\mathbf{m}_{\text{prog}} := \mathbf{F}_{\text{smooth}}(\tau_{\text{p}}, \mathbf{d}_i^{(\tau_{\text{p}})})$ which is $30\,\text{min}$ in advance of τ. Function $\mathbf{F}_{\text{mix}} : S_{\text{loop}} \times S_{\text{loop}} \times S_{\text{loop}} \times \mathbb{N} \longrightarrow S_{\text{obsv}}^4$ calculates the mixed value for the three measurement values $\mathbf{m}_{\text{curr}}$, $\mathbf{m}_{\text{hist}}$, and $\mathbf{m}_{\text{prog}}$, and any given index k. Equation 3.96 presents the definition of Function $\mathbf{F}_{\text{mix}}$. It implements the calculation method from [40, Sec. 3.5, Eq. 3.95] and uses the same value $\eta = 0.57$ as in the publication. The time horizon for the $30\,\text{min}$ prognosis is defined as $\tau_{\text{h}} := \tau_{\text{p}} - \tau := 30\Delta\tau$. Analogously to above, the resulting measurement value becomes available by composing the traffic observables into a measurement value $\mathbf{m}_{\text{mixed}}(\tau) := \mathbf{F}_{\text{mix}}(\mathbf{m}_{\text{curr}}, \mathbf{m}_{\text{hist}}, \mathbf{m}_{\text{prog}})$. Table 3.5.5 summarises the relevant functions, structure definitions, variables, and constants for this section.

$$\mathbf{F}_{\text{mix}}(\mathbf{m}_{\text{curr}}, \mathbf{m}_{\text{hist}}, \mathbf{m}_{\text{prog}}, k) := \mathbf{m}_{\text{prog}} \# b_k^{(t)} + \left\lceil \eta \left[1 - \frac{t_{\text{h}}}{t_{\text{hmax}}}\right] \left(\mathbf{m}_{\text{curr}} \# b_k^{(t)} - \mathbf{m}_{\text{hist}} \# b_k^{(t)}\right) \right\rceil . \qquad (3.96)$$

3.5.4 Invariant detectors

Similar to the loop and prognosis detectors, invariant detectors present detector data to the simulation. Thus, their definition is similar to those of the loop detectors. However, theoretically, they are not limited to the loop detector definition. Instead, they can adapt to any detector type. As an example

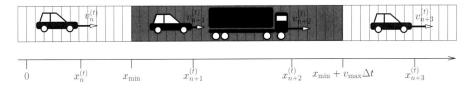

Figure 3.7: Local detector on a single lane with spatial extent $[x_{\min}; x_{\max} = x_{\min} + v_{\max}\Delta t]$.

for a possible definition, Equation 3.97 presents the one similar to the loop detectors. Equation 3.98 presents the definition for the invariant detector data.

$$S^4_{\text{invtype}*} := \langle \text{inv}*, \mathbf{j}_{\text{any}}, \mathbf{j}_{\text{car}}, \mathbf{j}_{\text{truck}}, \mathbf{v}_{\text{any}}, \mathbf{v}_{\text{car}}, \mathbf{v}_{\text{truck}}, \mathbf{p}_{\text{occ}} \rangle^{\text{dettype}} . \qquad (3.97)$$

$$S^4_{\text{inv}*} := \langle \text{inv}*, \tau, j^{(\tau)}_{\text{any}}, j^{(\tau)}_{\text{car}}, j^{(\tau)}_{\text{truck}}, v^{(\tau)}_{\text{any}}, v^{(\tau)}_{\text{car}}, v^{(\tau)}_{\text{truck}}, p^{(\tau)}_{\text{occ}} \rangle^{\text{measure}} . \qquad (3.98)$$

The simulation does not distinguish between stationary and spatio-temporal input traffic data. The behaviour of an invariant detector, and thus its data too, does not change over time. Periodic data is considered as invariant, too. Invariant detectors enable testing and verification of the simulation and its results. A typical scenario for the use of invariant detectors that has been published in [30]. It employs an invariant detector that presents a constant traffic flow and density target values to the simulation. The simulation with periodic boundary conditions tries to assimilate the values measured internally to those from to the invariant detectors. This way, the behaviour of the tuning elements can be studied. As constant detectors can be positioned at any place in the network, the constant detector approach is similar to the widely used origin/destination matrices. Functions that calculate measurement values are a generalisation of this approach. Equation 3.99 provides a test function for the effectiveness of tuning elements. It can be used in place of the aggregation operation and in combination with a constant measurement value $\mathbf{m}_{\text{const}} \in S^4_{\text{inv}*}$. It provides the measurement value from the aggregation operation periodically or zeros it for the interval of ω minutes. As a part of the test, the tuning elements have to reproduce the sharp slope somehow. More fine grained test functions could not zero out the measurement value but instead use several plateaus for the measurement value.

$$\mathbf{F}_{\text{aggregate}}(t, \mathbf{d}^{(t)}_j) := ((t \text{ div } 60\omega\Delta t) \text{ mod } 2) \cdot \mathbf{m}_{\text{const}} . \qquad (3.99)$$

3.5.5 Local detectors

Local detectors provide stationary traffic data as output from the simulation to the user and as feedback from the simulation back into the simulation again. In case of feeding data back in the simulation, the delay time is $60\,\text{s}$. The local detectors determine the traffic flow values based on events, i.e. as the number of vehicles that passed the detector during the last measurement cycle of one minute. This section presents two variants of the local detectors—one that calculates its measurements comparable to the loop detector data and another one comparable to the global detector data. Equation 3.100 presents the local detector type and Equation 3.101 the local detector data type. Except for the additional traffic observables the detector type name $\texttt{local}*$, the previous passing time $\mathbf{t}_{\text{pass}} := \langle \mathbf{s}^{\text{any}}_{\text{vehtype}}, 0.5\,\% \rangle^{\text{obsv}}$ in steps of $0.5\,\%$ of the measuring interval, the time headway $\mathbf{t}_{\text{hdw}} := \langle \mathbf{s}^{\text{any}}_{\text{vehtype}}, 100\,\text{ms} \rangle^{\text{obsv}}$, and the distance headway $\mathbf{x}_{\text{hdw}} := \langle \mathbf{s}^{\text{any}}_{\text{vehtype}}, 1\,\text{m} \rangle^{\text{obsv}}$, they are identical to the loop detectors. The traffic observable $\theta^{(t)}_{\text{pass}}$ stores the passing time of the vehicle that passed the detector as last in the previous time step.

$$S^4_{\text{lctype}*} := \langle \texttt{local}*, \mathbf{j}_{\text{any}}, \mathbf{j}_{\text{car}}, \mathbf{j}_{\text{truck}}, \mathbf{v}_{\text{any}}, \mathbf{v}_{\text{car}}, \mathbf{v}_{\text{truck}}, \mathbf{p}_{\text{occ}}, \mathbf{t}_{\text{pass}}, \mathbf{t}_{\text{hdw}}, \mathbf{x}_{\text{hdw}} \rangle^{\text{dettype}} . \qquad (3.100)$$

Table 3.5.6: Summary of relevant functions, structures, variables, and constants for Section 3.5.5.

definition	description	domain	range
$\mathfrak{T}_{\mathrm{milli}}$	milliseconds		$\{i\,\mathrm{ms}\mid i\in\mathbb{Z}\}$
$\mathbf{F}_{\mathrm{local}}$	vehicle area filter	$\mathfrak{T}_{\mathrm{secs}}\times S^4_{\mathrm{occ}}\times S^4_{\mathrm{det}}$	$\mathcal{P}^4_{\mathrm{veh}}(S^4_{\mathrm{veh}})$, see Equation 3.103
$\mathbf{F}_{\mathrm{passed}}$	triggering vehicles	$\mathfrak{T}_{\mathrm{secs}}\times S^4_{\mathrm{occ}}\times S^4_{\mathrm{det}}$	$\mathcal{P}^4_{\mathrm{veh}}(S^4_{\mathrm{veh}})$, see Equation 3.104
$\mathbf{F}_{\mathrm{vehtype}}$	vehicle type filter	$\mathcal{P}^4_{\mathrm{veh}}(S^4_{\mathrm{veh}})\times \mathfrak{L}^4_{\mathrm{vehtype}}$	$\mathcal{P}^4_{\mathrm{veh}}(S^4_{\mathrm{veh}})$, see Equation 3.105
$\mathbf{F}_{\mathrm{ext-local}}$	measure method	S^4_{local}	$S^4_{\mathrm{local*}}$, see Equation 3.110
$\langle\tau,a,b,c,\dots\rangle^{\mathrm{local*}}$	local* detector data type	$\mathfrak{T}_{\mathrm{mins}}\times(S^4_{\mathrm{obsv}})^{10}$	$S^4_{\mathrm{local*}}$, see Equation 3.101
$\mathbf{d}_i^{(t)}\#\theta_{\mathrm{pass}}^{(t)}$	previous passing time	$\mathbf{d}_i^{(t)}\in S^4_{\mathrm{local*}}$	S^4_{obsv}
$\mathbf{d}_i^{(t)}\#\theta_{\mathrm{hdw}}^{(t)}$	time headway	$\mathbf{d}_i^{(t)}\in S^4_{\mathrm{local*}}$	S^4_{obsv}
$\mathbf{d}_i^{(t)}\#x_{\mathrm{hdx}}^{(t)}$	distance headway	$\mathbf{d}_i^{(t)}\in S^4_{\mathrm{local*}}$	S^4_{obsv}

$$S^4_{\mathrm{local*}} := \langle\mathtt{local*},t,j_{\mathrm{any}}^{(t)},j_{\mathrm{car}}^{(t)},j_{\mathrm{truck}}^{(t)},v_{\mathrm{any}}^{(t)},v_{\mathrm{car}}^{(t)},v_{\mathrm{truck}}^{(t)},p_{\mathrm{occ}}^{(t)},\theta_{\mathrm{pass}}^{(t)},\theta_{\mathrm{hdw}}^{(t)},x_{\mathrm{hdw}}^{(t)}\rangle^{\mathrm{measure}} . \qquad (3.101)$$

The loop detectors in reality have zero spatial extent. However, the local detectors require a spatial extent due to the space and time discrete of the simulation. Thus, OLSIMv4 configures local detectors with a default spatial extent which is proportional to the maximum velocity of the $\mathtt{any}$ vehicle type. Figure 3.7 portraits the spatial extent of a local detector. Equation 3.102 specifies the default spatial extent $\mathbf{r}_{\mathrm{area}}$ of a local detector $\mathbf{d}_j^{(t)}\in S^4_{\mathrm{det}}$ on lane $y_{\min}$ and position $x_{\min}$.

$$\mathbf{r}_{\mathrm{area}} := \langle y_{\min},y_{\min},x_{\min},1+x_{\min}+\mathbf{s}_{\mathrm{vehtype}}\#V_{\max}\cdot\Delta t\rangle^{\mathrm{area}} . \qquad (3.102)$$

This section describes at first the local detectors' measuring method and thereafter the aggregate method. The measuring method $\mathbf{F}_{\mathrm{measure}}$ relies on a Function $\mathbf{F}_{\mathrm{local}}:\mathfrak{T}_{\mathrm{secs}}\times S^4_{\mathrm{occ}}\times S^4_{\mathrm{det}}\longrightarrow\mathcal{P}^4_{\mathrm{veh}}(S^4_{\mathrm{veh}})$ that in each time step filters all vehicles of an occupation table for those that are located on the spatial extent of the local detector. Equation 3.103 defines the filter function $\mathbf{F}_{\mathrm{local}}$. It uses the filter function $\mathbf{F}_{\mathrm{area}}^{<,>}$ from the context of Equation 3.30 in Section 3.3.5.

$$\mathbf{F}_{\mathrm{local}}(t,\mathbf{o}_r^{(t)},\mathbf{d}_j^{(t)}) := \{\mathbf{k}_i^{(t)}\in\mathbf{F}_{\mathrm{area}}^{<,>}(\mathbf{o}_r^{(t)}\#\mathbf{k}^{(t)},\mathbf{d}_j^{(t)}\#\mathbf{r}_{\mathrm{area}})\} . \qquad (3.103)$$

The vehicles that passed a local detector in the current timestamp trigger the event based measuring method. The measuring method performs all of the steps described in the following for each vehicle $\mathbf{k}_n^{(t)}$ that passed the detector $\mathbf{d}_j^{(t)}$ in the current time step t. Thus, the vehicles inside the spatial extent of a local detector require further filtering. Only those should remain that entered the spatial extent in the last time step and that did not enter the lane by a lane change in the last time step, i.e. for a vehicle the lane change velocity $\Delta y_n^{(t)} := 1/\Delta t\cdot((\mathbf{k}_n^{(t)}\#\mathbf{q}^{(t)})\#y-(\mathbf{k}_n^{(t)}\#\mathbf{q}^{(t-\Delta t)})\#y)$ has to equal o lanes/s in the current time step. The lane change velocity has already been discussed in the context of Equation 3.5.

Equation 3.104 defines Function $\mathbf{F}_{\mathrm{passed}}:\mathfrak{T}_{\mathrm{secs}}\times S^4_{\mathrm{occ}}\times S^4_{\mathrm{det}}\longrightarrow\mathcal{P}^4_{\mathrm{veh}}(S^4_{\mathrm{veh}})$ that filters the vehicles of the occupation table first by function $\mathbf{F}_{\mathrm{local}}$ and thereafter examines the remaining vehicles for the described no lane change criteria. The usual abbreviations such as $x_n^{(t)} := (\mathbf{k}_n^{(t)}\#\mathbf{q}^{(t)})\#x$ and $L_{\mathrm{n}} := \mathbf{k}_n^{(t)}\#L$ apply as far as they are unique.

$$\mathbf{F}_{\mathrm{passed}}(t,\mathbf{o}_r^{(t)},\mathbf{d}_j^{(t)}) := \{\mathbf{k}_n^{(t)}\in\mathbf{F}_{\mathrm{local}}(t,\mathbf{o}_r^{(t)},\mathbf{d}_j^{(t)})\mid \Delta t\Delta y_n^{(t)}=0 \wedge x_n^{(t)}-L_{\mathrm{n}}-v_n^{(t)}\cdot\Delta t<x_{\min}\} . \quad (3.104)$$

Additionally, the measuring method relies on a Function $\mathbf{F}_{\mathrm{vehtype}}:\mathcal{P}^4_{\mathrm{veh}}(S^4_{\mathrm{veh}})\times\mathfrak{L}^4_{\mathrm{vehtype}}\longrightarrow\mathcal{P}^4_{\mathrm{veh}}(S^4_{\mathrm{veh}})$ that filters the vehicles $\mathbf{k}^{(t)}\in\mathcal{P}^4_{\mathrm{veh}}(S^4_{\mathrm{veh}})$ for a specific vehicle type name $\mathfrak{t}_{\mathrm{name}}^{\mathrm{vehtype}}\in\mathfrak{L}^4_{\mathrm{vehtype}}$. Equation 3.105 presents the definition of that function.

$$\mathbf{F}_{\mathrm{vehtype}}(\mathbf{k}^{(t)},\mathfrak{t}_{\mathrm{name}}^{\mathrm{vehtype}}) := \{\mathbf{k}_n^{(t)}\in\mathbf{k}^{(t)}\mid(\mathbf{k}_n^{(t)}\#\mathbf{s}_{\mathrm{vehtype}})\#\mathfrak{t}_{\mathrm{name}}^{\mathrm{vehtype}}=\mathfrak{t}_{\mathrm{name}}^{\mathrm{vehtype}}\} . \qquad (3.105)$$

Table 3.5.7: Level-of-service values.

description	colour	LOS
congested traffic	red	LOS 6
viscous traffic	yellow	LOS 5
dense traffic	dark green	LOS 4
vital free flow traffic	green	LOS 3
free flow traffic	green	LOS 2
no data available	grey	LOS 1

When having all the vehicles that passed the detector in the current time step, the measure operation determines the vehicle type $t_{\text{name}}^{\text{vehtype,n}} := (\mathbf{k}_n^{(t)} \# \mathbf{s}_{\text{vehtype}}) \# t_{\text{name}}^{\text{vehtype}}$, the length L_n, and the velocity $v_n^{(t)}$ for each of the vehicles $\mathbf{k}_n^{(t)}$. It then calculates the time headway $t_{\text{headway},n} \in \mathfrak{T}_{\text{milli}} := \{i\,\text{ms} \mid i \in \mathbb{Z}\}$ with the help of the passing time $t_{\text{passed},n}$ that depends on the current time $t \in \mathfrak{T}_{\text{secs}}$, the detector $\mathbf{d}_j^{(t)}$, and the vehicle $\mathbf{k}_n^{(t)}$ that passed the detector. Equation 3.106 specifies how to calculate the passing time according to [164, Chap. 2].

$$t_{\text{passed},n} := \frac{x_{\min} - x_n^{(t)}}{v_n^{(t)}} + t \, . \tag{3.106}$$

The criteria that the calculation considers only those vehicles that passed the detector during the current time step asserts that $v_n^{(t)} > 0\,\text{cells/s}$ holds for the vehicle velocity. As the fraction term in the formula for the calculation of the passing time divides by the current velocity $v_n^{(t)}$, this assertion is important. The fraction term in Equation 3.106 is always negative, because the vehicle position $x_n^{(t)}$ that passed the detector is behind the detector's starting position $x_{\min}$ and the velocity is always positive. The resulting passing time $t_{\text{passed},n}$ is thus always a fraction smaller than the current time t. The local detectors store a variant of the passing time in the traffic observable $\theta_{\text{pass}}^{(t)}$. As the domain of this variable is in $0.5\,\%$, the measure method assigns the value $\theta_{\text{pass}}^{(t)} := \lceil 200 \cdot 1/\Delta t | t_{\text{passed},n} - t | \rceil$ each time a vehicle passes the local detector. Thus, the traffic observable $\theta_{\text{pass}}^{(t)}$ contains for each measurement $\mathbf{m}_t \in S_{\text{local}*}^4$ always the passing time of the vehicle that passed the local detector as last in the previous time step. However, the measuring method assigns the value only after calculating the time headway. For each fresh instance of the measurement value, the measuring method assigns the passing time value of the previous measurement as the default value, i.e. $\theta_{\text{pass}}^{(t)} := \theta_{\text{pass}}^{(t-\Delta t)}$.

With the passing time function the time headway $t_{\text{headway},n}$ between two subsequent vehicles $\mathbf{k}_n^{(t)}$, $\mathbf{k}_{n+1}^{(t)}$ equals the difference of their passing times. The local detectors operate based on vehicle events. However, the calculation of the time headway depends on a sequence, i.e. a pair, of vehicles. Thus, the local detectors have to maintain a state for the calculation of the headway. The traffic observable $\theta_{\text{pass}}^{(t)}$ has been chosen for that purpose. Equation 3.107 specifies the time headway between a pair of subsequent vehicles according to [164, Chap. 2].

$$t_{\text{headway},n} := \begin{cases} 1/\Delta t(t_{\text{passed},n} - (t - 1/200\Delta t \cdot \theta_{\text{pass}}^{(t)})) & \text{, if } \theta_{\text{pass}}^{(t)} \leqslant 200 \, , \\ 1/\Delta t(t - t_{\text{passed},n}) & \text{, otherwise.} \end{cases} \tag{3.107}$$

After the calculation of the time headway, the measuring method assigns first the passing time of the current vehicle to the passing time traffic observable $\theta_{\text{pass}}^{(t)}$. Thereafter, it stores the time headways of the vehicles of the current time step for the purpose of averaging them later. The distance headways $x_{\text{headway},n}$ follow from the time headways by applying Equation 3.108 to a vehicle $\mathbf{k}_n^{(t)}$.

$$x_{\text{headway},n} := v_n^{(t)} \cdot t_{\text{headway},n} - L_n \, . \tag{3.108}$$

The local detectors determine in each time step the occupation time $p_{\text{occ},n}$ from the vehicles lengths and velocities with the formula in Equation 3.109. In the first step, the detectors sum the occupation times, i.e. the duration within the detector was covered by a vehicle for each vehicle that passed the detector in the current time step. In the second step, they convert the sum into percent of the measuring interval.

$$p_{\text{occ},n} := \frac{L_{\text{n}}}{v_n^{(t)}} \, . \tag{3.109}$$

The value for the traffic flow is the number of vehicles that passed the detector in the measuring interval. Each vehicle that passed the detector contains already the length and the velocity for the traffic observable value. The velocity value $(\mathbf{k}_n^{(t)} \# \mathbf{q}_n^{(t)}) \# v$ of the vehicle, however, represents the vehicles velocity in the time step when it passes the detector. When the passing of the detector takes several time steps, this velocity might be inappropriate. Choosing a more appropriate velocity value, however, results in a different detector behaviour which is more similar to the global detectors. It is discussed below in more detail.

With all these values, the traffic observables for a time step are complete. In each time step, the measuring method computes the sums per vehicle type for the traffic flows $j_{\text{any}}^{(\tau)}$, $j_{\text{car}}^{(\tau)}$, and $j_{\text{truck}}^{(\tau)}$ and the average values for the velocities $\overline{v}_{\text{any}}^{(\tau)}$, $\overline{v}_{\text{car}}^{(\tau)}$, and $\overline{v}_{\text{truck}}^{(\tau)}$, the occupation times $\overline{p}_{\text{occ}}^{(\tau)}$, and the time headways $\overline{\theta}_{\text{hdw}}^{(t)}$, and the distance headways $\overline{x}_{\text{hdw}}^{(t)}$. The traffic observables result then in a measurement value per time step in the form of Equation 3.110.

$$\mathbf{F}_{\text{ext}-\text{local}}(\mathbf{m}^{(\tau)}) := \langle \texttt{local}*, \tau, j_{\text{any}}^{(\tau)}, j_{\text{car}}^{(\tau)}, j_{\text{truck}}^{(\tau)}, \overline{v}_{\text{any}}^{(\tau)}, \overline{v}_{\text{car}}^{(\tau)}, \overline{v}_{\text{truck}}^{(\tau)}, \overline{p}_{\text{occ}}^{(\tau)}, \theta_{\text{pass}}^{(t)}, \overline{\theta}_{\text{hdw}}^{(t)}, \overline{x}_{\text{hdw}}^{(t)} \rangle^{\text{measure}} \, . \tag{3.110}$$

The aggregate operation $\mathbf{F}_{\text{aggregate}*} : \mathfrak{T}_{\text{mins}} \times S_{\text{det}}^4 \longrightarrow S_{\text{measure}*}^4$, the detector sums and averages all measurements as it has been demonstrated using the average function $\mathbf{F}_{\text{avg}}$ in the context of Equation 3.92 in the prognosis detector section. The aggregate operation results then also in a value in the form of Equation 3.110.

Until now, all presented operations apply only for the virtual loop detectors. The reason is that the velocity recorded by the measure operation uses the velocity of the last time step in which the vehicle passed the detector. For velocities with $v_n^{(t)} \Delta t \leqslant L_{\text{n}}$, the time for a vehicle to pass the detector involves more than one time step. This approximisation of the velocity is not very accurate for low velocities. Thus, the fundamental diagrams "tend to underestimate the density in the congested regime" [183, Sec. 6.2] as discussed in Section 2.2.2.

However, it is possible to fix the "underestimating behaviour" by adapting the calculation of the velocity and thereafter all other traffic variables that depend on the modified value of the average velocity. Equation 3.111 defines how to calculate the velocity. The formula averages the velocity over all time steps and lists the conditions that each vehicle that covers the detector has to meet. Finer and even more exact approaches take the detector passing time for the first and the last time step into account.

$$\overline{v}_n^{(\tau)} := \frac{1}{N} \sum_{t=T_0+\Delta t}^{\tau} (\mathbf{k}_n^{(t)} \# \mathbf{q}_n^{(t)}) \# v \, , \text{ with } \tau := T_0 + N\Delta t \, ,$$

$$\Delta y_n^{(t)} := \text{o lanes/s} \, , \forall t \in [T_0 + \Delta t; T_0 + N\Delta t] \, , \tag{3.111}$$

$$x_n^{(t)} \geqslant x_{\text{min}} > x_n^{(t)} - v_n^{(t)} \Delta t \, , \text{ for } t = T_0 + \Delta t \, , \text{ and}$$

$$x_n^{(t)} - L_{\text{n}} \geqslant x_{\text{min}} > x_n^{(t)} - L_{\text{n}} - v_n^{(t)} \Delta t \, , \text{ for } t = T_0 + N\Delta t \, .$$

Calculating the velocity in this way requires to adapt the vehicle filtering by $\mathbf{F}_{\text{local}}$ from Equation 3.103. Vehicles that cover the detector have a position greater than or equal to the detector position. The simplest way to filter also for these kind of vehicles is to temporary manipulate the detector's area structure element $\mathbf{k}_n^{(t)} \# \mathbf{r}_{\text{area}}$ with the spatial extent according to Equation 3.112 and reuse the filtering function $\mathbf{F}_{\text{local}}$ from Equation 3.103 with the manipulated area structure element. The filtering in this

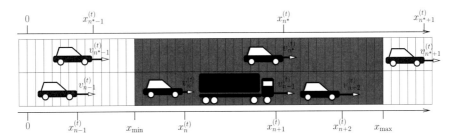

Figure 3.8: Spatial extent $[x_{\mathrm{min}}; x_{\mathrm{max}}]$ over two lanes of a global detector.

way results in a set that also contains the vehicles that still cover the detector. The first vehicle of the–
by position–ordered set of vehicles then contains eventually a vehicle that covers the detector. Such a
vehicle can then be fed into Equation 3.111.

$$\mathbf{d}_j^{(t)} := \mathbf{F}_{\mathrm{clone}}(\mathbf{d}_j^{(t)}, \mathbf{r}_{\mathrm{area}}, \langle y_{\mathrm{min}}, y_{\mathrm{max}}, x_{\mathrm{min}} - 1 \text{ cells}, x_{\mathrm{max}} \rangle^{\mathrm{area}}) \,. \tag{3.112}$$

The calculation of the passing time uses the updated average velocity directly instead of the velocity of
the last time step that was used previously when passing the detector. The adapted passing time results
finally in a more accurate time headway. The different behaviour of the local detector compared to the
virtual loop detector results then in comparable measurements to those of a global detector. Table 3.5.6
summarises the relevant functions, structure definitions, variables, and constants for this section.

3.5.6 Global detectors

Global detectors provide spatio-temporal traffic information as output data from the simulation. The
data refers to tracks, roadway segments, or complete roadways. Similar to the local detectors, global
detectors examine the vehicles on their spatial extent. Figure 3.8 illustrates by the colored region the
spatial extent of a global detector with some vehicles in it. Maybe the greatest advantage of the global
detectors can be seen from the figure. The density of the global detector matches the number of cells
covered by a vehicle divided by the total number of cells of the global detector's spatial extent. As the
roadways in OLSIMv4 have maximum extent, overlapping detectors for varying digital maps provide
an exact way to distribute the traffic information over several digital maps that vary in the sectioning
of the roadway sections. The straight forward calculation of the spatio-temporal traffic information
density, travel-time, and level-of-service are the main advantages of global detectors [164, Sec. 2.1].
The concept of the global detector and the calculation methods have been taken from [164, Sec. 2.1],
too. However, Pottmeier's publication [164, Sec. 2.1] does not clearly distinguish between the measure
and the aggregation operation. Additionally, it does not take closed lanes and time dependent network
structure into account. In case of feeding data back into the simulation the delay time is 60 s in the
context of OLSIMv4.

A global detector provides the same traffic observables as the local detectors except for the passing
times and the time and distance headways. Instead of the latter, they provide the three spatio-temporal
traffic observables density $\rho_{\mathrm{any}} := \langle \mathbf{s}_{\mathrm{vehtype}}^{\mathrm{any}}, 1 \text{ vehs/km} \rangle^{\mathrm{obsv}}$, the travel-time $\tau_{\mathrm{any}} := \langle \mathbf{s}_{\mathrm{vehtype}}^{\mathrm{any}}, 1 \text{ min} \rangle^{\mathrm{obsv}}$,
and the level-of-service $\sigma_{\mathrm{any}} := \langle \mathbf{s}_{\mathrm{vehtype}}^{\mathrm{any}}, 1 \text{ SL} \rangle^{\mathrm{obsv}}$. Table 3.5.7 lists the possible values for the level-of-
service data. Each of the observables refers to the `any` vehicle type and to its associated road segment.
Equation 3.100 presents the global detector type while Equation 3.101 presents the global detector
data.

$$S_{\mathrm{gltype}*}^4 := \langle \texttt{global}*, \mathbf{j}_{\mathrm{any}}, \mathbf{j}_{\mathrm{car}}, \mathbf{j}_{\mathrm{truck}}, \mathbf{v}_{\mathrm{any}}, \mathbf{v}_{\mathrm{car}}, \mathbf{v}_{\mathrm{truck}}, \mathbf{p}_{\mathrm{occ}}, \rho_{\mathrm{any}}, \tau_{\mathrm{any}}, \sigma_{\mathrm{any}} \rangle^{\mathrm{dettype}} \,. \tag{3.113}$$

Table 3.5.8: Parameters for the classification of the level-of-service with the scheme of Figure 3.9(a).

maximum velocity	domain	name	lanes	value	lanes	value
> 100 km/h	density	d6	2	> 60 vehs/km	3	> 75 vehs/km
> 100 km/h	density	d5	2	> 55 vehs/km	3	> 70 vehs/km
> 100 km/h	density	d4	2	> 30 vehs/km	3	> 45 vehs/km
> 100 km/h	density	d3	2	> 15 vehs/km	3	> 20 vehs/km
> 100 km/h	velocity	v6	⩾2	30 km/h		
> 100 km/h	velocity	v5	⩾2	80 km/h		

$$S^4_{\text{global}*} := \langle \texttt{global}*, t, j^{(t)}_{\text{any}}, j^{(t)}_{\text{car}}, j^{(t)}_{\text{truck}}, v^{(t)}_{\text{any}}, v^{(t)}_{\text{car}}, v^{(t)}_{\text{truck}}, p^{(t)}_{\text{occ}}, \rho^{(t)}_{\text{any}}, \tau^{(t)}_{\text{any}}, \sigma^{(t)}_{\text{any}} \rangle^{\text{measure}} . \tag{3.114}$$

Except for the name of the filtering function, a global detector filters for the vehicles in its spatial extent in the same way as the local detectors do by the use of Equation 3.103. To distinguish the filtering function from the one for the local detectors, the name of the filter function for the global detector is $\mathbf{F}_{\text{global}}$ in the following. Further filtering per vehicle type may also take place in the same way as described by Equation 3.104. However, for the global detector, the "measured" number $N^{(t)}_j := |\mathbf{k}^{(t)}|$ of vehicles $\mathbf{k}^{(t)} := \mathbf{F}_{\text{global}}(t, \mathbf{o}^{(t)}_r, \mathbf{d}^{(t)}_j)$ in each time step $t \in \mathfrak{T}_{\text{secs}}$ that are located in the occupation table $\mathbf{o}^{(t)}_r \in S^4_{\text{occ}}$ on the spatial extent of the global detector $\mathbf{d}^{(t)}_j$ does not match the traffic flow. The computation of the traffic flow observable requires instead the value for the total number of non-closed cells $L^{(t)}_j$, the average length $\bar{\mathrm{l}}^{(t)}_j$, and the average velocity $\bar{\mathrm{v}}^{(t)}_j$. In the following, the computation of these traffic observables is described.

With the vehicles on the detector's spatial extent, the measure operation can determine in each time step the average velocity $\bar{\mathrm{v}}^{(t)}_j$. In the case of the local detector, the calculation of the traffic observables were event-driven because only a few vehicles per time step pass the local detector. The spatial extent of a global detector and consequently the number of vehicles on it per time step, however, is likely to be larger. As the measure operation in each time step of the traffic observables traffic flow, level-of-service, and travel time depends on the average velocity of the vehicles, the vehicle velocities require averaging in each time step, too. Equation 3.115 defines the averaging of the vehicle velocities for a time step t and a global detector $\mathbf{d}^{(t)}_j$ that measures on occupation table $\mathbf{o}^{(t)}_r$. The usual abbreviations such as $x^{(t)}_n := (\mathbf{k}^{(t)}_n \# \mathbf{q}^{(t)}) \# x$ and $L_{\mathrm{n}} := \mathbf{k}^{(t)}_{\mathrm{n}} \# L$ apply as far as they are unique.

$$\bar{\mathrm{v}}^{(t)}_j := \begin{cases} 1/N^{(t)}_j \sum_{\mathbf{k}^{(t)}_i \in \mathbf{k}^{(t)}} (\mathbf{k}^{(t)}_i \# \mathbf{q}^{(t)}_i) \# v \text{ , where } \mathbf{k}^{(t)} := \mathbf{F}_{\text{global}}(t, \mathbf{o}^{(t)}_r, \mathbf{d}^{(t)}_j) & \text{, if } N^{(t)}_j > 0 \text{ ,} \\ 0 \text{ cells/s} & \text{, otherwise.} \end{cases} \tag{3.115}$$

Equation 3.116 defines the average length $\bar{\mathrm{l}}^{(t)}_j$ for all the vehicles on the spatial extent of the detector. The calculation proceeds analogous to the calculation of the average velocity. The value for the average traffic flow and the average occupation time which are introduced in the following. They depend on the average length and the average velocity, respectively.

$$\bar{\mathrm{l}}^{(t)}_j := \begin{cases} 1/N^{(t)}_j \sum_{\mathbf{k}^{(t)}_i \in \mathbf{k}^{(t)}} \mathbf{k}^{(t)}_i \# L \text{ , where } \mathbf{k}^{(t)} := \mathbf{F}_{\text{global}}(t, \mathbf{o}^{(t)}_r, \mathbf{d}^{(t)}_j) & \text{, if } N^{(t)}_j > 0 \text{ ,} \\ 0 \text{ cells} & \text{, otherwise.} \end{cases} \tag{3.116}$$

In addition to the number of vehicles, the computation of the traffic observables requires the number of non-closed cells $L^{(t)}_j$ in the spatial extent of the global detector $\mathbf{d}^{(t)}_j$, too. With the help of Function $\mathbf{F}^{\text{cell}}_{\text{Y} \times \text{X}}$ from Equation 3.22, counting the total number of non-closed cells is possible. Some cells may be closed in the roadway and thus in the rectangular area of the global detector

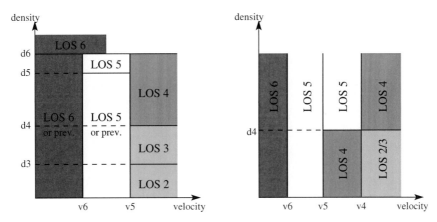

(a) Mappings for maximum velocity above 100 km/h . (b) Mappings for 60 km/h , 80 km/h , or 100 km/h .
Mappings from [108] and published with permission.

Figure 3.9: LOS classification schemes with (3.9(b)) and without (3.9(a)) velocity restrictions.

due to temporary network changes such as road works, deviations, and lane closings because of po-
lice traffic regulations. The roadway element $\mathbf{o}_r^{(t)}\#\mathbf{w}^i$ of the occupation table $\mathbf{o}_r^{(t)}$ depends on the
time. It contains the kind of information whether a cell is closed or not. The Function $\mathbf{F}_{\text{non-closed}}$:
$S_{\text{occ}}^4 \times S_{\text{area}}^4 \times S_{\text{vehtype}}^4 \longrightarrow \mathbb{N}_0$ counts the number of non-closed cells L_j per vehicle type by evaluating
$L_j^{(t)} = \mathbf{F}_{\text{non-closed}}(\mathbf{o}_r^{(t)}, \mathbf{d}_j^{(t)}\#\mathbf{r}_{\text{area}}, t_{\text{name}}^{\text{vehtype}})$. With the total number of non-closed cells $L_j^{(t)}$, the average
velocity $\overline{\mathbf{v}}_j^{(t)}$, and the total number of vehicles $N_j^{(t)}$ the prerequisites to compute the total traffic flow $\overline{\mathbf{j}}_j^{(t)}$
for time step t are complete. Equation 3.117 specifies the relation between the three traffic observables
and the traffic flow.

$$\overline{\mathbf{j}}_j^{(t)} := \frac{N_j^{(t)}\overline{\mathbf{v}}_j^{(t)}}{L_j^{(t)}} \ . \tag{3.117}$$

The average occupation $p_{\text{occ},j}^{(t)}$ in percent for each time step t becomes available with the total number
of non-closed cells $L_j^{(t)}$ similar to Equation 3.117. The fraction of the total length that is covered by
vehicles and the total length of the area covered by the detector with respect to the vehicle type form
the average occupation in percent. Equation 3.118 presents the fraction.

$$p_{\text{occ},j}^{(t)} = 100\frac{N_j^{(t)}\overline{\mathbf{l}}_j^{(t)}}{L_j^{(t)}} \ . \tag{3.118}$$

The computation of the average density contrasts the one of the occupation in percent by simply
omitting the average vehicle length in the denominator of the fraction. The density $\rho_j^{(t)}$ for the time
step t matches the fraction of the number of vehicles to the number of non-closed cells. Equation 3.119
specifies the calculation of the density value.

$$\rho_j^{(t)} := \frac{N_j^{(t)}}{L_j^{(t)}} \ . \tag{3.119}$$

A first estimation for the travel time $\tau_j^{(t)}$ per vehicle type $t_{\text{name}}^{\text{vehtype}}$ in each time step t is then given by
Equation 3.120. The calculation determines the maximum number of non-closed cells in a single-lane

Table 3.5.9: Summary of relevant functions, structures, variables, and constants for Section 3.5.6.

definition	description	domain	range
$\mathbf{F}_{\text{global}}$	vehicle area filter	$\mathfrak{T}_{\text{secs}} \times S_{\text{occ}}^4 \times S_{\text{det}}^4$	$\mathcal{P}_{\text{veh}}^4(S_{\text{veh}}^4)$, see Equation 3.103
$\mathbf{F}_{\text{ext−global}}$	measure method	S_{global}^4	$S_{\text{global}*}^4$
$\mathbf{F}_{\text{aggregate}*}$	aggregate operation	$\mathfrak{T}_{\text{mins}} \times S_{\text{det}}^4$	$S_{\text{glocal}*}^4$
$\langle \tau, a, b, c, \ldots \rangle^{\text{global}*}$	global* detector data type	$\mathfrak{T}_{\text{mins}} \times (S_{\text{obsv}}^4)^9$	$S_{\text{global}*}^4$, see Equation 3.114
$\mathbf{d}_i^{(t)} \# \rho_{\text{any}}^{(t)}$	density	$\mathbf{d}_i^{(t)} \in S_{\text{global}*}^4$	S_{obsv}^4
$\mathbf{d}_i^{(t)} \# \tau_{\text{any}}^{(t)}$	travel time	$\mathbf{d}_i^{(t)} \in S_{\text{global}*}^4$	S_{obsv}^4
$\mathbf{d}_i^{(t)} \# \sigma_{\text{any}}^{(t)}$	level-of-service	$\mathbf{d}_i^{(t)} \in S_{\text{global}*}^4$	S_{obsv}^4

by applying $\mathbf{F}_{\text{non−closed}}(\mathbf{o}_r^{(t)}, \mathbf{r}_{\text{area},m}, \mathbf{t}_{\text{name}}^{\text{vehtype}})$ over all lanes. It divides them by the average velocity $\bar{\mathbf{v}}_j^{(t)}$.

$$\tau_j^{(t)} := \frac{\max \left\{ \mathbf{F}_{\text{non−closed}}(\mathbf{o}_r^{(t)}, \mathbf{r}_{\text{area},m}, \mathbf{t}_{\text{name}}^{\text{vehtype}}) \right\}}{\bar{\mathbf{v}}_j^{(t)}} \text{ , where}$$

$$\mathbf{r}_{\text{area},m} = \langle y_i, y_i, x_{\min}, x_{\max} \rangle^{\text{area}} \text{ and } y_i \in \{y_{\min}, \ldots, y_{\max}\} \text{ .}$$

(3.120)

With all these values, the traffic observables for a time step τ are complete. In each time step, the measuring method computes the values for the traffic flows $j_{\text{any}}^{(\tau)}$, $j_{\text{car}}^{(\tau)}$, and $j_{\text{truck}}^{(\tau)}$ and the average values for the velocities $\bar{\mathbf{v}}_{\text{any}}^{(\tau)}$, $\bar{\mathbf{v}}_{\text{car}}^{(\tau)}$, and $\bar{\mathbf{v}}_{\text{truck}}^{(\tau)}$, the travel times $tau_j^{(\tau)}$, and the density $\rho_j^{(t)}$. The traffic observables result then in a measurement value per time step in the form of Equation 3.121.

$$\mathbf{F}_{\text{ext−global}}(\mathbf{m}^{(\tau)}) := \langle \text{global}*, \tau, j_{\text{j,any}}^{(\tau)}, j_{\text{j,car}}^{(\tau)}, j_{\text{j,truck}}^{(\tau)}, \bar{\mathbf{v}}_{\text{j,any}}^{(\tau)}, \bar{\mathbf{v}}_{\text{j,car}}^{(\tau)}, \bar{\mathbf{v}}_{\text{j,truck}}^{(\tau)}, \rho_j^{(t)}, \tau_j^{(t)}, \sigma_j^{(t)} \rangle^{\text{measure}} .$$

(3.121)

The aggregate operation $\mathbf{F}_{\text{aggregate}*} : \mathfrak{T}_{\text{mins}} \times S_{\text{det}}^4 \longrightarrow S_{\text{measure}*}^4$, the detector sums and averages all measurements as it has been demonstrated using the average function $\mathbf{F}_{\text{avg}}$ in the context of Equation 3.92 in the prognosis detector section. However, there is one exception to this. The global detectors determine the level-of-service value during the aggregate operation. With the level-of-service value, the aggregate operation results then also in a value in the form of Equation 3.121.

The aggregation operation calculates the level-of-service $\sigma_j^{(t)}$ once in a minute and using the values for the average velocity and the average density. Table 3.5.7 lists the six level-of-service (LOS) mappings from discrete values to colours as well as their interpretation as shown by Figure 2.1. OLSIMv3 and OLSIMv4 used the LOS mappings of [108, Sec. 5.4.1.5] that matches Figure 3.9(a). As a disadvantage of this scheme, the classification did not take the maximum velocity restrictions into account. The many road works, variable traffic signs, and static speed limits cause a lot of maximum velocity restrictions. Thus, Mazur, Zaksek, and the author of this thesis introduced a classification scheme for OLSIMv4 that takes the maximum velocity restrictions into account [52]. It will be briefly described in the following.

The diagrams in Figure 3.9 show the relation between velocity, density, and the level-of-service value. Without any maximum velocity restriction, the computation of the level-of-service value uses the scheme that the diagram in Figure 3.9(a) describes. Whenever a velocity restriction of 100 km/h , 80 km/h , or 60 km/h is in effect, the scheme described by Figure 3.9(b) is used. In the case where the classification leads to two level-of-service values, the higher value is chosen.

The classification schemes in Figure 3.9 rely on the density parameters d6, d5, d4, d3 and the velocity parameters v6, v5, and v4. Table 3.5.8 presents values for the parameters for the determination of the LOS value. The choice of the right parameter depends on the number of lanes. Table 3.5.9 summarises the relevant functions, structure definitions, variables, and constants for this section.

Table 3.5.10: Parameters for the classification of the level-of-service with the scheme of Figure 3.9(b).

maximum velocity	domain	name	lanes	value	lanes	value
100 km/h	density	d4	2	> 50 vehs/km	3	> 75 vehs/km
100 km/h	velocity	v6	≥2	< 30 km/h		
100 km/h	velocity	v5	≥2	< 65 km/h		
100 km/h	velocity	v4	≥2	< 90 km/h		
80 km/h	density	d4	2	> 25 vehs/km	3	> 40 vehs/km
80 km/h	velocity	v6	≥2	< 30 km/h		
80 km/h	velocity	v5	≥2	< 60 km/h		
80 km/h	velocity	v4	≥2	< 75 km/h		
60 km/h	density	d4	2	> 25 vehs/km	3	> 35 vehs/km
60 km/h	velocity	v6	≥2	< 30 km/h		
60 km/h	velocity	v5	≥2	< 45 km/h		
60 km/h	velocity	v4	≥2	< 55 km/h		

3.6 Tuning model

OLSIMv3 used lane-wise checkpoints that combine loop detectors with a tuning method. The tuning method implemented a tuning strategy that depends on the comparison per vehicle type of the number of vehicles measured by the loop detectors and counted at the checkpoints during the interval of one minute. The checkpoint compared the number of vehicles from the loop detector to the number of vehicles counted in the simulation at the checkpoint's position. The checkpoints inserted vehicles whenever the number of vehicle balance was positive. In case of a negative balance they removed vehicles. The insertion strategy involved searching for a gap that is sufficiently large for an additional vehicle. The removal strategy involved searching for a vehicle with the right vehicle type that has a position between the checkpoint and its downstream successor. Searching for a gap and inserting a vehicle involves searching across several tracks and adjusting the neighbouring vehicle references.

The strategy, as described so far, is a rough simplification of the details that emerge with the tuning methods. Kaumann et al. [94] introduced several tuning methods by examining their fitness systematically. The tuning methods vary between the sink-and-source strategy, the flow-tuning, and the tuning of the mean gap. The sink-and-source strategy cuts the road-segments at the checkpoints' position into even shorter segments and places a source at the beginning of each segment as well as a sink at the end of each segment. Although the results of this approach "are quite good" [94], the dynamics of the system breaks with this strategy. The flow-tuning strategy tries to adapt the traffic flow values between an inductive loop detector and the value measured in the simulation either by adding or removing of vehicles. The tuning-of-the-mean-gap strategy tries to insert vehicles without disturbing the system dynamics too much. However, all approaches are far to simple to deal with the problems that occur in real-world systems such as OLSIMv4. The problems that occur include the following ones. Inserting large amounts of vehicles in case of increased traffic volume is not possible with the mean gap approach. It has already been discussed in sections 2.1.4.2, 2.1.4.3, and 3.4. Resolving the ambiguity as introduced in Section 2.2.2 that the loop detector data contains, requires to recognise and reproduce the synchronised and the jammed traffic patterns. None of the above approaches recognises and reproduces traffic patterns appropriately.

Additionally, OLSIMv3 also had the following difficulty. The checkpoints and the loop detectors had always the same position because the checkpoints had no associated detectors. Instead, they counted the vehicles and summed their velocities that pass them in the simulation. This mimics partly the behaviour of a local detector. However, complex decisions such as resolving the ambiguity as introduced by the loop and the checkpoint local detector data requires a more global view on the traffic situation in the near of a checkpoint. As an example, OLSIMv3 did not tolerate fluctuations in the vehicle type distributions across the track between the loop detector and the checkpoint value. As another example, fluctuations in the beginning and ending of regions of dense traffic inside a track and that cover a loop detector resulted in the inserting or removing of vehicles by the checkpoint. However, that might be inappropriate when viewed from the global perspective. Moreover, even though the checkpoints detected and deactivated themselves whenever their loop detectors did not receive any data, the OLSIMv3 checkpoints did not support switching them on or off at runtime. Closely neighboured subsequent loop detectors lead to configurations where the checkpoints interfered with each other. OLSIMv3 neither supported flexible positioning nor lane-wise cross-section bundling of the checkpoints. These shortcomings demanded for a more general approach. OLSIMv4 generalizes the checkpoints concept with the accumulators and the tuning elements.

OLSIMv4 improves the situation in the following way. First, choosing the right strategy requires reliable detector data of any lane and position from the simulation and with any perspective. As the tuning strategy has an impact on the traffic state of the simulation system [15] in [30], reliable traffic data is extremely important. OLSIMv4 supports this by means of the detectors from Section 3.5 and the accumulators that Section 3.6.1 will present. Accumulators allow to calculate balances over any set of detector measurements. They free the tuning elements to deal with details of analysing, grouping, and calculating traffic observables. Consequently, they enable tuning elements to focus on their main task which is adapting actual with should-be values. Second, OLSIMv4 supports positioning the tuning elements at any spatial extent in the network. Third, OLSIMv4 supports runtime configuration of the tuning elements. Section 3.6.2 introduces the tuning elements. A description of the exact behaviour of the OLSIMv4 tuning elements is, however, not in the scope of this thesis. Thus, Section 3.6.2 alternatively presents the tuning elements in their general form.

3.6.1 Accumulators

As noted above, accumulators allow to calculate balances over any set of detector measurements. Consequently, their structure definition reflects this property. It consists of the unique accumulator name $\mathfrak{t}_{\text{name}}^{\text{accu}} \in S_{\text{accu}}^4$, an accumulator type name $\mathfrak{t}_{\text{name}}^{\text{actype}} \in S_{\text{actype}}^4$, the aggregated measurement $\mathbf{M}^{(t)} \in S_{\text{measure}}^4$ for the time step t, and the detector names $\mathfrak{t}_{\text{name},1}^{\text{detector}}, \ldots, \mathfrak{t}_{\text{name},n}^{\text{detector}} \in \mathfrak{L}_{\text{det}}^4$. The accumulator type name helps to distinguish and implement several accumulating methods. Example accumulator types are cross-section accumulators and balancing accumulators. The accumulator name is a unique name to identify a single accumulator in a list. The accumulator stores its current aggregated measurement as $\mathbf{M}^{(t)} \in S_{\text{measure}}^4$. The detector names specify the detectors that are associated with the accumulator. In addition to the usual accessors, the accumulator structure provides the accessor $\mathbf{a}_i^{(t)} \# \mathfrak{t}_{\text{name}}^{\text{detector}}$ for the sequence of all detector names. Equation 3.122 introduces the definition of the accumulator structure.

$$S_{\text{accu}}^4 := \langle \mathfrak{t}_{\text{name}}^{\text{actype}}, \mathfrak{t}_{\text{name}}^{\text{accu}}, \mathbf{M}^{(t)}, \mathfrak{t}_{\text{name},1}^{\text{detector}}, \ldots, \mathfrak{t}_{\text{name},n}^{\text{detector}} \rangle^{\text{accu}} . \tag{3.122}$$

Accumulators aggregate the measurements of their associated detectors in each time step using the aggregate function $\mathbf{F}_{\text{accumulate}} : \mathfrak{T}_{\text{secs}} \times S_{\text{accu}}^4 \times \mathcal{P}_{\text{det}}(S_{\text{det}}^4) \longrightarrow S_{\text{accu}}^4$. All detectors provide output traffic information or traffic data input as a feedback information into the simulation. OLSIMv4 uses the cross-section aggregating of measurement values. The accumulate function filters the set of detectors to match with the detector name their corresponding set of detector names $\mathbf{a}_i^{(t)} \# \mathfrak{t}_{\text{name}}^{\text{detector}}$. The averaging method for this kind of aggregating has already been introduced in Section 3.5.3. Table 3.6.1 summarises all relevant functions, structure definitions, variables, and constants for this section.

3.6.2 Tuning elements

The tuning elements insert and remove vehicles into or from, respectively, the occupation tables according to their actual should-be value comparison. The balance of the actual should-be value comparison serves, however, only as a first indicator for the right strategy to choose. Depending on the long-term balance and on the traffic state, different strategies might be chosen by the tuning elements. For the needs of the real-world microscopic traffic simulation OLSIMv4, Habel developed and implemented the

Table 3.6.1: Summary of relevant functions, structures, variables, and constants for Section 3.6.1.

definition	description	domain	range
$\mathfrak{L}^4_{\text{actype}}$	accum. type names		$\{\text{xsect}, \dots\}$
$\mathfrak{L}^4_{\text{accu}}$	accumulator names		$\{\text{a00001}, \dots\}$
$\mathbf{F}_{\text{accumulate}}$	accumulate	$\mathcal{T}_{\text{secs}} \times S^4_{\text{accu}} \times \mathcal{P}_{\text{det}}(S^4_{\text{det}})$	S^4_{accu}
$\langle a, b, c, d_1, \dots, d_n \rangle^{\text{accu}}$	accumulators	$\mathfrak{L}^4_{\text{actype}} \times \mathfrak{L}^4_{\text{accu}} \times S_{\text{msre}} \times \mathcal{P}^4_{\text{det}}(\mathfrak{L}^4_{\text{det}})$	S^4_{accu}, see Eq. 3.122
$\mathbf{a}_i^{(t)} \# \mathbf{t}_{\text{name}}^{\text{actype}}$	acc. type name	$\mathbf{a}_i^{(t)} \in S^4_{\text{accu}}$	$\mathfrak{L}^4_{\text{actype}}$
$\mathbf{a}_i^{(t)} \# \mathbf{t}_{\text{name}}^{\text{accu}}$	accumulator name	$\mathbf{a}_i^{(t)} \in S^4_{\text{accu}}$	$\mathfrak{L}^4_{\text{accu}}$
$\mathbf{a}_i^{(t)} \# \mathbf{M}^{(t)}$	aggr. measurement.	$\mathbf{a}_i^{(t)} \in S^4_{\text{accu}}$	S^4_{measure}
$\mathbf{a}_i^{(t)} \# \mathbf{t}_{\text{name},i}^{\text{detector}}$	detector name	$\mathbf{a}_i^{(t)} \in S^4_{\text{accu}}$	$\mathfrak{L}^4_{\text{det}}$
$\mathbf{a}_i^{(t)} \# \mathbf{t}_{\text{name}}^{\text{detector}}$	detector names	$\mathbf{a}_i^{(t)} \in S^4_{\text{accu}}$	$\mathcal{P}^4_{\text{det}}(\mathfrak{L}^4_{\text{det}})$

tuning algorithm. It employs a flow based tuning strategy in combination with the so-called tuning of the mean gap [94]. The author of this thesis contributed only the idea of the tuning elements itself and their use of the `lc-safe?` operation to the improved tuning for OLSIMv4. The operation ensures that the traffic model's lane change safety rules hold. In the case of the lane change model by Habel, the rules from Equation 3.65 have to hold. However, depending on the traffic state and the particular tuning strategy, the tuning elements may also violate the safety rules by temporarily increasing the value for D_{tol}. Employing an `lc-safe?` operation results in the benefit that the tuning model does not depend on any concrete traffic model as it can use the abstraction. As part of the enhancement effort from [30], the author introduced the `SafeInsert` operation that constructs a vehicle inside a bulk of neighbouring vehicles as such that it has highly adapted position and velocity attributes. Instead of preparing the vehicle for insertion and testing the safety of the insertion with `lc-safe?`, the safe insert function $\mathbf{F}_{\text{safe}-\text{ins}}$ adjusts the vehicle attributes to be in harmony with the neighbouring vehicles. The tuning elements may use this function when inserting vehicles.

Tuning elements employ a pair of accumulators each to sum or balance the measurements from their associated detectors. The pair consists of an actual $\mathbf{a}_{\text{actual}}^{(t)} \in S^4_{\text{accu}}$ and a should-be $\mathbf{a}_{\text{should}-\text{be}}^{(t)} \in S^4_{\text{accu}}$ accumulator. The actual accumulator $\mathbf{a}_{\text{actual}}^{(t)} \in S^4_{\text{accu}}$ as an element from Equation 3.122 adopts and generalizes the behaviour from OLSIMv3 where the checkpoints counted the number of vehicles that passed the checkpoint in order to compare it to the one of the loop detector. The should-be accumulator $\mathbf{a}_{\text{should}-\text{be}}^{(t)} \in S^4_{\text{accu}}$ replaces the previously integrated local type detector by detectors with the more general model from Section 3.5. A checkpoint in OLSIMv3 covered always only a single lane. The accumulators in OLSIMv4 bundle detectors cross-section wise, segment-wise, or even across several roadways for the calculation of sums or balances of the measurements.

Equation 3.123 presents the tuning element structure S^4_{tun}. A tuning element type name $\mathbf{t}_{\text{name}}^{\text{tuntype}} \in \mathfrak{L}^4_{\text{ttpe}}$ enables the tuning elements to distinguish varying behaviour. The unique tuning element name $\mathbf{t}_{\text{name}}^{\text{tuning}} \in \mathfrak{L}^4_{\text{tuning}}$ helps to identify tuning elements by their name. The first generalisation for the OLSIMv4 tuning elements affects the positioning of the tuning elements. That is no longer required to be the same as for the loop detectors. As has been the case with the detectors, too, the tuning element positions use the area structure S^4_{area} from Equation 3.13 to specify their spatial extent $\mathbf{r}_{\text{area}}$. The actual accumulator $\mathbf{a}_{\text{actual}}^{(t)} \in S^4_{\text{accu}}$ and the should-be $\mathbf{a}_{\text{should}-\text{be}}^{(t)} \in S^4_{\text{accu}}$ accumulator form the next two components of the tuning element structure. The remaining components are the $j + 1$ measurements $\mathbf{m}^{(t)}, \mathbf{m}^{(t-\Delta t)}, \dots, \mathbf{m}^{(t-j\Delta t)} \in S^4_{\text{measure}}$. The tuning elements maintain this sequence to account how many vehicles they have inserted or removed already.

$$S^4_{\text{tun}} := \langle \mathbf{t}_{\text{name}}^{\text{tuntype}}, \mathbf{t}_{\text{name}}^{\text{tuning}}, \mathbf{r}_{\text{area}}, \mathbf{a}_{\text{actual}}^{(t)}, \mathbf{a}_{\text{should}-\text{be}}^{(t)}, \mathbf{m}^{(t)}, \mathbf{m}^{(t-\Delta t)}, \dots, \mathbf{m}^{(t-j\Delta t)} \rangle^{\text{tun}} . \quad (3.123)$$

While detectors read or analyse the traffic situation on a roadway, the tuning elements modify or adjust the traffic situation on a roadway. Tuning elements therefore provide a tuning function $\mathbf{F}_{\text{tune}}$:

Table 3.6.2: Summary of relevant functions, structures, variables, and constants for Section 3.6.2.

definition	description	domain	range
$\mathcal{L}^4_{\text{ttpe}}, \mathcal{L}^4_{\text{ttpe}}$	tun. type names		$\{\text{act/shbe}, \dots\}$
$\mathcal{L}^4_{\text{tuning}}$	tun. elem. names		$\{\text{tn00001}, \dots\}$
$\mathbf{F}_{\text{tune}}$	recursive tune	$\mathfrak{T}_{\text{secs}} \times S^4_{\text{occ}} \times S^4_{\text{tun}}$	$S^4_{\text{occ}} \times S^4_{\text{tun}}$
$\mathbf{F}_{\text{tune}*}$	single tune	$\mathfrak{T}_{\text{secs}} \times S^4_{\text{occ}} \times S^4_{\text{tun}}$	$S^4_{\text{occ}} \times S^4_{\text{tun}}$, Eq. 3.124
$\langle a, b, c, d_1, \dots, d_n \rangle^{\text{tun}}$	tuning elements	$\mathcal{L}^4_{\text{ttpe}} \times \mathcal{L}^4_{\text{tun}} \times S^4_{\text{area}}$ $\times (S^4_{\text{accu}})^2 \times \mathcal{P}^4_{\text{tun}}(S_{\text{msre}})$	S^4_{tun}, Eq. 3.123
$\mathbf{e}^{(t)}_i \, \#\mathbf{t}^{\text{tuntype}}_{\text{name}}$	tun. type name	$\mathbf{e}^{(t)}_i \in S^4_{\text{tun}}$	$\mathcal{L}^4_{\text{ttpe}}$
$\mathbf{e}^{(t)}_i \, \#\mathbf{t}^{\text{tuning}}_{\text{name}}$	tun. elem. name	$\mathbf{e}^{(t)}_i \in S^4_{\text{tun}}$	$\mathcal{L}^4_{\text{tun}}$
$\mathbf{e}^{(t)}_i \, \#\mathbf{r}_{\text{area}}$	spatial extent	$\mathbf{e}^{(t)}_i \in S^4_{\text{tun}}$	$\mathcal{L}^4_{\text{det}}$
$\mathbf{e}^{(t)}_i \, \#\mathbf{a}^{(t)}_{\text{actual}}$	actual accum.	$\mathbf{e}^{(t)}_i \in S^4_{\text{tun}}$	S^4_{measure}
$\mathbf{e}^{(t)}_i \, \#\mathbf{a}^{(t)}_{\text{should}-\text{be}}$	should-be accum.	$\mathbf{e}^{(t)}_i \in S^4_{\text{tun}}$	S^4_{measure}
$\mathbf{e}^{(t)}_i \, \#\mathbf{m}_j, \mathbf{e}^{(t)}_i \, \#\mathbf{m}^{(t-j\Delta t)}$	measurement	$\mathbf{e}^{(t)}_i \in S^4_{\text{tun}}$	S^4_{measure}

$\mathfrak{T}_{\text{secs}} \times S^4_{\text{occ}} \times S^4_{\text{tun}} \longrightarrow S^4_{\text{occ}} \times S^4_{\text{tun}}$. The signature of the tune function ranges over a pair of an occupation table and a tuning element. The pair is necessary, as the tune operation is designed to be able to modify the occupation tables. At the same time, the tune operation has to maintain its internal state. Thus, the range consists of a pair.

The general procedure implemented by the tune operation $\mathbf{F}_{\text{tune}}$ is as follows. The tune operation starts with determining the current actual should-be value comparison. The most simplest approach uses the traffic flow values of the associated detectors and does not distinguish between the various vehicle types. It is sufficient to demonstrate the general functioning and is, thus, done in the following.

The simulation update procedure evaluates tune operation for a time $t \in \mathfrak{T}_{\text{secs}}$, an occupation table $\mathbf{o}^{(t)}_r \in S^4_{\text{occ}}$, and a tuning element $\mathbf{e}^{(t)}_j \in S^4_{\text{tun}}$. The accumulators provide the measurements as $\mathbf{M}^{(t)}_{\text{actual}} :=$ $(\mathbf{e}^{(t)}_i \#\mathbf{a}^{(t)}_{\text{actual}})\#\mathbf{M}^{(t)}$ for the actual accumulator and as $\mathbf{M}^{(t)}_{\text{should}-\text{be}} := (\mathbf{e}^{(t)}_i \#\mathbf{a}^{(t)}_{\text{should}-\text{bel}})\#\mathbf{M}^{(t)}$ for the should-be accumulator. The tuning elements determine, for example, the balance of the traffic flows with $\Delta j^{(t)} := \mathbf{M}^{(t)}_{\text{actual}}\#j^{(t)}_{\text{any}} - \mathbf{M}^{(t)}_{\text{should}-\text{be}}\#j^{(t)}_{\text{any}}$. Depending on the value of $\Delta j^{(t)}$, they decide, in the general case, whether to insert or remove vehicles into their occupation table. A positive value indicates inserting and a negative value indicates removing. As the tuning elements keep track of the total number of vehicles inserted or remove since the last minute $J^{(t)} := \sum_{k=0}^{\tau} (\mathbf{e}^{(t)}_i \#\mathbf{m}^{(t-k\Delta t)})\#j^{(t)}_{\text{any}}$ where $k \in \mathbb{N}^{59}$. In the more realistic case, an aggregate function similar to the already presented ones is well-suited for this task. With the two values, $J^{(t)}$ and $\Delta j^{(t)}$, the decision whether to insert or remove is unique. Equation 3.124 presents the cases for the simplified tune operation $\mathbf{F}_{\text{tune}*}$ that depend one the two values. In case of inserting a vehicle into the occupation table, the tuning elements wrap the insert function $\mathbf{F}_{\text{insert}}$ from Equation 3.28 and the remove function $\mathbf{F}_{\text{remove}}$ from Equation 3.29, respectively. The functions use the measurement value for the next time step which is $\mathbf{m}*^{(t)} := \mathbf{F}_{\text{clone}}(\mathbf{m}^{(t)}, j^{(t)}_{\text{any}}, j^{(t)}_{\text{any}} + 1\,\text{vehs/min})$ and and $\mathbf{m}*^{(t)} := \mathbf{M}^{(t)}_{\text{zero}}$ initially for the insertion case. For the removal case the measurement value is $\mathbf{m}*^{(t)} := \mathbf{F}_{\text{clone}}(\mathbf{m}^{(t)}, j^{(t)}_{\text{any}}, j^{(t)}_{\text{any}} - 1\,\text{vehs/min})$ and $\mathbf{m}*^{(t)} := \mathbf{M}^{(t)}_{\text{zero}}$ initially. The vehicle $\mathbf{k}^{(t)}_n$ to be inserted or to be removed into the occupation table is determined as follows.

$$\mathbf{F}_{\text{tune}*}(t, \mathbf{o}^{(t)}_r, \mathbf{e}^{(t)}_j) := \begin{cases} (\mathbf{F}_{\text{insert}}(\mathbf{o}^{(t)}_r, \mathbf{k}^{(t)}_n), \mathbf{F}_{\text{clone}}(\mathbf{e}^{(t)}_j, \mathbf{m}^{(t)}, \mathbf{m}*^{(t)})) & , \text{if } \Delta j^{(t)} > J^{(t)} > 0, \\ (\mathbf{F}_{\text{remove}}(\mathbf{o}^{(t)}_r, \mathbf{k}^{(t)}_n), \mathbf{F}_{\text{clone}}(\mathbf{e}^{(t)}_j, \mathbf{m}^{(t)}, \mathbf{m}*^{(t)})) & , \text{if } \Delta j^{(t)} < J^{(t)} < 0, \\ (\mathbf{o}^{(t)}_r, \mathbf{e}^{(t)}_j) & , \text{otherwise.} \end{cases} \quad (3.124)$$

The $\mathbf{F}_{\text{tune}}$ functions relies on a function $\mathbf{F}_{\text{gap}} : S^4_{\text{occ}} \times S^4_{\text{area}} \longrightarrow S^4_{\text{veh}} \times \mathcal{X}^4_{\text{veh}}$ that searches for the largest headway between each two successive vehicles on the lanes of the area given as an argument to the function. It returns the vehicle at the front of the headway. Its counterpart is the vehicle search function $\mathbf{F}_{\mathrm{veh}} : S_{\mathrm{occ}}^4 \times S_{\mathrm{area}}^4 \longrightarrow S_{\mathrm{veh}}^4$ that searches for a vehicle in the spatial extent of the tuning element. With the vehicle, the lane, and the position, the tuning elements can construct a vehicle $\mathbf{k}_i^{(t_0)} \in S_{\mathrm{veh}}^4$ with the initial coordinates $\langle y_{\mathrm{gap}}, x_{\mathrm{gap}}, 0\,\mathrm{cells/s}, 0 \rangle^{\mathrm{coord}}$. The tuning element chooses the vehicle's vehicle type attribute depending on the comparison of the actual and should-be values. The tuning element derives the remaining vehicle attributes from the vehicle type. Before inserting the vehicle into the occupation table $\mathbf{o}^{(t)}$, the vehicle requires adjustment by the safe insert function from the context of Equation 3.48 $\mathbf{u}^{(t)} := (s_{\mathrm{rand}}', \mathbf{k}_i^{(t)}) = \mathbf{F}_{\mathrm{safe-ins}}(\mathbf{o}^{(t)} \# s_{\mathrm{rand}}, \mathbf{k}_i^{(t_0)}, \mathbf{p}, \mathbf{b})$ with $\mathbf{p} = \mathbf{F}_{\mathrm{topo}}(\mathbf{o}^{(t)} \# \mathbf{w}_i, \mathbf{k}_i^{(t_0)})$ and $\mathbf{b} = \mathbf{F}_{\mathrm{nbrs}*}(\mathbf{o}^{(t)} \# \mathbf{w}_i, \mathbf{k}_i^{(t_0)})$.

The simplified tune operation $\mathbf{F}_{\mathrm{tune}*}$ inserts or removes only a single vehicle per evaluation. Thus, the real tune function $\mathbf{F}_{\mathrm{tune}}$ evaluates the simplified version recursively. The balance between the number of vehicles to insert or remove per time step and the number of vehicles insert in the current time step form the termination criteria for the recursive function. One of the main differences of the presented tune function to the real-world implementation in OLSIMv4 by Habel exists with the implementation of the $\mathbf{F}_{\mathrm{gap}}$ function. The real-world counterpart determines the current mean time headway in advance to searching for the most appropriate gap. If possible, it also inserts the vehicle with a time and distance headway that does not differ too much from the mean value. For the special case of the OLSIMv4 tuning elements in combination with the traffic model by Lee et al., Pottmeier, and Habel as described in Section 3.4, there is a non-obvious difficulty that proved to be hard to solve. The model distinguishes between optimistic and pessimistic driving behaviour and during simulation clusters of homogeneous driving behaviours evolve. When inserting a vehicle into a cluster of optimistic vehicles, the whole cluster toggles from optimistic into pessimistic driving behaviour. When trying to insert several vehicles, this effect intensifies the chosen strategy of increasing the traffic flow in order to decrease the velocity. In the case where this effect is not part of the strategy, the insertion procedure has to maintain the distance headways appropriately. Fortunately, the situations where no intensification is aimed are limited to free flow traffic where appropriate gaps are always available.

Instead of modelling on- and off-ramps and let the vehicles lane change onto the main roadways in the simulation, the tuning elements of OLSIMv4 provide the way to calculate a balance according to which a tuning element inserts or removes vehicles into or from the simulation. Tuning elements also serve to model traffic lights in intersections scenarios. Whenever the traffic light turns to red, a tuning element inserts special dummy vehicles that drive with an extreme braking manoeuvre in order to slow down successive vehicles. Various other kinds of tuning elements may be of interest. A flow based tuning element uses the balance between the number of vehicles per measuring interval of the should-be and the actual accumulator to decide whether to insert or remove vehicles from the simulation. Density based tuning introduces an alternative approach to the flow based tuning. It is especially useful for scenarios with periodic boundary conditions where a constant density is aimed at. Such a scenario was used to verify the correct behaviour of the tuning elements in [30].

Maybe the greatest advantage of the flexible tuning element approach results from the potential to feedback traffic information from a detector of a certain roadway element into a tuning element of another roadway element. Turning behaviour that would otherwise have to be modelled by lane changing across several roadways. The tuning element solution decouples the roadway-wise simulation systems. They are only coupled through their input/output data but not through their state. The following section discusses this in more detail. Finally, Table 3.6.2 summarises all relevant functions, structure definitions, variables, and constants for Section 3.6.2.

3.7 Simulation model

Section 2.2.1 classified the microscopic traffic simulations with stochastic cellular automaton models as discrete time systems due to their parallel update mechanism. The simulations as discrete time models and their systems "assume a stepwise mode of execution"[216, Sec. 3.2]. For these kind of simulations the *Discrete Time System Specification* (DTSS) applies. Section 3.7 explains the latter and applies it to the components of this chapter's conceptual design.

Even though OLSIMv3 lacked a formal specification, mapping of OLSIMv3 onto the DTSS structure formalism is possible in principle. Section 3.7.1 introduces this formalism. The formalism is, however, the one that provides the least system knowledge when applied to the NRW highway network as a whole. Additionally, it has no parallel interpretation. It is thus desirable to decompose the system into several smaller subsystems such as a single DTSS for each roadway or each track.

The second level of system specification is the *multicomponent Discrete Time System Specification* (multiDTSS). When applied to the microscopic traffic simulation system OLSIMv4, it decomposes the system into a network of components per roadway or per track and defines a state transition function per component. The state transition function depends on the current state of the component itself as well as the states of its influencers, i.e. the neighbouring tracks or the connected roadways. This decomposition method might be applicable to other simulations, too. However, its application requires extreme care and special preparation of the component state transition functions in order to fulfill the multiDTSS formalism. In general, this also applies to all systems that rely on section based road traffic networks such as SUMO [119] and AIMSUN where a "leaving vehicles entity process" [196, Figs. 275, 276] redirects the leading vehicles of a section [31]. Section 3.7.2 discusses the multiDTSS formalism.

The next level of system specification decomposes a system into a network of smaller, less complex subsystems. This specification results in a structure that has a highly parallel interpretation. It is thus very desirable to apply it to OLSIM. In principle, it it applies to OLSIMv4. However, it does not apply to systems such as OLSIMv3 due to the following reasons. "Networks must obey certain constraints to fulfill closure under coupling. The most important constraint is that the network must be free of any delayless loops" [216, p. 164]. Consequently, a network with delayless loops cannot be decomposed of less complex subsystems. Unfortunately, the longitudinally connected tracks introduce delayless cyclic dependencies when modelled as a network of subsystems. A leading vehicle's velocity and position for the next time step depend on the properties of the subsequent vehicle(s). However, the subsequent vehicle(s) may be positioned as the last vehicle(s) on the subsequent track. Thus, the state for the next time step of the subsystem that represents the subsequent track depends on the update result of the leading vehicle of the previous track. Consequently, the update result would have to propagate immediately onto the subsequent track. Thus, modelling these kind of systems as a network of subsystems is not possible. Section 3.7.3 details the network of systems specification (DTSN).

The specification as a network of systems poses the question of how to model the dynamic character of the road traffic network. The cell properties can vary over time and are designed to reflect the dynamic character of the network. As a consequence, the network specification might require adaptions. Section 3.7.4 discusses potential adaptions by the dynamic structure system network (DSSN) approach. Finally, Section 3.7.5 thinks about further decomposition extensions. The nomenclature in this section follows the one used by Zeigler, Kim and Praehofer [216]. Figure 3.10 depicts the hierarchy of system specifications that are used in this section.

3.7.1 System specification

According to [216, Sec. 6.4] the DTSS is a structure S_{DTSS} in the form of Equation 3.125 and with the components as described in the following.

$$S_{\mathrm{DTSS}} := \langle X, Y, Q, \delta, \lambda, c \rangle \ . \tag{3.125}$$

The component X denotes the set of inputs, Y the set of outputs, Q the set of states, $\delta : Q \times X \longrightarrow Q$ the state transition function, $\lambda : Q \times X \longrightarrow Y$ the Mealy-type or $\lambda : Q \longrightarrow Y$ the Moore-type output function, and $c = \Delta t$ the time base. According to Zeigler, Kim and Praehofer [216, Sec. 6.4], the DTSS specifies a structure $S_{\mathrm{DTSStruct}}$ in the form of Equation 3.126 that consists of the following components. T is the time base, X, Q, and Y are arbitrary sets, Ω represents the set of admissible input segments, i.e. the set of all sequences over X and T, and the state and output trajectories are sequences over Q, T, and Y.

$$S_{\text{DTSStruct}} := \langle T, X, \Omega, Y, Q, \Delta, \Lambda \rangle^{\text{DTSStruct}} . \tag{3.126}$$

$$\Delta(q, \omega) := \begin{cases} q & \text{, if } t_1 = t_2 \\ \delta(q, \omega(t_1)) & \text{, if } t_1 = t_2 - \Delta t \\ \delta(\Delta(q, \omega(t_2 - \Delta t)), \omega(t_2 - \Delta t)) & \text{, otherwise.} \end{cases} \tag{3.127}$$

Still according to [216, Sec. 6.4], a DTSStruct defines the dynamic behaviour by the two functions Δ and Λ both introduced in the following. The book by Zeigler, Kim and Praehofer [216, Sec. 6.4] defines the global state transition function Δ of the dynamic system to a given input segment $\omega : [t_1, t_2) \longrightarrow X$ and an initial state q at time t_1 in the form of Equation 3.127. The global state transition function Δ describes the complete state transition that results from starting with the initial state q and applying the state transition function $\delta : Q \times X \longrightarrow Q$ recursively to the state of time t for each time step $t \in [t_1, t_2)$. The left hand side of Equation 3.127 does not consume the time t as argument. As the function definition specifies a state transition for the complete interval of $[t_1, t_2)$, the time variable is superfluous. The global state transition from the initial state q for time t_1 to the final state for time t_2 as described by $\Delta(q, \omega)$ results in a set of nested per time step state transitions for the time steps between t_1 and t_2. The third clause in Equation 3.127 produces the nested set of state transitions by descending recursively into the next inner level of state transitions for each next time step $t - \Delta t$. The resulting state transition is then achieved by applying δ first to the initial state and in every step thereafter to the step-wise previous state achieved.

The specification of the dynamic system's output function $\Lambda : Q \times X \longrightarrow Y$ or $\Lambda : Q \longrightarrow Y$ completes the description of the dynamic behaviour. For Moore-type systems the output function $\Lambda(q, x) = \lambda(q)$ is appropriate while Mealy-type systems employ $\Lambda(q, x) = \lambda(q, x)$. Assigning the input and output functions ω and λ and the state transition function δ onto the functions as presented in the previous sections, completes the discrete time system specification for the microscopic traffic simulation in the context of OLSIM and is done in the following. While assigning the input and output functions ω and λ onto the $\mathbf{F}_{\text{input}}$ and $\mathbf{F}_{\text{output}}$ functions is relatively straight forward, the assignment of the state transition function δ is much more complex and at the same time the most interesting part of it as it provides insights into the core of a simulation.

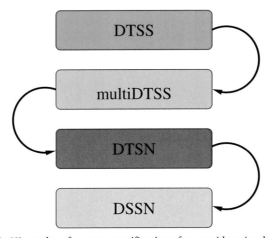

Figure 3.10: Hierarchy of system specifications for consideration by OLSIMv4.

Table 3.7.1: Summary of relevant functions, structures, variables, and constants for Section 3.7.1.

definition	description	domain	range
$\langle a, b, c, d, e, f, g, h \rangle^{\text{sys}}$	discrete time system	$(\mathfrak{T}_{\text{secs}}^4)^2 \times (\mathbf{F})^2 \times \mathcal{P}_{\text{occ}}^4(S_{\text{occ}}^4) \times \mathcal{P}_{\text{det}}^4(S_{\text{det}}^4)$ $\times \mathcal{P}_{\text{tun}}^4(S_{\text{tun}}^4) \times \mathcal{P}_{\text{swarea}}^4(S_{\text{swarea}}^4)$	S_{sys}^4, Eq. 3.128
$\mathbf{F}_{\text{output}}$	output function	S_{sys}^4	$\mathcal{P}_{\text{measure}}^4(S_{\text{measure}}^4)$
$\mathbf{F}_{\text{measure}}$	measure function	$S_{\text{sys}}^4 \times \mathcal{P}_{\text{measure}}^4(S_{\text{measure}}^4)$	S_{sys}^4, Eq. 3.129
$\mathbf{F}_{\text{accu}}$	accumulate function	S_{sys}^4	S_{sys}^4, Eq. 3.130
$\mathbf{F}_{\text{tune}}$	tune function	S_{sys}^4	S_{sys}^4, Eq. 3.131
$\mathbf{F}_{\text{drive}}$	drive update	S_{sys}^4	S_{sys}^4, Eq. 3.132
$\mathbf{F}_{\text{chlane}}$	lane change update	S_{sys}^4	S_{sys}^4, Eq. 3.133
$\mathbf{F}_{\text{inc}-t}$	increase time	S_{sys}^4	S_{sys}^4, Eq. 3.134
$\mathbf{F}_{\text{update}}$	state transition	$S_{\text{sys}}^4 \times \mathcal{P}_{\text{measure}}^4(S_{\text{measure}}^4)$	S_{sys}^4, Eq. 3.135

The OLSIMv4 simulation system structure in the non-decomposed form of Equation 3.128 may have been similar to the form of OLSIMv3. The structure S_{sys}^4 in this form consists of the constant time interval size Δt, the current time $t \in \mathfrak{T}_{\text{secs}}$, the two vehicle update procedures $\mathbf{F}_{\text{drive}}$ and $\mathbf{F}_{\text{chlane}}$, the list of occupation tables $\mathbf{L}_{\text{occ}} \subset \mathcal{P}(S_{\text{occ}}^4)$ that represent tracks or roadways, the list of detectors $\mathbf{L}_{\text{det}} \subset \mathbf{L}_{\text{det}}$, the list of tuning elements $\mathbf{L}_{\text{tun}}$, and the list of switching areas $\mathbf{L}_{\text{swarea}} \subset \mathcal{P}(S_{\text{swarea}}^4)$. A list herein, such as the list $\mathbf{l}^{(t)} := \mathbf{L}_{\text{det}} \subset \mathcal{P}(S_{\text{det}}^4) = \{\mathbf{d}_1^{(t)}, \ldots, \mathbf{d}_n^{(t)} \in S_{\text{det}}^4\}$ of detectors S_{det}^4, refers to an ordered set of elements, detectors $\mathbf{d}_1^{(t)}, \ldots, \mathbf{d}_n^{(t)}$ in the example, that are uniquely accessible by an index referer such as, for example, $\mathbf{l}_i^{(t)}$ where $i \in \mathbb{N}$ refers to the ith-element. The vehicle update functions thereby have the signatures $\mathbf{F}_{\text{drive}} : \mathbf{L}_{\text{occ}} \times \mathbf{L}_{\text{swarea}} \longrightarrow \mathbf{L}_{\text{occ}}$ and $\mathbf{F}_{\text{chlane}} : \mathbf{L}_{\text{occ}} \times \mathbf{L}_{\text{swarea}} \longrightarrow \mathbf{L}_{\text{occ}}$.

$$S_{\text{sys}}^4 := \langle t, \Delta t, \mathbf{F}_{\text{drive}}, \mathbf{F}_{\text{chlane}}, \mathbf{L}_{\text{occ}}, \mathbf{L}_{\text{det}}, \mathbf{L}_{\text{tun}}, \mathbf{L}_{\text{swarea}} \rangle^{\text{sys}}, \text{ with } S_{\text{sys}}^4 \subset Q. \tag{3.128}$$

The structure S_{sys}^4 contains every component required to specify the simulation system and to classify it as a discrete time simulation system in the sense of Equation 3.126. The classification requires to assign the components of the S_{sys}^4 structure to the ones of structure $S_{\text{DTSStruct}}^4$. Thereby, the per time step state transition function consumes the main part in finding the corresponding components. With the components of a system state structure $\mathbf{s}^{(t)} \in S_{\text{sys}}^4$ the input function can be defined as $\omega(t) := \{\mathbf{F}_{\text{input}}(t, \mathbf{d}_i^{(t)}) \mid \mathbf{d}_i^{(t)} \in \mathbf{s}^{(t)} \# \mathbf{L}_{\text{det}}\}$ with $\mathbf{F}_{\text{input}}$ from Equation 3.83. Applying the definition to a certain time would, however, result in a set of measurements $\omega(t) \subset S_{\text{measure}}^4$ that would have no references to the detectors. The conceptual design uses, thus, the different definition that inputs the measurements as part of the detectors' measure operation from Equation 3.85. The difference does, however, not break the discrete time system specification for the simulation as the detector definition could easily be extended to associate the input and output measurements with the corresponding detector name. The detectors' input function $\mathbf{F}_{\text{input}}$ could then be adjusted to sort out –according to the detector name– the detectors input data from the set of input measurements for the current timestep. Similar holds true for the output function. The latter can be defined as $\lambda(q) = \lambda(\mathbf{s}^{(t)}) := \{\mathbf{F}_{\text{output}}(\mathbf{s}^{(t)} \# t, \mathbf{d}_i^{(t)}) \mid \mathbf{d}_i^{(t)} \in \mathbf{s}^{(t)} \# \mathbf{L}_{\text{det}}\} \subset \mathcal{P}_{\text{measure}}(S_{\text{measure}}^4)$ with $\mathbf{F}_{\text{output}} : S_{\text{sys}}^4 \longrightarrow \mathcal{P}_{\text{measure}}(S_{\text{measure}}^4)$ from Equation 3.80. The state transition function involves, of course, the update function of the simulation system which is assigned in the following.

The update of a simulation in the discrete time system specification $\mathbf{s}^{(t)} \in S_{\text{sys}}^4$ starts with the output function from above. The measure operation follows. It is implemented by the measure function $\mathbf{F}_{\text{measure}} : S_{\text{sys}}^4 \times \mathcal{P}_{\text{measure}}(S_{\text{measure}}^4) \longrightarrow S_{\text{sys}}^4$ from Equation 3.129. The function definition in Equation 3.129 takes a system state $\mathbf{s}^{(t)} \in S_{\text{sys}}^4$ and, for the context of OLSIMv4, the set of loop detector measurements $\mathfrak{M}_{\text{loop}*}^4 \subset X$ as arguments and relies on the additional constraints $\mathbf{F}_{\text{rway}=?}(\mathbf{o}_r^{(t)}, \mathbf{d}_i^{(t)}) := \mathbf{d}_i^{(t)} \# t_{\text{name}}^{\text{roadway}} = \mathbf{o}_r^{(t)} \# t_{\text{name}}^{\text{roadway}}$ and $\mathbf{F}_{\text{elem-of}?}(\mathbf{s}^{(t)}, \mathbf{o}_r^{(t)}, \mathbf{d}_i^{(t)}) := \mathbf{d}_i^{(t)} \in \mathbf{s}^{(t)} \# \mathbf{L}_{\text{det}} \wedge \mathbf{o}_r^{(t)} \in \mathbf{s}^{(t)} \# \mathbf{L}_{\text{occ}}$ for the occupation table $\mathbf{o}_r^{(t)}$, the detector $\mathbf{d}_i^{(t)}$, and the system state $\mathbf{s}^{(t)}$. It generalises the definition of the equally named function from Equation 3.82 of the detectors section by mapping it over each occupation table and each detector.

$$\mathbf{F}_{\text{measure}}(\mathbf{s}^{(t)}, \mathfrak{M}_{\text{loop}*}^4) := \mathbf{F}_{\text{clone}}(\mathbf{s}^{(t)}, \mathbf{L}_{\text{det}}, \mathbf{L}_{\text{det}}^*) \text{ , with}$$

$$\mathbf{L}_{\text{det}}^* = \{\mathbf{F}_{\text{measure}}(\mathbf{s}^{(t)} \# t, \mathbf{o}_r^{(t)}, \mathbf{d}_i^{(t)}) \mid \mathbf{F}_{\text{elem-of?}}(\mathbf{s}^{(t)}, \mathbf{o}_r^{(t)}, \mathbf{d}_i^{(t)}) \wedge \mathbf{F}_{\text{rway=?}}(\mathbf{o}_r^{(t)}, \mathbf{d}_i^{(t)})\} \quad (3.129)$$

The next step of the system update involves accumulating the detector measurements for the accumulators. The accumulation depends only on the current system state $\mathbf{s}^{(t)} \in S_{\text{sys}}^4$. Function $\mathbf{F}_{\text{accu}}$: $S_{\text{sys}}^4 \longrightarrow S_{\text{sys}}^4$ from Equation 3.130 performs the accumulators' update.

$$\mathbf{F}_{\text{accu}}(\mathbf{s}^{(t)}) := \mathbf{F}_{\text{clone}}(\mathbf{s}^{(t)}, \mathbf{L}_{\text{tun}}, \mathbf{L}_{\text{tun}}^*) \text{ , with}$$

$$\mathbf{L}_{\text{tun}}^* = \{\mathbf{F}_{\text{acc2}}(\mathbf{F}_{\text{acc1}}(\mathbf{e}_i^{(t)}, \mathbf{s}^{(t)} \# \mathbf{L}_{\text{det}}), \mathbf{s}^{(t)} \# \mathbf{L}_{\text{det}}) \mid \mathbf{e}_i^{(t)} \in \mathbf{s}^{(t)} \# \mathbf{L}_{\text{tun}}\} \text{ ,}$$

$$\mathbf{F}_{\text{acc1}}(t, \mathbf{e}_i^{(t)}, \mathbf{L}_{\text{det}}) = \mathbf{F}_{\text{clone}}(\mathbf{e}_i^{(t)}, \mathbf{a}_{\text{actual}}^{(t)}, \mathbf{F}_{\text{accumulate}}(t, \mathbf{e}^{(t)} \# \mathbf{a}_{\text{actual}}^{(t)}, \mathbf{L}_{\text{det}})) \text{ , and}$$

$$\mathbf{F}_{\text{acc2}}(t, \mathbf{e}_i^{(t)}, \mathbf{L}_{\text{det}}) = \mathbf{F}_{\text{clone}}(\mathbf{e}_i^{(t)}, \mathbf{a}_{\text{should-be}}^{(t)}, \mathbf{F}_{\text{accumulate}}(t, \mathbf{e}^{(t)} \# \mathbf{a}_{\text{should-be}}^{(t)}, \mathbf{L}_{\text{det}})) \text{ .}$$

$$(3.130)$$

After the accumulate operation, the tune operation $\mathbf{F}_{\text{tune}} : S_{\text{sys}}^4 \longrightarrow S_{\text{sys}}^4$ is applied to the simulation system. The tuning depends only on the current system state $\mathbf{s}^{(t)} \in S_{\text{sys}}^4$. Function from Equation 3.131 performs the update of the tuning elements and the occupation tables. The function definition in Equation 3.131 takes a system state $\mathbf{s}^{(t)} \in S_{\text{sys}}^4$ as argument and relies on the additional constraint $\mathbf{F}_{\text{rway=?}}(\mathbf{o}_r^{(t)}, \mathbf{e}_j^{(t)}) := \mathbf{e}_j^{(t)} \# \mathbf{t}_{\text{name}}^{\text{roadway}} = \mathbf{o}_r^{(t)} \# \mathbf{t}_{\text{name}}^{\text{roadway}}$ for the occupation table $\mathbf{o}_r^{(t)}$ and the tuning element $\mathbf{e}_j^{(t)}$. The tune update from Equation 3.131 applies the tune operation from Equation 3.124 to all tuning elements for any occupation table in the system state. As applying the tune operation results in a pair $(\mathbf{o}_r^{(t')}, \mathbf{e}_j^{(t')}) = \mathbf{F}_{\text{tune}}(t, \mathbf{o}_r^{(t)}, \mathbf{e}_j^{(t)})$ that consists of the updated occupation table $\mathbf{o}_r^{(t')}$ and the updated tuning element $\mathbf{e}_j^{(t')}$, the tune update for a single occupation table is of sequential or recursive nature.

$$\mathbf{F}_{\text{tune}}(\mathbf{s}^{(t)}) := \mathbf{F}_{\text{clone}}(\mathbf{F}_{\text{clone}}(\mathbf{s}^{(t)}, \mathbf{L}_{\text{tun}}, \mathbf{L}_{\text{tun}}^*), \mathbf{L}_{\text{occ}}, \mathbf{L}_{\text{occ}}^*) \text{ , where}$$

$$\mathbf{L}_{\text{occ}}^* = \bigcup\{\mathbf{o}_r^{(t')}\} \qquad \text{and} \qquad \mathbf{L}_{\text{tun}}^* = \bigcup \mathbf{L}_{\text{tun}}' \text{ , with}$$

$$(\mathbf{o}_r^{(t')}, \mathbf{L}_{\text{tun}}') = \mathbf{F}_{\text{occ-tune}}(t, \mathbf{o}_r^{(t)}, \mathbf{s}^{(t)} \# \mathbf{L}_{\text{tun}}, \{\}) \text{ , } \forall \mathbf{o}_r^{(t)} \in \mathbf{s}^{(t)} \# \mathbf{L}_{\text{occ}}$$

$$\mathbf{F}_{\text{occ-tune}}(t, \mathbf{o}_r^{(t)}, \mathbf{L}_{\text{tun}}, \mathbf{L}_{\text{tun}}') = \begin{cases} \mathbf{F}_{\text{occ-tune}}(t, \mathbf{o}_r^{(t')}, \mathbf{L}_{\text{tun}} \setminus \{\mathbf{e}_j^{(t)}\}, \{\mathbf{e}_j^{(t')}\} \cup \mathbf{L}_{\text{tun}}') \text{ , if} \\ \quad \exists \mathbf{e}_j^{(t)} \in \mathbf{L}_{\text{tun}} \mid \mathbf{F}_{\text{rway=?}}(\mathbf{o}_r^{(t)}, \mathbf{e}_j^{(t)}) \text{ with} \\ \quad (\mathbf{o}_r^{(t')}, \mathbf{e}_j^{(t')}) = \mathbf{F}_{\text{tune}}(t, \mathbf{o}_r^{(t)}, \mathbf{e}_j^{(t)}) \text{ , or} \\ (\mathbf{o}_r^{(t)}, \mathbf{L}_{\text{tun}}') \text{ , otherwise.} \end{cases}$$

$$(3.131)$$

As the second last step of the system update, the vehicle update operations $\mathbf{F}_{\text{drive}} : S_{\text{sys}}^4 \longrightarrow S_{\text{sys}}^4$ and $\mathbf{F}_{\text{chlane}} : S_{\text{sys}}^4 \longrightarrow S_{\text{sys}}^4$ are applied. As the roadways or tracks in the discrete time system specification may be inter-connected laterally and longitudinally, the vehicle update functions do not only depend on the state of the occupation table as is the case in Equation 3.33. The occupation tables, thus, cannot be updated in parallel. Additionally, the occupation table update function from Equation 3.33 does not apply either. Instead the vehicle update functions depend on the list of occupation tables and the list of switching areas and, consequently, they have the signature $\mathbf{F}_{\text{drive}} : \mathfrak{T}_{\text{secs}} \times \mathcal{P}_{\text{occ}}(S_{\text{occ}}^4) \times \mathcal{P}_{\text{swarea}}(S_{\text{swarea}}^4) \longrightarrow \mathcal{P}_{\text{occ}}(S_{\text{occ}}^4)$ and $\mathbf{F}_{\text{chlane}} : \mathfrak{T}_{\text{secs}} \times \mathcal{P}_{\text{occ}}(S_{\text{occ}}^4) \times \mathcal{P}_{\text{swarea}}(S_{\text{swarea}}^4) \longrightarrow \mathcal{P}_{\text{occ}}(S_{\text{occ}}^4)$. Equations 3.132 and 3.133 provide the function definitions. In general, the order of applying the drive and the lane change update should be irrelevant. However, as the lane change update may potentially disturb the coupling between the vehicles, lane changing first may result in a more consistent coupling among the vehicles with respect to the measure and the tune operations of the next cycle which is why this order is preferred herein. The signatures for the vehicle update functions demonstrates that the set of occupation tables or tracks is atomic with respect to the vehicle update. The vehicle update procedures operate on the non-decomposed and complex set of the occupation tables that represents the entire network by a set. The vehicle update procedures cannot be decomposed trivially into procedures that operate on a single track or a single roadway as the vehicle update depends also on the states of the neighboured tracks. As a consequence, testing or verification of an implementation of the vehicle update always involves operating on a complete network. Using smaller or simpler networks bears the risk of unrealistic test cases. The system modelling provides, thus, very little knowledge about the system S_{sys}^4.

$$\mathbf{F}_{\text{drive}}(\mathbf{s}^{(t)}) := \mathbf{F}_{\text{clone}}(\mathbf{s}^{(t)}, \mathbf{L}_{\text{occ}}, \mathbf{s}^{(t)} \# \mathbf{F}_{\text{drive}}(\mathbf{s}^{(t)} \# t, \mathbf{s}^{(t)} \# \mathbf{L}_{\text{occ}}, \mathbf{s}^{(t)} \# \mathbf{L}_{\text{swarea}})) \tag{3.132}$$

$$\mathbf{F}_{\text{chlane}}(\mathbf{s}^{(t)}) := \mathbf{F}_{\text{clone}}(\mathbf{s}^{(t)}, \mathbf{L}_{\text{occ}}, \mathbf{s}^{(t)} \# \mathbf{F}_{\text{chlane}}(\mathbf{s}^{(t)} \# t, \mathbf{s}^{(t)} \# \mathbf{L}_{\text{occ}}, \mathbf{s}^{(t)} \# \mathbf{L}_{\text{swarea}})) \tag{3.133}$$

The update operation of the discrete time simulation system ends with the time update function $\mathbf{F}_{\text{inc}-\text{t}} : S_{\text{sys}}^4 \longrightarrow S_{\text{sys}}^4$ with the definition from Equation 3.134.

$$\mathbf{F}_{\text{inc}-\text{t}} := \mathbf{F}_{\text{clone}}(\mathbf{s}^{(t)}, t, t + \Delta t) \tag{3.134}$$

The system update finally results in a function $\mathbf{F}_{\text{update}} : S_{\text{sys}}^4 \times \mathcal{P}_{\text{measure}}(S_{\text{measure}}^4) \longrightarrow S_{\text{sys}}^4$ with the definition from Equation 3.135. The resulting update function composes the update functions for the particular update steps accumulate, tune, change lane, drive, and increase time and applies the composed function to the state updated by the measure function.

$$\mathbf{F}_{\text{update}}(\mathbf{s}^{(t)}, \mathfrak{M}_{\text{loop}*}^4) := \mathbf{F}_{\text{inc}-\text{t}} \circ \mathbf{F}_{\text{drive}} \circ \mathbf{F}_{\text{chlane}} \circ \mathbf{F}_{\text{tune}} \circ \mathbf{F}_{\text{accu}} \circ \mathbf{F}_{\text{measure}}(\mathbf{s}^{(t)}, \mathfrak{M}_{\text{loop}*}^4) \tag{3.135}$$

With $\mathbf{F}_{\text{update}} : S_{\text{sys}}^4 \times \mathcal{P}_{\text{measure}}^4(S_{\text{measure}}^4) \longrightarrow S_{\text{sys}}^4$ from Equation 3.135, the definition of the state transition function can be defined as $\delta(q, x) := \mathbf{F}_{\text{update}}(\mathbf{s}^{(t)}, \mathfrak{M}_{\text{loop}*}^4)$. The definition of the state transition function completes the discrete time system specification S_{DTSS}. A more promising way to handle complex systems exists with decomposing them into less complex ones [18]. The network model of OLSIMv3 with the partitioned roadways prevented the simulation system from decomposition. Thus, the OLSIMv4 network model prepares for a potential decomposition. Section 3.3 introduced the OLSIMv4 network model, already. Having a decomposition with closure under coupling results in having a formalism that has a parallel interpretation. A parallel interpretation gives the benefit of making the system "amenable for a parallel implementation" [18].

Table 3.7.1 summarises the relevant functions, structure definitions, variables, and constants of this section.

3.7.2 Multicomponent specification

The discrete time system specification of the last section treats the road traffic network as an atomic entity. The simulation system results, thus, in a relatively complex system that provides only minimal granularity and no or at least unknown parallel interpretation. The system specification level of the last section, therefore, matches the I/O system level. It is, however, possible to break down the complex system into smaller and less complex system components. The next specification level that applies for this chapter's simulation system therefore relies on the Multicomponent Discrete Time System Formalism [216, Sec. 7.4] which is explained in the following. The components compound the system state per roadway and their update depends on the state of the connected components. The multicomponent specification, in contrast to their system specification counterpart, supports, however, component-wise parallel update. Consequently, the update also supports component-wise testing and verification. These two improvements and the increased system knowledge makes the multicomponent specification preferable over the system specification.

Equation 3.136 defines the specification (multiDTSS) structure $S_{\text{multiDTSS}}^4$ for the formalism that consists of an arbitrary set of input values X_{N}, the time interval h_{N}, the index set D, and the set of components $\{M_{\text{d}}\}$.

$$S_{\text{multiDTSS}}^4 := \langle X_{\text{N}}, D, \{M_{\text{d}}\}, h_{\text{N}} \rangle^{\text{multiDTSS}} . \tag{3.136}$$

Table 3.7.2: Summary of relevant functions, structures, variables, and constants for Section 3.7.2.

definition	description	domain	range
$\langle a,b,c,d,e,f,g,h\rangle^{\text{road}}$	component system	$(\mathfrak{T}_{\text{secs}}^4)^2 \times (\mathbf{F})^2 \times S_{\text{occ}}^4 \times \mathcal{P}_{\text{det}}^4(S_{\text{det}}^4)$ $\times \mathcal{P}_{\text{tun}}^4(S_{\text{tun}}^4) \times \mathcal{P}_{\text{swarea}}^4(S_{\text{swarea}}^4)$	S_{road}^4, Eq. 3.140
$\mathbf{F}_{\text{output}}(s)$	output function	S_{road}^4	$\mathcal{P}_{\text{measure}}^4(S_{\text{measure}}^4)$
$\mathbf{F}_{\text{measure}}(s,m)$	measure function	$S_{\text{road}}^4 \times \mathcal{P}_{\text{measure}}^4(S_{\text{measure}}^4)$	S_{road}^4, Eq. 3.141
$\mathbf{F}_{\text{accu}}(s,d)$	accumulate function	$S_{\text{road}}^4 \times \mathcal{P}_{\text{det}}^4(S_{\text{det}}^4)$	S_{road}^4, Eq. 3.142
$\mathbf{F}_{\text{tune}}(s)$	tune function	S_{road}^4	S_{road}^4, Eq. 3.143
$\mathbf{F}_{\text{swarea-drive}}(s,o)$	swarea drive upd.	$S_{\text{road}}^4 \times \mathcal{P}_{\text{occ}}^4(S_{\text{occ}}^4)$, see Eq. 3.146	$S_{\text{road}}^4 \times \mathcal{P}_{\text{veh}}^4(S_{\text{veh}}^4)$
$\mathbf{F}_{\text{drive}}(s,v)$	drive update	$S_{\text{road}}^4 \times \mathcal{P}_{\text{veh}}^4(S_{\text{veh}}^4)$	S_{road}^4, Eq. 3.147
$\mathbf{F}_{\text{swarea-chlane}}(s,o)$	swarea lane change	$S_{\text{road}}^4 \times \mathcal{P}_{\text{occ}}^4(S_{\text{occ}}^4)$, see Eq. 3.146	$S_{\text{road}}^4 \times \mathcal{P}_{\text{veh}}^4(S_{\text{veh}}^4)$
$\mathbf{F}_{\text{chlane}}(s,v)$	lane change update	$S_{\text{road}}^4 \times \mathcal{P}_{\text{veh}}^4(S_{\text{veh}}^4)$	S_{road}^4, Eq. 3.147
$\mathbf{F}_{\text{inc-t}}(s)$	increase time	S_{road}^4	S_{road}^4, Eq. 3.148
$\mathbf{F}_{\text{update}}(s,m,d,o)$	state transition	$S_{\text{road}}^4 \times \mathcal{P}_{\text{measure}}(S_{\text{measure}}^4)$ $\times \mathcal{P}_{\text{det}}(S_{\text{det}}^4) \times \mathcal{P}_{\text{occ}}(S_{\text{occ}}^4)$	S_{road}^4, Eq. 3.149

Additionally, for each $d \in D$ the component M_d is specified as $M_d = \langle Q_d, Y_d, I_d, \delta_d, \lambda_d \rangle$ [216, Sec. 7.4]. Thereby Q_d is an arbitrary set of states, Y_d is an arbitrary set of output values, $I_d \in D$ is a set of influencers of d, Signature 3.137 defines the state transition function δ_d of d, and Signature 3.138 defines the output function λ_d of d. Formally, any multicomponent discrete time system specifications specifies a discrete time system specification built from the crossproduct of the components [216, Sec. 7.4]. As a consequence, all multicomponent systems that use the multiDTSS specification have a corresponding DTSS specification, too. As another consequence, it is sufficient to define the local state transition function for the components.

$$\delta_d : \underset{i \in I_d}{\times} Q_i \times X \longrightarrow Q_d \tag{3.137}$$

$$\lambda_d : \underset{i \in I_d}{\times} Q_i \times X \longrightarrow Y_d \tag{3.138}$$

In order to map a microscopic traffic simulation onto the multiDTSS formalism, the road traffic network is divided into structures S_{road}^4 from Equation 3.136 that contain one occupation table per roadway as well as all detectors, tuning elements, and switching areas of the occupation table. As the switching areas connect the roadways with each other, they require special treatment. Because the vehicles in a road occupation table can switch from one occupation table into the other, the component state transition function might want to transfer a vehicle from one component state into the other. Assuming a straightforward definition of the state transition function, such an update would end in modifying the states of the two affected occupation tables synchronously. Consequently, the signature of a minimal function that implements the synchronous operation would be in the form of Equation 3.139. In the given specification, however, the state transition functions require the form of the multicomponent approach from Equation 3.137 which limits the dependency to reading the state of neighboured components only. The form of Equation 3.139 requires, however, two write updates instead of the otherwise one write update per component. More complex variations might involve even more than two write updates.

$$\delta_{d,d*} : \underset{i \in I_d}{\times} Q_i \times X \longrightarrow Q_d \times Q_{d*} \tag{3.139}$$

As a result, such an approach disqualifies not only the specification from conforming to the multicomponent formalism but also prevents it from having an efficient parallel interpretation. This kind of specification has been implemented in microscopic traffic simulation systems that rely on road-segment-wise components such as OLSIMv3 and SUMO. Consequently, these kind of systems require an almost sequential evaluation strategy as the component state transition function has the property to strengthen the coupling among the simulation components. As another consequence, the simulation system is not a multicomponent system because the signature of the state transition function $\delta_{\mathrm{d,d}*}$ from Equation 3.139 does not match to the one of function δ_{d} from Equation 3.137. Thus, applying the strictly conforming multicomponent formalism to OLSIM results not only in a higher level of system knowledge but also in a better parallel interpretation and an improved verification capabilities. As having a better parallel interpretation provides a high practical benefit, studying whether and how the multicomponent formalism could apply to microscopic traffic simulations is worthwile and, thus, done as part of the vehicle update discussion.

As already pointed out, the component's state transition functions take the form of $\delta_{\mathrm{d,d}*}$ in the straightforward implementation. However, it is required and possible to simplify the signature in such a way that it conforms strictly to δ_{d} from Equation 3.137. The simplification concerns only the vehicle update process and exploits the principles of deterministic behaviour and reproducibility. Pseudo-random number generators generate and distribute numbers in such a way that they match to a given stochastic distribution. However, they provide their numbers in fact deterministically. Thus, two independent simulations that start with the same initial state and operate sequentially in the same order and the same vehicle update function will produce always the same result. As the components instantiate the occupation tables of the switching areas for the vehicle update in each time step, they require only to start the instantiation with the same state of the random number generator. As long as the two instantiations of the occupation tables for a vehicle update hold the two aforementioned constraints, they will always result in the same state after the vehicle update process. The two components holding the random number generator states, however, interpret their identical results of the vehicle update process individually. They mutually trust their corresponding component that its behaviour matches the specification. The component of the 'from' part chooses all vehicles on the 'from' side of the joined occupation table. The 'to' part behaves analogously.

The simplification involves the following provisions. First, it requires to provide each switching area definition twice—one for the 'from' part component and the other one for the 'to' part component. It is important, that both instances of the initially identical switching areas have identical states before and after each update step for the whole time of the simulation. Second, it requires to instantiate the joined occupation tables twice. Additionally, it requires to perform the vehicle update independently on both joined occupation tables. The components interpret then the updated joined occupation tables individually by choosing only those vehicles of their corresponding part. The components merge then the chosen and updated vehicles back into their originating occupation table. Concluding from the algorithm as described above, the simplification maps the concept of the switching areas on the concept of the input/output ports of a component. In the following the algorithm is described in more detail and in the overall context of the component update function.

The structure $S_{\mathrm{road}}^{4} \subset Q_{\mathrm{d}}$ from Equation 3.140 represents a component system state in the multicomponent specification. The structure is similar to the one of the system specification from Section 3.7.1 with the exception that the former contains a single occupation table only and the remaining components belong all to the roadway of occupation table $\mathbf{o}_{d}^{(t)}$. The elements of this structure $\mathbf{s}_{d}^{(t)} \in S_{\mathrm{road}}^{4}$ consist of the single occupation table $\mathbf{o}_{d}^{(t)} \in S_{\mathrm{occ}}^{4}$, the set of detectors $\mathbf{L}_{\mathrm{det,d}} \in \mathcal{P}_{\mathrm{det}}(S_{\mathrm{det}}^{4})$, the set of tuning elements $\mathbf{L}_{\mathrm{tun,d}} \in \mathcal{P}_{\mathrm{tun}}(S_{\mathrm{tun}}^{4})$, and the set of switching areas $\mathbf{L}_{\mathrm{swarea,d}} \in \mathcal{P}_{\mathrm{swarea}}(S_{\mathrm{swarea}}^{4})$. Additionally, the elements of the road structure $\mathbf{s}_{d}^{(t)} \in S_{\mathrm{road}}^{4}$ contain the current time t, the size of the time discretisation Δt, the functions for the drive $\mathbf{F}_{\mathrm{drive}}$, and the lane change update $\mathbf{F}_{\mathrm{chlane}}$.

$$S_{\mathrm{road}}^{4} := \langle t, \Delta t, \mathbf{F}_{\mathrm{drive}}, \mathbf{F}_{\mathrm{chlane}}, \mathbf{o}_{d}^{(t)}, \mathbf{L}_{\mathrm{det,d}}, \mathbf{L}_{\mathrm{tun,d}}, \mathbf{L}_{\mathrm{swarea,d}} \rangle^{\mathrm{road}}. \qquad (3.140)$$

As has been the case in Equation 3.129 for the system specification, the update of a system component $\mathbf{s}_{d}^{(t)} \in S_{\mathrm{road}}^{4}$ starts with the output function as $\lambda(q) = \lambda(\mathbf{s}_{d}^{(t)}) := \{\mathbf{F}_{\mathrm{output}}(\mathbf{s}_{d}^{(t)} \# t, \mathbf{d}_{i}^{(t)}) \mid \mathbf{d}_{i}^{(t)} \in \mathbf{s}_{d}^{(t)} \# \mathbf{L}_{\mathrm{det,d}}\} \subset \mathcal{P}_{\mathrm{measure}}(S_{\mathrm{measure}}^{4})$ with $\mathbf{F}_{\mathrm{output}} : S_{\mathrm{road}}^{4} \longrightarrow \mathcal{P}_{\mathrm{measure}}(S_{\mathrm{measure}}^{4})$ from Equation 3.80. The measure operation follows. It is implemented by the measure function $\mathbf{F}_{\mathrm{measure}} : S_{\mathrm{road}}^{4} \times \mathcal{P}_{\mathrm{measure}}(S_{\mathrm{measure}}^{4}) \longrightarrow S_{\mathrm{road}}^{4}$ from Equation 3.141. The function definition in Equation 3.141 takes a component system state $\mathbf{s}_{d}^{(t)} \in S_{\mathrm{road}}^{4}$ and, for the context of OLSIMv4, the set of loop detector

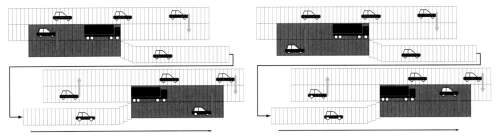

(a) Network state before the update. (b) Network state after the switching area update.

Figure 3.11: Isolated lane change update for the switching areas.

measurements $\mathfrak{M}^4_{\text{loop}*} \subset X_N$ as arguments and relies on the additional constraints $\mathbf{F}_{\text{rway}=?}(\mathbf{o}^{(t)}_r, \mathbf{d}^{(t)}_i) :=$ $\mathbf{d}^{(t)}_i \# \mathbf{t}^{\text{roadway}}_{\text{name}} = \mathbf{o}^{(t)}_r \# \mathbf{t}^{\text{roadway}}_{\text{name}}$ and $\mathbf{F}_{\text{elem}-\text{of}?}(\mathbf{s}^{(t)}_d, \mathbf{o}^{(t)}_r, \mathbf{d}^{(t)}_i) := \mathbf{d}^{(t)}_i \in \mathbf{s}^{(t)}_d \# \mathbf{L}_{\text{det},d} \wedge \mathbf{o}^{(t)}_r = \mathbf{s}^{(t)}_d \# \mathbf{o}^{(t)}_d$ for the occupation table $\mathbf{o}^{(t)}_r$, the detector $\mathbf{d}^{(t)}_i$, and the component system state $\mathbf{s}^{(t)}_d$. It generalizes the definition of the equally named function from Equation 3.82 of the detectors section by mapping it over each of the components' occupation table and its detectors.

$$\mathbf{F}_{\text{measure}}(\mathbf{s}^{(t)}_d, \mathfrak{M}^4_{\text{loop}*}) := \mathbf{F}_{\text{clone}}(\mathbf{s}^{(t)}_d, \mathbf{L}_{\text{det},d}, \mathbf{L}^*_{\text{det},d}) , \text{ with}$$
$$\mathbf{L}^*_{\text{det},d} = \{\mathbf{F}_{\text{measure}}(\mathbf{s}^{(t)}_d \# t, \mathbf{o}^{(t)}_r, \mathbf{d}^{(t)}_i) \mid \mathbf{F}_{\text{elem}-\text{of}?}(\mathbf{s}^{(t)}_d, \mathbf{o}^{(t)}_r, \mathbf{d}^{(t)}_i) \wedge \mathbf{F}_{\text{rway}=?}(\mathbf{o}^{(t)}_r, \mathbf{d}^{(t)}_i)\} \quad (3.141)$$

The next step of the component system update involves accumulating the detector measurements for the accumulators. Function $\mathbf{F}_{\text{accu}} : S^4_{\text{road}} \longrightarrow S^4_{\text{road}}$ from Equation 3.142 performs the accumulators' update. In addition to the current component system state $\mathbf{s}^{(t)}_d \in S^4_{\text{road}}$, the accumulation depends also on the subset $\mathbf{L}_{\text{det}} := \bigcup_{k \in D} \mathbf{s}^{(t)}_k \# \mathbf{L}_{\text{det},k}$ of all global detectors. The accumulation involves the detector subset $\mathbf{L}_{\text{det}}$ and the detector names $\mathbf{L}_{\text{dnames}}$ from Equation 3.142. The set contains all detectors that the component's accumulators reference by name. The special case where $\mathbf{L}_{\text{det}} = \mathbf{s}^{(t)}_d \# \mathbf{L}_{\text{det},d}$ results in a less complex accumulation operation. It does not involve detectors that are not part of the component system state. As it does not require synchronisation with other components, it provides, hence, a better parallelism than in the general case. The special case is likely to occur in the multicomponent model as the components depend already on the state of other connected components and the set of influencers I_d represents the latter already for component d. Alternatively, to force this special case, the detector measurements can be provided as part of the detectors input data and as discussed in the context of Equation 3.153 from Section 3.7.3.

$$\mathbf{F}_{\text{accu}}(\mathbf{s}^{(t)}_d, \mathbf{L}_{\text{det}}) := \mathbf{F}_{\text{clone}}(\mathbf{s}^{(t)}_d, \mathbf{L}_{\text{tun},d}, \mathbf{L}^*_{\text{tun},d}) , \text{ with}$$
$$\mathbf{L}^*_{\text{tun},d} = \{\mathbf{F}_{\text{acc2}}(\mathbf{s}^{(t)}_d \# t, \mathbf{F}_{\text{acc1}}(\mathbf{s}^{(t)}_d \# t, \mathbf{e}^{(t)}_i, \mathbf{L}_{\text{det}}), \mathbf{L}_{\text{det}}) \mid \mathbf{e}^{(t)}_i \in \mathbf{s}^{(t)}_d \# \mathbf{L}_{\text{tun},d}\} ,$$
$$\mathbf{L}_{\text{det}} = \{\mathbf{d}^{(t)}_j \in \bigcup_{k \in D} \mathbf{s}^{(t)}_k \# \mathbf{L}_{\text{det},k} \mid \mathbf{d}^{(t)}_j \# \mathbf{t}^{\text{detector}}_{\text{name}} \in \mathbf{L}_{\text{dnames}}\} ,$$
$$\mathbf{F}_{\text{acc1}}(t, \mathbf{e}^{(t)}_i, \mathbf{L}_{\text{det}}) = \mathbf{F}_{\text{clone}}(\mathbf{e}^{(t)}_i, \mathbf{a}^{(t)}_{\text{actual}}, \mathbf{F}_{\text{accumulate}}(t, \mathbf{e}^{(t)}_i \# \mathbf{a}^{(t)}_{\text{actual}}, \mathbf{L}_{\text{det}})) , \quad (3.142)$$
$$\mathbf{F}_{\text{acc2}}(t, \mathbf{e}^{(t)}_i, \mathbf{L}_{\text{det}}) = \mathbf{F}_{\text{clone}}(\mathbf{e}^{(t)}_i, \mathbf{a}^{(t)}_{\text{should}-\text{be}}, \mathbf{F}_{\text{accumulate}}(t, \mathbf{e}^{(t)} \# \mathbf{a}^{(t)}_{\text{should}-\text{be}}, \mathbf{L}_{\text{det}})) , \text{ and}$$
$$\mathbf{L}_{\text{dnames}} = \bigcup_{\mathbf{e}^{(t)}_i \in \mathbf{s}^{(t)}_d \# \mathbf{L}_{\text{tun},d}} \left((\mathbf{e}^{(t)}_i \# \mathbf{a}^{(t)}_{\text{actual}}) \# \mathbf{t}^{\text{detector}}_{\text{name}} \cup (\mathbf{e}^{(t)}_i \# \mathbf{a}^{(t)}_{\text{should}-\text{be}}) \# \mathbf{t}^{\text{detector}}_{\text{name}}\right) .$$

After the accumulate operation, the tune operation is applied to the component system. The tuning depends only on the current system state $\mathbf{s}^{(t)}_d \in S^4_{\text{road}}$. Function $\mathbf{F}_{\text{tune}} : S^4_{\text{road}} \longrightarrow S^4_{\text{road}}$ from Equation

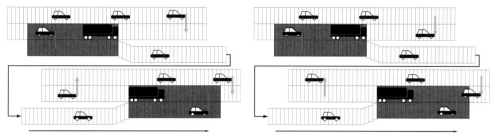

(a) Network state before the update.   (b) Network state after the isolated lane change update.

Figure 3.12: Isolated lane change update without the impacts of the switching area update.

3.143 performs the update of the tuning elements and of the occupation tables. The function definition in Equation 3.143 takes a system state $s_d^{(t)} \in S_{\text{road}}^4$ as argument and relies on the additional constraint $\mathbf{F}_{\text{rway}=?}(\mathbf{o}_r^{(t)}, \mathbf{e}_j^{(t)}) := \mathbf{e}_j^{(t)} \# \mathbf{t}_{\text{name}}^{\text{roadway}} = \mathbf{o}_r^{(t)} \# \mathbf{t}_{\text{name}}^{\text{roadway}}$ for the occupation table $\mathbf{o}_r^{(t)}$ and the tuning element $\mathbf{e}_j^{(t)}$. The tune update from Equation 3.143 applies the tune operation from Equation 3.124 to all tuning elements $\mathbf{s}^{(t)} \# \mathbf{L}_{\text{tun},d}$ on the same roadway as the occupation table $\mathbf{s}^{(t)} \# \mathbf{o}_d^{(t)}$ in the component system state. As has been the case for the system specification, applying the tune operation results in an update with recursive nature. As the component structure elements contain only a single occupation table, the tune update is less complex for the multicomponent model than for the system specification case.

$$\mathbf{F}_{\text{tune}}(\mathbf{s}_d^{(t)}) := \mathbf{F}_{\text{clone}}(\mathbf{F}_{\text{clone}}(\mathbf{s}_d^{(t)}, \mathbf{L}_{\text{tun}}, \mathbf{L}_{\text{tun}}^*), \mathbf{o}_d^{(t)}, \mathbf{o}_d^{(t*)}) , \text{ where}$$

$$(\mathbf{o}_d^{(t*)}, \mathbf{L}_{\text{tun}}^*) = \mathbf{F}_{\text{otune}}(t, \mathbf{s}^{(t)} \# \mathbf{o}_d^{(t)}, \mathbf{s}^{(t)} \# \mathbf{L}_{\text{tun},d}, \{\}) ,$$

$$\mathbf{F}_{\text{otune}}(t, \mathbf{o}_r^{(t)}, \mathbf{L}_{\text{tun}}, \mathbf{L}_{\text{tun}}') = \begin{cases} \mathbf{F}_{\text{otune}}(t, \mathbf{o}_r^{(t')}, \mathbf{L}_{\text{tun}} \backslash \{\mathbf{e}_j^{(t)}\}, \{\mathbf{e}_j^{(t')}\} \cup \mathbf{L}_{\text{tun}}') , \text{ with} \\ \quad (\mathbf{o}_r^{(t')}, \mathbf{e}_j^{(t')}) = \mathbf{F}_{\text{tune}}(t, \mathbf{o}_r^{(t)}, \mathbf{e}_j^{(t)}) , \text{ if } \exists \mathbf{e}_j^{(t)} \in \mathbf{L}_{\text{tun}} \mid \mathbf{F}_{\text{rway}=?}(\mathbf{o}_r^{(t)}, \mathbf{e}_j^{(t)}) , \\ (\mathbf{o}_r^{(t)}, \mathbf{L}_{\text{tun}}') , \text{ otherwise.} \end{cases}$$

(3.143)

The so far presented update steps did not respect the set of influencers I_d as the latter affect only the vehicle update steps for the case of the microscopic traffic simulation. The set of occupation tables $\mathbf{L}_{\text{occ}} \subset I_d$ as connected by the components' switching areas $\mathbf{s}_d^{(t)} \# \mathbf{L}_{\text{swarea},d}$ represents the set of influencers I_d. Equation 3.144 defines the set of occupation tables for component $\mathbf{s}_d^{(t)}$ that relies on a function $\mathbf{F}_{\text{neighbour?}} : S_{\text{occ}}^4 \times S_{\text{swarea}}^4 \times S_{\text{occ}}^4 \longrightarrow \mathfrak{B}_{\text{occ}}^4$ from Equation 3.145 which is used in the following.

$$\mathbf{L}_{\text{occ}}(d) = \bigcup_{\mathbf{a}_j \in \mathbf{s}_d^{(t)} \# \mathbf{L}_{\text{swarea},d}} \mathbf{L}_{\text{occ}}(d, \mathbf{a}_j) = \{\mathbf{o}_k^{(t)} \in \bigcup_{k \in D, k \neq d} \mathbf{s}_k^{(t)} \# \mathbf{o}_k^{(t)} \mid \mathbf{F}_{\text{neighbour?}}(\mathbf{s}_d^{(t)} \# \mathbf{o}_d^{(t)}, \mathbf{a}_j, \mathbf{o}_k^{(t)})\} , \text{ with}$$

(3.144)

$$\mathbf{F}_{\text{neighbour?}}(\mathbf{o}_d^{(t)}, \mathbf{a}_j, \mathbf{o}_k^{(t)}) := (\mathbf{a}_j \# \mathbf{f}_{\text{name}}^{\text{roadway}} = (\mathbf{o}_d^{(t)} \# \mathbf{w}^{(t)}) \# \mathbf{t}_{\text{name}}^{\text{roadway}} \wedge \mathbf{a}_j \# \mathbf{t}_{\text{name}}^{\text{roadway}} = (\mathbf{o}_k^{(t)} \# \mathbf{w}^{(t)}) \# \mathbf{t}_{\text{name}}^{\text{roadway}})$$
$$\vee (\mathbf{a}_j \# \mathbf{t}_{\text{name}}^{\text{roadway}} = (\mathbf{o}_d^{(t)} \# \mathbf{w}^{(t)}) \# \mathbf{t}_{\text{name}}^{\text{roadway}} \wedge \mathbf{a}_j \# \mathbf{f}_{\text{name}}^{\text{roadway}} = (\mathbf{o}_k^{(t)} \# \mathbf{w}^{(t)}) \# \mathbf{t}_{\text{name}}^{\text{roadway}}) .$$

(3.145)

As the vehicle update of a component $\mathbf{s}_d^{(t)}$ depends on the state of the connected components, it requires an intermediate step to update the vehicles on the areas affected by the connected components. The intermediate update steps involve updating the switching areas of the occupation tables before each of the two update steps, namely the drive update and the lane change update step. The functions for the intermediate update steps employ the vehicle update function for the occupation tables as, for example,

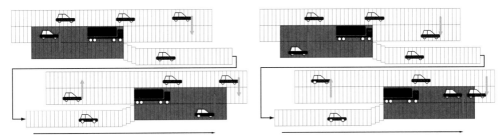

(a) Network state before the update.   (b) Network state after the completed lane change update.

Figure 3.13: Complete lane change update that reflects the intermediate update of the switching areas as well as the lane change update.

the drive update function of Equation 3.33 from Section 3.3.5. To make use of the latter, they instantiate a fresh occupation table with the dimensions of the switching area in each time step. Equation 3.35 provided a function that instantiates an area structure element with the appropriate dimensions. The switching area update function instantiates the fresh occupation table with the vehicles of the two connected occupation tables and the random number generator state as provided by the switching area element.

$$\mathbf{F}_{\text{swarea}-\text{drive}}(\mathbf{s}_d^{(t)}, \mathbf{L}_{\text{occ}}) := (\mathbf{s}_d^{(t*)}, \mathbf{L}_{\text{veh}}^*) = (\mathbf{F}_{\text{clone}}(\mathbf{s}_d^{(t)}, \mathbf{L}_{\text{swarea}}, \mathbf{L}_{\text{swarea}}'), \mathbf{L}_{\text{veh}}') \text{ with}$$

$$(\mathbf{L}_{\text{swarea}}', \mathbf{L}_{\text{veh}}') = \mathbf{F}_{\text{swdrive}}(\mathbf{s}_d^{(t)} \# t, \mathbf{s}_d^{(t)} \# \mathbf{o}_d^{(t)}, \{\}, \mathbf{s}_d^{(t)} \# \mathbf{L}_{\text{swarea},d}, \{\}),$$

$$\mathbf{F}_{\text{swdrive}}(t, \mathbf{o}_d^{(t)}, \mathbf{L}_{\text{v}}, \mathbf{L}_{\text{s}}, \mathbf{L}_{\text{s}}') = \begin{cases} \mathbf{F}_{\text{swdrive}}(t, \mathbf{o}_d^{(t)}, \mathbf{L}_{\text{v}} \cup \mathbf{L}_{\text{v}}', \mathbf{L}_{\text{s}} \backslash \{\mathbf{a}_{\text{j}}\}, \{\mathbf{a'}_{\text{j}}\} \cup \mathbf{L}_{\text{s}}'), \text{ with} \\ \quad (\mathbf{a'}_{\text{j}}, \mathbf{L}_{\text{v}}') = \mathbf{F}_{\text{swa}-\text{drive}}(t, \mathbf{o}_d^{(t)}, \mathbf{a}_{\text{j}}, \mathbf{o}_k^{(t)}), \text{ if} \\ \quad \exists \mathbf{a}_{\text{j}} \in \mathbf{L}_{\text{s}}, \mathbf{o}_k^{(t)} \in \mathbf{L}_{\text{occ}} \mid \mathbf{F}_{\text{neighbour?}}(\mathbf{o}_d^{(t)}, \mathbf{a}_{\text{j}}, \mathbf{o}_k^{(t)}), \\ (\mathbf{L}_{\text{s}}', \mathbf{L}_{\text{v}}), \text{ otherwise.} \end{cases} \quad (3.146)$$

After the vehicle update for the freshly instantiated occupation table, the switching area update merges the vehicles on the component's part of the occupation table back into the originating occupation table or, depending on the accuracy of the implementation, holds them for merging them with the vehicles after the vehicle update step. The switching area update function $\mathbf{F}_{\text{swdrive}} : S_{\text{road}}^4 \times \mathcal{P}_{\text{occ}}(S_{\text{occ}}^4) \longrightarrow S_{\text{road}}^4 \times \mathcal{P}_{\text{veh}}(S_{\text{veh}}^4)$ results then in a pair consisting of the updated component state and the set of updated vehicles. Equation 3.146 provides a definition for Function $\mathbf{F}_{\text{swdrive}}$. Figure 3.11 illustrates this intermediate update step for the lateral motion case. This step takes place in parallel to the isolated lane change update.

As a component's switching areas do not overlap, the switching area update can be extended as in Function $\mathbf{F}_{\text{swdrive}}$ to process a component's entire set of switching areas. The processing results then in a set of vehicles that contains all updated vehicles of all the component's switching areas. The subsequent vehicle update steps will then respect the already updated vehicles. After the switching area updates, the component state reflects the updated random number generator states of the switching areas. However, it does not reflect any updates on the component's occupation table. To keep the random number generator states synchronous to the corresponding ones of the connected components, the vehicle updates on the joined occupation tables should be performed in a global order such as always perform the lane change update in the order of the lane numbers. After the switching area update step, the update continues with the vehicle update step. Instead of the general occupation table vehicle update from Equation 3.33, the vehicle update step requires, however, to use the variants $\mathbf{F}_{\text{drive}} : \mathcal{T}_{\text{secs}} \times S_{\text{occ}}^4 \times \mathcal{P}_{\text{swarea}}(S_{\text{swarea}}^4) \longrightarrow S_{\text{occ}}^4$ of the vehicle update functions as introduced at the end of Subsection 3.3.5. The latter are used to exclude vehicles on switching areas from the general vehicle update as the switching area update function updated them already. Equation 3.147 demonstrates the use of the variant for the drive update. The latter merges the vehicles provided by the switching area update with the updated vehicles from the drive update in a final step. The procedure for the lane change update is analogous to the one of the drive update. An implementation should, however, take care about potential conflicts that may arise from lane changes in scenarios with more than two lanes. Additionally, there may be simulation scenarios where each of the vehicle updates requires its own set of coupled occupation tables. This may be the case when vehicles switch roadways first via the lane change update and second via the drive update in the same time step. Figure 3.12 illustrates the isolated lane change update. The latter operates on the same initial state as the intermediate switching area update step as illustrated by Figure 3.11. The final state, finally, results from the merging of both the updates. Figure 3.13 illustrates the states before and after the update.

$$
\begin{aligned}
\mathbf{F}_{\text{drive}}(\mathbf{s}_d^{(t)}, \mathbf{L}_{\text{occ}}) &= \mathbf{F}_{\text{sw-drive}}(\mathbf{s}_d^{(t*)}, \mathbf{L}_{\text{veh}}^*) \text{ , with } (\mathbf{s}_d^{(t*)}, \mathbf{L}_{\text{veh}}^*) = \mathbf{F}_{\text{swarea-drive}}(\mathbf{s}_d^{(t)}) \text{ ,} \\
\mathbf{F}_{\text{sw-drive}}(\mathbf{s}_d^{(t)}, \mathbf{L}_{\text{veh}}) &:= \mathbf{F}_{\text{clone}}(\mathbf{s}_d^{(t)}, \mathbf{o}_d^{(t)}, \mathbf{o}_d^{(t*)}) \text{ ,} \\
\mathbf{o}_d^{(t*)} &= \mathbf{F}_{\text{clone}}(\mathbf{F}_{\text{clone}}(\mathbf{o}_d^{(t)}, \mathbf{k}_d^{(t)}, \mathbf{L}_{\text{veh}} \cup \mathbf{o}_d^{(t')} \# \mathbf{k}_d^{(t)}), s_{\text{rand}}, \mathbf{o}_d^{(t')} \# s_{\text{rand}}) \text{ , and} \\
\mathbf{o}_d^{(t')} &= \mathbf{s}_d^{(t)} \# \mathbf{F}_{\text{drive}}(\mathbf{s}_d^{(t)} \# t, \mathbf{s}_d^{(t)} \# \mathbf{o}_d^{(t)}, \mathbf{s}_d^{(t)} \# \mathbf{L}_{\text{swarea,d}})
\end{aligned}
\tag{3.147}
$$

The component's update operation ends with the time update function $\mathbf{F}_{\text{inc-t}} : S_{\text{road}}^4 \longrightarrow S_{\text{road}}^4$ with the definition from Equation 3.148.

$$
\mathbf{F}_{\text{inc-t}} := \mathbf{F}_{\text{clone}}(\mathbf{s}_d^{(t)}, t, t + \Delta t)
\tag{3.148}
$$

The system update finally results in a function $\mathbf{F}_{\text{update}} : S_{\text{road}}^4 \times \mathcal{P}_{\text{measure}}(S_{\text{measure}}^4) \times \mathcal{P}_{\text{det}}(S_{\text{det}}^4) \times \mathcal{P}_{\text{occ}}(S_{\text{occ}}^4) \longrightarrow S_{\text{road}}^4$ with the definition from Equation 3.149. The resulting component update function applies first the measure function to the initial state $\mathbf{s}_d^{(t)}$ and the set of measurements $\mathfrak{M}_{\text{loop}*}^4$. The accumulate function continues updating the state in combination with the set of detectors $\mathbf{L}_{\text{det}}$ from Equation 3.142 that may also contain detectors from other components. The tune function takes over the updated state from the accumulation and passes its update to the next update steps which are the vehicle update functions for the drive and the lane change update. The vehicle update functions operate on the component state but depend also on the set of connected occupation tables $\mathbf{L}_{\text{occ}}(d) \subset I_d$ from Equation 3.144, namely the set of influencers I_d. Finally, the time update function advances the time by one time step.

$$
\begin{aligned}
\mathbf{F}_{\text{update}}(\mathbf{s}_d^{(t)}, \mathfrak{M}_{\text{loop}*}^4, \mathbf{L}_{\text{det}}, \mathbf{L}_{\text{occ}}) := \\
\mathbf{F}_{\text{inc-t}} \circ \mathbf{F}_{\text{drive}}(\mathbf{F}_{\text{chlane}}(\mathbf{F}_{\text{tune}} \circ \mathbf{F}_{\text{accu}}(\mathbf{F}_{\text{measure}}(\mathbf{s}_d^{(t)}, \mathfrak{M}_{\text{loop}*}^4), \mathbf{L}_{\text{det}}), \mathbf{L}_{\text{occ}}), \mathbf{L}_{\text{occ}}) \text{ .}
\end{aligned}
\tag{3.149}
$$

With the update function $\mathbf{F}_{\text{update}} : S_{\text{road}}^4 \times \mathcal{P}_{\text{measure}}(S_{\text{measure}}^4) \longrightarrow S_{\text{road}}^4$ from Equation 3.149, the definition of the multicomponent model state transition function can be defined as $\delta(q, q_1, \ldots, q_n, x) := \mathbf{F}_{\text{update}}(\mathbf{s}_d^{(t)}, \mathfrak{M}_{\text{loop}*}^4, \mathbf{L}_{\text{det}}, \mathbf{L}_{\text{occ}})$. Thereby, the set of detectors $\mathbf{L}_{\text{det}}$ and the set of occupation tables $\mathbf{L}_{\text{occ}}$ can be derived in each time step from the set of component states $q_1, \ldots, q_n$. Table 3.7.2 summarises the relevant functions, structure definitions, variables, and constants of this section.

3.7.3 Network specification

The multicomponent discrete time system specification of the previous section provided the state transition functions δ_d of Equation 3.137 by Function $\mathbf{F}_{\text{update}}$ from Equation 3.149 for the components M_d. Their consecutive states $\mathbf{s}_d^{(t)}, \mathbf{s}_d^{(t+\Delta t)}, \ldots$ depend on the states of the connected components $Q_i := \mathbf{s}_i^{(t)}$ with $i \in I_d$. In particular, the accumulate step and the vehicle update steps depend on the states of the connected components. The definition of $\mathbf{F}_{\text{update}}$ reflects the dependency by passing the set of detectors and the set of occupation tables as arguments to the accumulate function and the vehicle update function, respectively. In an implementation, this dependency requires to read the state of the connected components in each update step. Reading the connected components' states, however, requires synchronisation in a parallel interpretation. Thus, it is desirable to avoid this limiting dependency. In the context of OLSIMv4, the on- and off-ramps introduce this dependency. Additionally and as another aspect that questions the gain of modelling on- and off-ramps for a microscopic traffic simulation such as OLSIMv4, the number of vehicles that the traffic model supports to enter the main roadway via on-ramps is relatively low compared with realistic values. This has been discussed already in sections 2.1.4 and 2.2.5. Considering the fact, that modelling on- and off-ramps leads to less parallelisation, the more promising solution is not to consider them and rely only on the tuning of the network. The tuning elements, however, may require detector data originating from other components. The multicomponent model of the previous section, therefore, made the accumulate operation depend on other components' detectors. The dependency, however, involves an unnecessary synchronisation as the detector data can also be made available for the tuning elements as part of the measure operation. The output function $\mathbf{F}_{\text{output}} : S^4_{\text{net}} \longrightarrow \mathcal{P}^4_{\text{measure}}(S^4_{\text{measure}})$ of the simulation provides the detector data, thus, as part of the input detector data.

Without the dependency on other components' states and without the unnecessary synchronisation during the accumulate operation, the state transition function simplifies significantly. As a consequence, the system becomes subject to the the next higher level of system specification which is the Network of System Specification Formalism [216, Sec. 5.10]. The specification couples components of input/output systems modularly by connecting their input and output interfaces [216, Sec. 5.10]. The specification is also called a coupled system specification or network of system specifications [216, Sec. 5.10] and it is introduced in the following. The components are themselves systems with their own input and output interfaces [216, Sec. 5.10]. Consequently, the coupled systems for the case of the microscopic traffic simulations have a parallel interpretation with only one synchronisation per state transition.

$$S^4_{\text{net}} = N = \langle T, X_{\text{N}}, Y_{\text{N}}, D, \{M_{\text{d}} \mid d \in D\}, \{I_{\text{d}} \mid d \in D \cup \{N\}\}, \{Z_{\text{d}} \mid d \in D \cup \{N\}\} \rangle^{\text{net}} . \quad (3.150)$$

Equation 3.150 defines the coupled system specification structure S^4_{net}. It consists of the components' time base T, the set of inputs of the network X_{N}, the set of outputs of the network Y_{N}, and the set of component references D. Its further components are the interface map

$$Z_{\text{d}} : \underset{i \in I_{\text{d}}}{\times} YX_{\text{i}} \longrightarrow XY_{\text{d}}$$

for the cases from Equations 3.151 and 3.152, the I/O systems $M_{\text{d}} = \langle T, X_{\text{d}}, Y_{\text{d}}, \Omega, Q, \Delta, \Lambda \rangle^{\text{sys}}$ with $\forall d \in D$, and the set of influencers $I_{\text{d}} \subseteq D \cup N$. Figure 3.14 illustrates an example of a coupled system specification.

$$YX_{\text{i}} = \begin{cases} X_{\text{i}} & \text{, if } i = N , \\ Y_{\text{i}} & \text{, if } i \neq N , \end{cases} \quad (3.151)$$

$$XY_{\text{d}} = \begin{cases} X_{\text{d}} & \text{, if } d = N , \\ Y_{\text{d}} & \text{, if } d \neq N . \end{cases} \quad (3.152)$$

A first approach in mapping this chapter's microscopic traffic simulation onto the Network of System Specification Formalism exists with reusing the definition of Equation 3.140 from the Subsection 3.7.2 and the state transition function from Equation 3.149 with an empty set of switching areas $s_d^{(t)} \# \mathbf{L}_{\text{swarea}} = \{\}$ and the additional limitation that the accumulators may only refer to detectors of the same component $s_d^{(t)}$. The empty set of switching areas results in an likewise empty set of the component's influencers represented by the dependent occupation tables $\mathbf{L}_{\text{occ}}(s_d^{(t)}) = \{\}$ from Equation 3.144. Consequently, the vehicle update operations no longer require synchronisation. According to Equation 3.142, The limitation for the accumulators results in a set of detectors that equals or is a subset of the component's set of detectors $\mathbf{L}_{\text{det}} \subseteq s_d^{(t)} \# \mathbf{L}_{\text{det,d}}$. Consequently, the accumulate operation does no longer require synchronisation. The main consequence from freeing the vehicle update of the neighbour component's state results, however, in a simulation scenario where the detectors have to provide the data that reflects the dependency among the components for the roadways. As the accumulators

Table 3.7.3: Summary of relevant functions, structures, variables, and constants for Section 3.7.3.

definition	description	domain	range
$\langle a, b, c, d, e, f, g \rangle^{\mathrm{rdsys}}$	network component	$(\mathfrak{T}_{\mathrm{secs}}^4)^2 \times (\mathbf{F})^2 \times S_{\mathrm{occ}}^4$ $\times \mathcal{P}_{\mathrm{det}}^4(S_{\mathrm{det}}^4) \times \mathcal{P}_{\mathrm{tun}}^4(S_{\mathrm{tun}}^4)$	S_{rdsys}^4, Eq. 3.154
$\mathbf{F}_{\mathrm{output}}(s)$	output function	S_{rdsys}^4	$\mathcal{P}_{\mathrm{measure}}^4(S_{\mathrm{measure}}^4)$
$\mathbf{F}_{\mathrm{measure}}(s, m)$	measure function	$S_{\mathrm{rdsys}}^4 \times \mathcal{P}_{\mathrm{measure}}^4(S_{\mathrm{measure}}^4)$	S_{rdsys}^4
$\mathbf{F}_{\mathrm{accu}}(s)$	accumulate function	S_{rdsys}^4	S_{rdsys}^4, Eq. 3.155
$\mathbf{F}_{\mathrm{tune}}(s)$	tune function	S_{rdsys}^4	S_{rdsys}^4
$\mathbf{F}_{\mathrm{drive}}(s)$	drive update	S_{rdsys}^4	S_{rdsys}^4, Eq. 3.156
$\mathbf{F}_{\mathrm{chlane}}(s)$	lane change update	S_{rdsys}^4	S_{rdsys}^4, Eq. 3.156
$\mathbf{F}_{\mathrm{inc-t}}(s)$	increase time	S_{sys}^4	S_{sys}^4, Eq. 3.157
$\mathbf{F}_{\mathrm{update}}(s, m)$	state transition	$S_{\mathrm{rdsys}}^4 \times \mathcal{P}_{\mathrm{measure}}^4(S_{\mathrm{measure}}^4)$	S_{rdsys}^4, Eq. 3.158

with the additional constraint no longer may depend on external detectors, the question arises how this can be achieved.

The solution is to provide the detector data as part of the input data to the measure operation. The latter differs from the previously used data providing as part of the accumulate operation in the following way. While the measurements provided to the accumulate operation are considered to be delayless, the measurements provided to the measure operation have at least one cycle delay. The exact delay depends on the detector type specific value of N_{aggvals} which is by default 60 for the OLSIMv4 local and global detectors. The additional constraint for the accumulate operation is not a big restriction because of the following reasons. Firstly, no detector types have been introduced that operate with a value of $N_{\mathrm{aggvals}} < 1$. Neither do the tuning elements rely on detector values other than the aggregated values. In fact, by now the accumulators already rely on the delay as defined by N_{aggvals}. Even in the case where this delay should be too restrictive, the following restriction should hold for all simulation scenarios. While the actual accumulators of the tuning elements may require delayless detectors, the should-be accumulators do not as they are always at least one cycle ahead.

The measure operation can, thus, input the detector measurements for the should-be accumulators. With the empty set of the component's influencers represented by the dependent occupation tables $\mathbf{L}_{\mathrm{occ}}(\mathbf{s}_d^{(t)}) = \{\}$ and the set of detectors $\mathbf{L}_{\mathrm{det}} \subseteq \mathbf{s}_d^{(t)} \# \mathbf{L}_{\mathrm{det,d}}$ for the accumulate operation, the update function simplifies significantly. The update function does no longer depend on other components and does no longer require synchronisation. The update function for the road structure elements from the previous section could, thus, be written as in Equation 3.153. The intermediate system state $\mathbf{s}_d^{(t*)}$ therein reflects the results of the measure operation already, i.e. $\mathbf{s}_d^{(t*)} := \mathbf{F}_{\mathrm{measure}}(\mathbf{s}_d^{(t)}, \mathbf{F}_{\mathrm{output}}(\mathbf{s}_d^{(t)}) \cup \mathfrak{M}_{\mathrm{loop*}}^4)$.

$$\mathbf{F}_{\mathrm{update}}(\mathbf{s}_d^{(t)}, \mathfrak{M}_{\mathrm{loop*}}^4) :=$$
$$\mathbf{F}_{\mathrm{inc-t}} \circ \mathbf{F}_{\mathrm{drive}}(\mathbf{F}_{\mathrm{chlane}}(\mathbf{F}_{\mathrm{tune}} \circ \mathbf{F}_{\mathrm{accu}}(\mathbf{F}_{\mathrm{measure}}(\mathbf{s}_d^{(t)}, \mathbf{F}_{\mathrm{output}}(\mathbf{s}_d^{(t)}) \cup \mathfrak{M}_{\mathrm{loop*}}^4), \mathbf{s}_d^{(t*)} \# \mathbf{L}_{\mathrm{det,d}}), \{\}), \{\}) \,.$$
$$(3.153)$$

However, this simplification is not the only one. As the component system state definition got rid of the switching areas, the system state specification also simplifies significantly. Mapping this chapter's microscopic traffic simulation onto the Network of System Specification Formalism requires then first to specify the OLSIMv4 components as input/output systems or even discrete time system specifications. Section 3.7.1 demonstrated this already for the whole network using the S_{sys}^4 structure from Equation 3.128. The structure has a component $\mathbf{L}_{\mathrm{occ}}$ for the set of occupation tables. As has been the case for the multicomponent model, in the special case of the network of system specification this set reduces to a single occupation table element. Equation 3.154 specifies the set of system states

$\mathbf{s}_d^{(t)}, \mathbf{s}_d^{(t+\Delta t)}, \ldots S_{\text{rdsys}}^4 \subseteq Q_d$ for this special case.

$$S_{\text{rdsys}}^4 = \langle t, \Delta t, \mathbf{F}_{\text{drive}}, \mathbf{F}_{\text{chlane}}, \mathbf{o}_d^{(t)}, \mathbf{L}_{\text{det,d}}, \mathbf{L}_{\text{tun,d}} \rangle^{\text{rdsys}}. \tag{3.154}$$

While it is sufficient to simply extend the output and the measure function definition onto the new domain kind S_{rdsys}^4, the accumulate function requires adjusting. The latter no longer depends on the set of component-external detectors. Function $\mathbf{F}_{\text{accu}} : S_{\text{rdsys}}^4 \longrightarrow S_{\text{rdsys}}^4$ from Equation 3.155 performs the accumulators' update. The accumulation depends only on the current component state $\mathbf{s}_d^{(t)} \in S_{\text{rdsys}}^4$.

$$
\begin{aligned}
\mathbf{F}_{\text{accu}}(\mathbf{s}_d^{(t)}) :=& \mathbf{F}_{\text{clone}}(\mathbf{s}_d^{(t)}, \mathbf{L}_{\text{tun,d}}, \mathbf{L}_{\text{tun,d}}^*), \text{ with} \\
\mathbf{L}_{\text{tun,d}}^* =& \{ \mathbf{F}_{\text{acc2}}(\mathbf{s}_d^{(t)} \# t, \mathbf{F}_{\text{acc1}}(\mathbf{s}_d^{(t)} \# t, \mathbf{e}_i^{(t)}, \mathbf{s}_d^{(t)} \# \mathbf{L}_{\text{det,d}}), \mathbf{s}_d^{(t)} \# \mathbf{L}_{\text{det,d}}) \mid \mathbf{e}_i^{(t)} \in \mathbf{s}_d^{(t)} \# \mathbf{L}_{\text{tun,d}} \}, \\
\mathbf{F}_{\text{acc1}}(t, \mathbf{e}_i^{(t)}, \mathbf{L}_{\text{det}}) =& \mathbf{F}_{\text{clone}}(\mathbf{e}_i^{(t)}, \mathbf{a}_{\text{actual}}^{(t)}, \mathbf{F}_{\text{accumulate}}(t, \mathbf{e}^{(t)} \# \mathbf{a}_{\text{actual}}^{(t)}, \mathbf{L}_{\text{det}})), \text{ and} \\
\mathbf{F}_{\text{acc2}}(t, \mathbf{e}_i^{(t)}, \mathbf{L}_{\text{det}}) =& \mathbf{F}_{\text{clone}}(\mathbf{e}_i^{(t)}, \mathbf{a}_{\text{should}-\text{be}}^{(t)}, \mathbf{F}_{\text{accumulate}}(t, \mathbf{e}^{(t)} \# \mathbf{a}_{\text{should}-\text{be}}^{(t)}, \mathbf{L}_{\text{det}})).
\end{aligned}
\tag{3.155}
$$

For the tune operation it is also sufficient to extend it onto the new domain kind S_{rdsys}^4. The vehicle update functions, however, require adjustment as they also become much less complex in the Network of System Specification. Because the vehicle update does no longer depend on other components' states, the intermediate step for the switching areas in the multicomponent model from the previous section is superfluous. The vehicle update functions can directly use the general occupation table vehicle update $\mathbf{F}_{\text{drive}} : \mathfrak{T}_{\text{secs}} \times S_{\text{occ}}^4 \times \longrightarrow S_{\text{occ}}^4$ from Equation 3.33. The lane change function $\mathbf{F}_{\text{drive}} : \mathfrak{T}_{\text{secs}} \times S_{\text{occ}}^4 \times \longrightarrow S_{\text{occ}}^4$ is defined analogously.

$$\mathbf{F}_{\text{drive}}(\mathbf{s}_d^{(t)}) = \mathbf{F}_{\text{clone}}(\mathbf{s}_d^{(t)}, \mathbf{o}_d^{(t)}, \mathbf{s}_d^{(t)} \# \mathbf{F}_{\text{drive}}(\mathbf{s}_d^{(t)} \# t, \mathbf{s}_d^{(t)} \# \mathbf{o}_d^{(t)})) \tag{3.156}$$

The network's update operation ends with the time update function $\mathbf{F}_{\text{inc}-\text{t}} : S_{\text{rd}}^4 \longrightarrow S_{\text{rd}}^4$ with the definition from Equation 3.157.

$$\mathbf{F}_{\text{inc}-\text{t}} := \mathbf{F}_{\text{clone}}(\mathbf{s}_d^{(t)}, t, t + \Delta t) \tag{3.157}$$

The update function which also matches the state transition function for the network of systems specification, finally, results in the definition from Equation 3.158. Depending on the number of aggregate values N_{aggvals} which is 60 by default for OLSIMv4, the output function evaluation might not be necessary in each time step. In this case, the synchronisation effort drastically reduces to the fraction $1/N_{\text{aggvals}}$. Table 3.7.3 summarises the relevant functions, structure definitions, variables, and constants of this section.

$$
\begin{aligned}
\mathbf{F}_{\text{update}}(\mathbf{s}_d^{(t)}, \mathfrak{M}_{\text{loop}*}^4) :=& \\
& \mathbf{F}_{\text{inc}-\text{t}} \circ \mathbf{F}_{\text{drive}} \circ \mathbf{F}_{\text{chlane}} \circ \mathbf{F}_{\text{tune}} \circ \mathbf{F}_{\text{accu}} \circ \mathbf{F}_{\text{measure}}(\mathbf{s}_d^{(t)}, \mathbf{F}_{\text{output}}(\mathbf{s}_d^{(t)}) \cup \mathfrak{M}_{\text{loop}*}^4).
\end{aligned}
\tag{3.158}
$$

With $X_d \subset \mathcal{P}_{\text{measure}}(S_{\text{measure}}^4)$ as the set of inputs, $Y_d \subset \mathcal{P}_{\text{measure}}(S_{\text{measure}}^4)$ as the set of outputs, $\delta_d(q, x) := \mathbf{F}_{\text{update}}(\mathbf{s}_d^{(t)}, \mathfrak{M}_{\text{loop}*}^4)$ from Equation 3.158 as the component state transition function, $c := \Delta t$, and $\lambda_d(q) := \mathbf{F}_{\text{output}}(\mathbf{s}_d^{(t)})$ as the output function, the discrete time system specification formalism S_{DTSS} from Equation 3.125 applies and is complete. The formalism specifies a structure that matches those of the input/output systems. Thus, the component M_d as specified by the DTSS formalism $S_{\text{DTSS}} = \langle \Delta t, X_d, Y_d, Q_d, \delta_d, \lambda_d \rangle^{\text{DTSS}}$ is a valid input/output system specification. The Discrete Time System Network (DTSN) Specification Formalism lacks then only the specification of the set of influencers I_d and those of the interface mappings Z_d. The sets $I_d = \{d\}$ and $Z_d = \{d\}$ are the configuration for OLSIMv4 and provide the simplest way to complete the DTSN specification. Depending on the density of the detector network and the detector positions, this simulation setup might already provide sufficiently accurate results. A more complex solution for the first part of the lacking specification exists with the switching areas that provide the information for the set of influencers. In this solution the set of influencers is chosen as the set containing every system component with a detector referenced by one or more of its accumulators, i.e. $I_\mathrm{d} := \{d\} \cup \{k \mid \mathbf{s}_k^{(t)} \in D \land \mathbf{L}_{\mathrm{dnames},k} \cap \mathbf{L}_{\mathrm{dnames},\mathrm{d}} \neq \varnothing\}$ with $\mathbf{L}_{\mathrm{dnames},k}$ and $\mathbf{L}_{\mathrm{dnames},\mathrm{d}}$ from Equation 3.159. The set of interface mappings equals then to the set

$$Z_\mathrm{d} = \{ \underset{i \in I_\mathrm{d}}{\times} Y_\mathrm{i} \longrightarrow X_\mathrm{d} \} \; .$$

$$
\begin{aligned}
\mathbf{L}_{\mathrm{dnames},\mathrm{d}} &= \bigcup_{\mathbf{e}_i^{(t)} \in \mathbf{s}_\mathrm{d}^{(t)} \# \mathbf{L}_{\mathrm{tun},\mathrm{d}}} \left((\mathbf{e}_i^{(t)} \# \mathbf{a}_{\mathrm{actual}}^{(t)}) \# \mathbf{t}_{\mathrm{name}}^{\mathrm{detector}} \cup (\mathbf{e}_i^{(t)} \# \mathbf{a}_{\mathrm{should-be}}^{(t)}) \# \mathbf{t}_{\mathrm{name}}^{\mathrm{detector}} \right) \\
\mathbf{L}_{\mathrm{dnames},k} &= \bigcup_{\mathbf{d}_i^{(t)} \in \mathbf{s}_k^{(t)} \# \mathbf{L}_{\mathrm{det},k}} \mathbf{d}_i^{(t)} \# \mathbf{t}_{\mathrm{name}}^{\mathrm{detector}}
\end{aligned}
\tag{3.159}
$$

The practical implication of this solution is to place a detector and a tuning element on each part of a switching area and let the tuning elements remove and insert the vehicles according to the balances of their actual and should-be accumulator values. In principle, two ways exist to implement such a switching area constellation. In the first scenario, the loop detectors are already placed appropriately—one in the near of the 'from' part of the switching area, the other one in the near of the 'to' part of the switching area. In that case, it is sufficient to tune the traffic flow onto the flow values of the corresponding loop detector. In the second scenario, a tuning element, an accumulator, and a detector are placed on each part of the switching area. The tuning element's and the accumulator's specific type depend on whether the connection is an on- or an off-ramp or an interchanging area. As an example, in the case of an on-ramp, the actual accumulator on the main roadway accumulates only the traffic flow values of the right most single-lane while the should-be accumulator accumulates the traffic flow values with those from the detector of the on-ramp. In general, the should-be accumulator traffic flow values are higher than the actual values. As a consequence, the tuning element on the main roadway will insert the number of vehicles from the balance.

3.7.4 Dynamic network

The Discrete Time System Network Specification focuses on static networks only. As road traffic networks may vary over time for a lot of reasons it is, thus, questionable whether the specification reflects

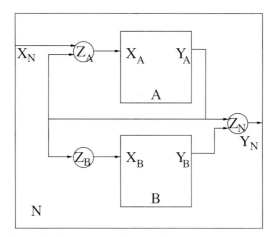

Figure 3.14: Example of a coupled system specification from [216, Fig. 16] with the set of influencers $I_A = \{N, A\}$, $I_B = \{A\}$, and $I_N = \{A, B\}$.

the dynamic character of the network appropriately. The question will, thus, be examined in the following. Road traffic networks may vary in structure and size over time because of several reasons. Lane closings and mergings, speed limits, and deviations due to road-works are a commonly known phenomena. Less known but also existent in some German states are the dynamic single-lanes. Normally, the right outer lane on German highways is reserved for vehicles with technical damage. However, in some German states, the traffic control centre may temporarily open the right outer most lane in high traffic flow situations. Police regulations to solve arising of unexpected traffic accidents may also affect the network in form of lane closings and mergings, speed limits, and deviations. As another source of changes in the traffic network, extreme weather situations such as massive snow may lead to spontaneous lane usage where vehicles drive in between of two single-lanes. Subsection 5.1.2 describes such a scenario. Last but not least, the dynamic traffic signs also have a dynamic impact on the traffic network.

Section 2.2.4 and Subsection 2.1.3 emphasised the dynamic character of the OLSIMv4 road traffic network and the importance for microscopic traffic simulations to adapt to a dynamic network already. The previous sections 3.7.1, 3.7.2, and 3.7.3 presented three different specification approaches. Among them, the network of systems specification provides the greatest benefits for the needs of the traffic simulation. It is, thus, the most important one in the context of OLSIMv4. However, the discrete time system specification applies to OLSIMv3 and probably a lot of other widely used microscopic traffic simulations systems. Alternatively and in the best case only, the multicomponent discrete time system specification applies. It requires a very careful system modelling and, thus, it is questionable whether any microscopic traffic simulation system has implemented it already.

Theoretically, the Dynamic Structure Discrete Time System Specification [18, Sec. 4.1] from the Modeling Formalisms for Dynamic Structure Systems publication applies for the dynamic network specification of OLSIMv4 as introduced in Section 3.7.3. The publication specifies a formalism for dynamic structure system networks, i.e. system networks with a structure that may change over time. Dynamic structures vary in the connections among elements of themselves over time. They vary from the coupled systems of the previous section in the form that they contain their coupling information as part of the system state. Thus, the state transition functions can change the coupling. For the purpose of this chapter's microscopic traffic simulations, the coupling will only change from an existing coupling to a closed or no longer existing coupling. Once a coupling between two components exists, it is possible to model it as closed without touching the coupling. Consequently, as the physical connections between the roadways of OLSIMv4 do not change over time, using this approach for OLSIMv4 is not necessary. As long as the state of a specification contains the network structure, the state "transition functions can change this state and, in consequence, change the structure of the network" [18, Sec. 3]. As the network model of OLSIMv4 provides the traffic rules in cells, the rules can vary over time. The rule variations can adapt to the required scenarios such as speed limits, lane change and overtaking restrictions, and lane closings. Thus, the dynamic structure system network specification is not necessary for the context of OLSIMv4.

3.7.5 Further decompositions

An even finer decomposition may result from modelling the vehicles inside of a roadway as a network of modular and autonomous systems. A specially crafted discrete event simulation could then update the components of the network, i.e. the vehicles. If the events pass the vehicle chain in the direction of travel, the result would be identical to the otherwise more common parallel update. The important point in crafting the discrete event simulation is not to make the result of an update immediately available as this would correspond to the merging of an updated vehicle back into occupation table and would also violate the parallel update philosophy. The relative distance headways and the relative velocities and accelerations could propagate as events through the network of systems and the systems calculate new velocities and distances only on changes.

The motivation for this relatively complex approach is based on the general assumption that implementations of discrete event simulation systems result in better performance than those of the corresponding discrete time simulation systems or a network of systems specification. In general, two reasons establish this assumption. Firstly, a discrete event system network updates only the relevant the parts

Table 3.7.4: Comparison of relevant system specification properties.

name	granularity	state trans. fun.	dependencies	knowledge	sync.-freq.
DTSS	monolithic	Eq. 3.135	$s_d^{(t)}, \mathfrak{M}_{loop*}^4$	poor	
multiDTSS	component	Eq. 3.149	$s_d^{(t)}, \mathfrak{M}_{loop*}^4, L_{det}, L_{occ}$	medium	$2/\Delta t - 3/\Delta t$
DTSN	network of systems	Eq. 3.158	$s_d^{(t)}, \mathfrak{M}_{loop*}^4$	good	$N_{aggval}1/\Delta t$

of the system or the network. The update of the latter depends on the component that changed its state after an event has been processed. Second, a discrete event structure system network has a finer grained component structure than the corresponding discrete time structure that would represent the whole network of vehicles per roadway. And, in general, a finer grained structure provides more parallelism. Thus, modelling a microscopic traffic simulation such as OLSIMv4 as a modular network of discrete event systems components that represent per roadway dynamic networks of vehicles looks very promising.

However, several aspects substantiate reasonable doubts for the effectiveness of this approach. First, as has been discussed in the previous section, the connections between a vehicle and its neighbouring vehicles have to be part of the state in order to change the structure of the system. While this is not difficult to fulfill, the connections between the vehicles may require updating in each time step which results in a non- ignorable. synchronisation effort that suppresses the parallelism benefit of this approach at least partly. Second, as already discussed before the vehicle update procedure has to process a parallel update. The implementation of a parallel drive update for a parallel discrete event structure system network could employ an ordered sequential update per roadway that updates each vehicle of a roadway and propagates in the direction of travel. Such an update would again update all vehicles on a roadway and, thus, not provide any benefit over the classical, parallel update in the case of a discrete time system simulation. Third, the dawdling in the vehicle update requires updating in each time step. However, this approach is a very interesting implementation option that could be investigated further. As OLSIMv4 is a productive system, it employs the more established discrete time structure network system. Table 3.7.4 summarises some of the relevant properties of system specifications.

3.8 Summary

This chapter presented a conceptual design in the form of a set of rules, function definitions, and equations as well as set definitions for every single type. Section 3.1.2 introduced a nomenclature for the latter. The rules, equations, and function definitions presented and described the heterogeneous models homogenously. The heterogeneous models consist of the vehicle models, the network model, the microscopic traffic models, the detector models, the tuning element models, and the simulation models.

Section 3.2 described the vehicle models and the vehicle type models that occur in various microscopic traffic simulations. Vehicle types Subsection 3.2.1 pointed out the contexts wherein vehicle types occur and may impact the simulation results. The contexts range from traffic rules and regulations over loop detectors and driving behaviour on highways to the various user perceptions. The current vehicle type definition distinguishes the vehicle types into passenger cars and trucks with the maximum velocity, the maximum length, and the strictness in obeying speed limits as their type related properties. As the vehicle model varies among simulation scenarios, Subsection 3.2.2 presented a rather minimal vehicle model as well as a vehicle model based on OLSIMv4 that is well-suited for a productive environment. While the former uses only the vehicle's coordinates of the last time step, the latter supports trajectories of several timesteps. The loop detectors can profit of the coordinates from subsequent time steps, as the calculation of the lane change velocity depends on at least two subsequent lane attributes.

Section 3.3 portrayed the network model. It consists of the cells, the sections, the roadways, the topological analysis, the occupation tables, and the switching areas. The presented network model improves several shortcomings of OLSIMv4 and earlier and widespread modelling techniques. Subsection 3.3.1 introduced the cells as collections of rules that provide the topological information for the cell's particular lane and position. As an enhancement with respect to OLSIMv4, the rules also distinguish between the various vehicle types properly. The sections from Subsection 3.3.2 represent single lane areas of maximum extent within which the topological information as provided by a prototype cell does not vary. Compared with the implementation from OLSIMv4, their definition is more flexible as the sections of neighbouring lanes are not required to start and end at identical positions. The roadways as described by Section 3.3.3 introduce probably the most significant enhancement that started with OLSIMv4. As they do not partition the roadways into segments, they simplify the vehicle update logic significantly and, at the same time, also increase the parallelism of the operation. The topological analysis from Subsection 3.3.4 determines and summarises the topological information that applies on a certain lane and position of a given vehicle. The resulting information reduces the topological information to a rather minimal set of values. The occupation tables from Section 3.3.5 combine the roadways with the list of vehicles that move on the roadway. They also provide an analogous function to the topological analysis that determines and summarises the vehicle neighbourhood constellation for a given vehicle into a single structure element. The occupation tables additionally provide the vehicle update of their entire list of vehicles which makes the occupation tables a more or less atomic unit. Simulation scenarios that do not model lateral merging between the roadways explicitly, may, thus, end up in a simulation model with more parallelism. Wherever vehicles should switch between occupation tables, the switching areas provide this functionality. Subsection 3.3.6 described them and attributed their updated back to the occupation table update.

Section 3.4 presented the interface to the microscopic traffic model as well as one of the integrated traffic model that implements the interface, namely the model by Lee *et al.* with multi-lane traffic rules by Pottmeier and asymmetric lane change extensions by Habel on German highways. The presentation compounded the longitudinal and the lateral part of the vehicle motion. As part of the longitudinal motion, Subsection 3.4.1 introduced the classical model in Paragraph 3.4.1.1 that requires optimisation for the implementation in a productive environment such as OLSIMv4. Subsection 3.4.1.2 detailed the optimisation in Paragraph 3.4.1.2 that turned out to require two to four times less computations. Further optimisations are possible and expressed in Paragraph 3.4.1.3. The latter presents also certain configurations with non-existent solutions for the vehicle update. It also discusses their role in the assumed collisions. Subsection 3.4.2 presents the lateral motion part of the vehicle motion.

Section 3.5 introduced the detector model along with various instantiations, namely the loop, prognosis, invariant, local, and global detectors. Subsection 3.5.1 specified the measure and the aggregate operation as supported by the detectors. One group of detectors provides their data to the simulation only. Others provide their data from the simulation to the user. Having comparable input and output data from the simulation has been the main modelling principle for combining both detector variants in a single model. The loop detectors and the prognosis detectors provide data from the inductive loop detectors to the simulation. Subsections 3.5.2 and 3.5.3 described them. The latter aggregate the data as historical time series of the former. The invariant detectors, the local detectors, and the global detectors are parts of the detector group that provides data from the simulation. Section 3.5.4 sketched the invariant detectors and showed that they are especially useful for testing. Local detectors Subsection 3.5.5 are the counterparts in the simulation to the loop detectors. While loop and local detectors provide stationary detector data, the global detectors provide spatio-temporal traffic data. Subsection 3.5.6 described the global detectors.

Section 3.6 sketched the tuning model that relies on accumulators and tuning elements. Subsection 3.6.1 presented the accumulators that facilitate to calculate balances over sets of detector measurements. As each tuning element employs an actual and a should-be accumulator to adapt the actual measurements to the should-be values, the accumulator provide a way to combine traffic flows from several roadways and also to distribute them across the roadways. The simulation models that employ these kind of tuning elements do therefore make the simulation's roadway-based components depend only on the detector output of neighboured components and, thus, free them of depending on the neighbour component's state. Subsection 3.6.2 introduced the tuning elements and their tune operation.

Section 3.7 presented several simulation models for applying them on an entire road traffic network. The simulation models specify systems where the specifications consist of a structure that describes the state, of an input and an output function, and a state transition function. The presentation also disclosed the interaction between road geometry and execution model. Additionally, it defined the state transition functions for each of the simulation models by means of the functions provided by the previous sections. Each state transition function involves the update steps output, measure, accumulate, tune, lane change, drive, and increase time. As the first of the simulation models, Subsection 3.7.1 presented the discrete time system specification that considers the road traffic network as an atomic network. The specification provides, thus, the least system knowledge and does not support a parallel interpretation. As a consequence of the atomicity, the specification supports result verification only when related to a concrete road traffic network for which the results are known and with a rectangular road segment being the only one. Subsection 3.7.2 detailed the multicomponent model which provides the next level of system knowledge. The components in the model represent the roadways and have a state transition function each but are not systems by themselves because they are coupled with other components. Consequently, the state transition function of a component depends on the states of the coupled component. However and unlike common to many simulation systems such as OLSIMv3 and SUMO, the state transition function does not require synchronous updating of two connected components. In addition to the dependency on the coupled components' states, the accumulate function depends on the detectors referenced by the accumulators including the ones from other components. This is in contrast to the next model where the state transition function provides the detector measurements for the accumulators as part of the input data for the measure operation. Subsection 3.7.3 detailed the next simulation model that implements the network of systems specification and consists of a network of input and output systems. The systems represent the roadways of the network and do not depend on other system states. They provide the next higher level of system knowledge as well as a high degree of parallelism. Instead of depending on the state of coupled components, the systems model the dependencies to other systems by appropriate tuning elements. Because the NRW highway network is a dynamic network, Subsection 3.7.4 discussed the need for a dynamic network of systems as part of the concluding thoughts. In addition to the network specification, the dynamic network specification models the input and output connections as part of the system state. However, the network of systems specification covers already all aspects necessary for a microscopic traffic simulation and, thus, the dynamic network specification is not necessary. Subsection 3.7.5 discussed an interesting idea for a further decomposition that models a roadway based system as a network of systems with a discrete event simulation system without relinquishing the parallel update property. It is, however, unlikely, that such a system specification will result in a more efficient implementation.

Implementation issues

This chapter describes some of the implementation issues that might occur when implementing the conceptual design of Chapter 3 in form of the simulation model and all its dependent sub-models. The structure of the conceptual design of that chapter suggested already a structure well-suited for an implementation by dividing the simulation system into the components that match the sectioning and by providing models for each of the presented components, namely the vehicles, the road traffic network, the traffic models, the detectors, the tuning elements, and the simulation as a state transition system. This chapter's implementation follows that component-based implementation design generally and, therefore, discusses the implementation issues per component or model, respectively.

The conceptual design of the previous chapter described all required components and models for a possible implementation of the simulation system in the context of OLSIMv4 and it also includes several enhancements over OLSIMv4. It serves, thus, as a semi-formal specification that describes all involved components and models to implement. Additionally, by providing a semantics for any possible implementation of it, the conceptual design answers the questions what to implement and to which part of the real-world the computation corresponds to. By providing several implementations for the simulation model, this chapter goes beyond the intention of the previous one. It extends the semi-formal specification into a formal one. Moreover, it answers the question what to implement more formally by providing a formal specification and a formal semantics for the conceptual design. Additionally, it poses and answers another related question, namely, how to implement the conceptual design for the requirements of a productive, real-time simulation system such as OLSIMv4 [31].

This chapter details one part of the implementation issues as a set of formal specifications that describe how to represent the semi-formal specification of the previous chapter formally in a declarative manner. The other part of the implementation issues discusses several aspects that arose during the implementation and the validation of OLSIMv4. The implementations employ the concepts and the logic provided by declarative and functional programming languages, namely the Maude declarative programming language [45, 135, 47] and the Bigloo Scheme dialect [189] that is based on the fifth Revised Report on the Algorithmic Language Scheme (R5RS) [1]. The abstract data types of the conceptual design in form of the semantic algebras correspond to abstract data types in declarative and functional programming languages. This chapter focuses, thus, on the implementation of the conceptual design in a declarative programming language such as Maude. Additionally, the "higher and more abstract level of programming" provided by the concepts, logic, and paradigms of declarative programming languages "leads to reliable and maintainable programs [...] mainly due to the fact that, in contrast to imperative programming, one does not describe *how* to obtain a solution to a problem by performing a sequence of steps but *what* are the properties of the problem and the expected solutions" [75]. The Maude implementation makes not only use of the higher and more abstract level of programming and provides not only the additional benefit of having an executable formal semantics for the conceptual design. It also prepares the simulation model for formal verification as demonstrated in the author's publication [30] and substantiates the usefulness and the correctness of this consideration. Additionally, this chapter relies heavily on the author's previous publication [30] that presents the Maude code and demonstrates how to formally verify the simulation model of the previous chapter.

In contrast to the properties that are only useful in a formal context, real-world programming languages such as Scheme extend purely functional languages by also providing typical imperative programming constructs such as I/O operations, assignments, and control constructs [179]. The Bigloo Scheme implementation, for example, enables "Scheme based programming style where C(++) is usually required" [189]. In addition to the real-world extensions, however, the referential transparency property, higher order functions, and generic programming equip the programmer with a clearly understandable programming logic. Lastly, the Bigloo Scheme programming language provides control structures for exploiting CPU thread-level parallelism. As OLSIMv4 requires a parallel interpretation of the simulation model, exploiting the hardware parallelism is essential for any implementation of it.

However, the presented implementations do not express the experiences that result from a direct and top-down implementation attempt of the previous chapter's simulation model. In contrary, the conceptual design, which may now also serve as a semi-formal specification, reflects the results of several enhancements over this chapter's implementation effort. This chapter discusses only those aspects of the implementation issues that are of some relevance in the context of OLSIM4. Among them dominate correctness in terms of verification and validation possibilities, readability and code maintainability, implementation difficulty, code size and complexity, code and computational scalability in particular under the aspect of a sequential or parallel interpretation, and performance. Moreover, as the term "conceptual design" suggests, the previous chapter provides only a semi-formal description. In general, only implementations in a machine verifiable interpretation provide a formal model. As a consequence, the conceptual design leaves several aspects unspecified. This chapter unveils them and discusses the implementation alternatives in each section.

The correctness of an implementation, the construction of a complete road network traffic, the real-time requirements, and the storage and access of the detector measurements as input to or output from the simulation form the greatest challenges in the implementation of the conceptual design. This chapter addresses them as follows. The analogy between the abstract data types as specified in form of the semantic algebras and their representation in a declarative and functional programming language forms a key to the verification of the correctness of an implementation. An abstraction layer between the road traffic network of the simulation engine and the database representation of the road-segment based digital maps in form of a graph library eases the construction of the complete network in the simulation engine as well as the merging of the traffic information between the various maps in the database. A database provides fast and reliable storage and access of the detector measurements and all other traffic information. Finally, a Concurrent ML library with a concurrency model based on message passing that has been extended to exploit CPU thread-level parallelism provides the functionality for the parallel simulation of the individual roadway based occupation tables.

The chapter starts with Section 4.1 that presents some general remarks around the implementation of the conceptual design. It reasons about the choices of the various involved programming languages. The section accompanies also each chosen programming language with a short introduction. After the general considerations, Section 4.2 presents some simulation engines as possible implementations of the simulation model from Section 3.7. The section also explains the differences to the simulation model that result from the enhancement effort. In addition to the implementation of the OLSIMv4 network model, the section also includes the implementation variations for the road traffic network model from Section 3.3. The simulation engine contains only abstract models. Thus, it does not contain any implementation for the vehicle model from Section 3.2, the microscopic traffic model from Section 3.4, the detector model from Section 3.5, and the tuning element model from Section 3.6. They follow with Sections 4.3, 4.4, 4.5, and 4.6, respectively.

4.1 General remarks

This section introduces the foundations of the simulation model implementations. The two dominating implementation variants are the one of the productive simulation system OLSIMv4 and the one intended for remodelling and reengineering the simulation system. The implementation of the productive simulation system uses the programming languages Bigloo Scheme [189] and PostgreSQL [68]. As members of Prof. Dr. Schreckenberg's staff of the Physics of Transport and Traffic Department of the University of Duisburg and Essen, mainly the author and two further colleagues, Habel and Zaksek, have written the OLSIMv4 software in Bigloo Scheme. The author has not only contributed to nearly all parts of the software but has also provided the architecture and the component design. However, only the team as a whole were able to write the entire software. Appendix A therefore acknowledges all team members. The author has further developed the simulation model for the purpose of the Maude implementation which aims to enhance the shortcomings of the OLSIMv4 implementation. The conceptual design as well as the implementation for the Maude part is the author's own work even though some parts of the Maude design are based on the prior OLSIMv4 design.

The development history of the implementations that will be presented in the following sections starts with the initial implementation of the simulation system for OLSIMv4. Viewed from the perspective of a software development process, the implementation is more like a proof of concept. Therefore, as part of a reflection over the proven concept and as a further development of the initial implementation, the author initiated the implementation in the declarative programming language Maude [45, 135, 47]. The implementation was designed to support a wider range of simulation models and to be portable across a lot of road traffic networks. It contains the redesigns from the lessons learned and, as soon as it also reflects the enhancement effort of the conceptual design, can serve as a formal specification for future implementations. As the last part of the cycle, the semi-formal specification builds upon the Maude implementation. Even though the development history contains that order of the implementations, the discussion of the implementation issues herein might suggest that it would have been the other way round. This is, however, not the case but will lead to a better understandable discussion of the implementation issues.

This section continues with some considerations about the software development process in Section 4.1.1. Thereafter, it discusses the choice of the various programming languages in Section 4.1.2. It concludes with the description of the implementation goals in Section 4.1.3.

4.1.1 Development process

Section 2.1.4 has already mentioned that OLSIMv3, as most other scientific research and development projects [36], too, lacked a formal software development process. Just like its predecessor, the implementation process of the simulation system for OLSIMv4 lacks also a formal software development process. The small budget and the researchers' unwillingness for committing themselves to a formal software development process are not the only reasons that lead to this lack. The missing specification distinguishes the research project from software industry projects and is the main reason for the informal software development process. Due to the nature of research projects, the requirements at the start of the OLSIMv4 implementation where only available as high-level requirements in the sense of the high-level requirements from Chapter 2 such as the improved traffic models from Section 2.2.3, feature requests from Section 2.1.3, and re-architecture demands from Section 2.1.4. Therefore, "when we come up against a poorly defined and understood problem and a highly volatile user need, we can hardly expect to output a full requirements specification at the start. In this case, we have to opt for longer and more complex life cycles, like the Spiral Model [24]" [95]. And, in fact, the development process for the OLSIMv4 implementation including the reengineering efforts in Maude by the author resulted in a model similar to the Spiral Model [24].

The lack of the software development process in most scientific and engineering projects is not surprisingly. In a case study about software engineering within the scientific and engineering computing domain [36], many of the studied projects began "as research projects without a definitive relationship among deliverables, schedule and resources" [36]. Several characteristics in the computational science and engineering domains (CSE) substantiate that the lack of software development and software quality ensuring processes has rational reasons rather than pure unwillingness of the researchers. Firstly, "CSE software often explores unknown science or engineering, so many requirements emerge during development and can't be known or adequately planned for a priori" [35]. Secondly and similarly, "while most scientific and engineering projects are ultimately based on the underlying laws of nature, which are fixed, the application of those laws to a specific problem is often unknown at the start of the project. Most requirements, beyond some obvious high-level ones, are discovered during the course of the project" [36]. However, "to make sure some solution solves a problem correctly, one must first state that problem correctly" [121]. Furthermore, "problem specifications are essential for designing, validating, documenting, communicating, reengineering, and reusing solutions. Formality helps in obtaining higher-quality specifications within such processes; it also provides the basis for their automated support" [121]. The lack of a specification turns out as a major implementation issue.

Thirdly, "another distinguishing characteristic is that the main driver for these projects is, not surprisingly, correct science or engineering, rather than ensuring software quality through the use of sound software engineering practices. In fact, many of the projects are not given adequate funding or support to implement even basic software engineering principles" [36]. Fourthly, "often the scientist or engineer doesn't know the expected results of the software's execution, making it difficult to perform traditional tests using test oracles" [35]. Fifthly, "the software-development life cycle is often either extremely long (tens of years in development and evolution) or short (code written quickly, executed once, and discarded); therefore, choosing an appropriate process to manage software development can be difficult" [35]. These five reasons may be summarised by a more general aspect, namely that "the development of CSE software doesn't follow traditional software engineering best practices" [35]. By the same token and as a last reason, "in many cases, existing software engineering best practices aren't applicable to CSE software without significantly tailoring and adapting them" [35].

However, the researchers' unwillingness to commit to a software development process is apparently existent. According to [36], "the developers tend to be averse to the "process"-oriented software development approaches which have been successfully used to manage risk on other types of software projects. This aversion is due in part to the lack of formal software engineering training and in part to the nature of the application domain. Many of the projects have long life-cycles that last for decades during which the model of science embodied in the software evolves as knowledge evolves. Developers tend to think they will have greater flexibility by not following rigid software development processes." In the context of OLSIMv4, the researchers were not only unwilling but also had neither the time, the funding, the infrastructure, nor the expertise to start a formal software development process. Additionally, the concrete dimensions of the simulation model, of the computational model with respect to parallelism, of the road traffic network data model in vehicles per second, and of the tuning model were unknown at the beginning of the OLSIMv4 implementation phase. Furthermore, it was generally unknown whether the traffic models will produce adequate results that will pass the validation tests.

Due to the mentioned reasons, the software finally took the course through the following stages. At first, the initial position as described by Chapter 2 contained several high-level requirements. The lack of a more concrete specification led to the bottom-up style development of the OLSIMv4 simulation system. The procedure is not astonishing as "specifications are never formal in the first place" [121]. During the development, several severe problems emerged that required an enormous effort to solve and led to a significant delay of the project. The computation time in the parallel interpretation did not meet the real-time requirements and the simulation results did not pass the validation criteria. Finally, however, the OLSIMv4 implementation met the real-time requirements and passed the validation tests. The OLSIMv4 can now serve as a proof of concept. Thus, the author remodelled, reengineered, and thereafter implemented the simulation model in the Maude term rewriting system. In contrast to the Bigloo Scheme language, the latter is free of mutable state as every expression results in a term. The reinvestigated simulation model can now serve as a formal specification. In combination with the semi-formal specification and the semantics of Chapter 3, a full specification is now available. The semantics is an important part of the formal specification as "formal specifications are meaningless without a precise, informal definition of how to interpret them in the domain considered" [121]. The development progress corresponds one cycle in the Spiral Model development process [24].

OLSIMv4 has been validated "using a section control based method and as described in the following. On two subsequent bridges across a highway, digital cameras have taken traffic photos in multi-shot mode. Subtracting associated time-stamps of the photos for a unique set of vehicle attributes such as parts of the license plate, color, and manufacturer yielded to travel times that the simulation results had to match" [30]. However, validation relies on correct computation and takes place at a very late phase of the implementation process. Additionally, "programming errors in the computer code, deficiencies in the numerical algorithms, or inaccuracies in the numerical solution, for example, may cancel one another in specific validation calculations and give the illusion of an accurate representation of the experimental measurements" [154]. Thus, it is essential for the development process of microscopic traffic simulations to not only rely on validation but also on verification. Additionally, as part of the lessons learned, these processes require a significant importance in terms of time and effort in the software development process of microscopic traffic simulations.

Formal verification without a specification and without a verifiable simulation model is almost impossible for microscopic traffic simulations as "for real-world simulation scenarios and apart from free flow traffic, the results of a formal simulation model are still unknown. Thus, it is not possible to develop a set of formal criteria that simulation results have to meet" [30]. This is, however, not limited to the domain of microscopic traffic simulations because "often the scientist or engineer doesn't know the expected results of the software's execution, making it difficult to perform traditional tests using test oracles" [35]. Verification and validation is, thus, generally "very difficult in this environment" [36, Sec. 5.1]. Again according to the author's publication [30], "the only way to verify the correctness of an implementation of a microscopic traffic simulation is by first verifying the correctness of each of the part-taking models and then by simplifying the simulation scenario into a one with periodic boundary conditions whose results have to reproduce the results known from literature." The goal of the verification process is to provide "evidence, or substantiation, that the mathematical model, which is derived from the conceptual model, is solved correctly by the computer code that is being assessed" [154, 153]. Verification activities include code verification, formal verification, software quality and result verification, benchmarks, and accuracy assessment [12].

As "validation activities presume that the computational model result is an accurate solution of the mathematical model" [154, Chap. 2, 153], the validation activities come next after verification activities in the software development process. In contrast to [12], or beyond verification, it is the goal of the validation process to determine "how accurately the computational model simulates the real world for system responses of interest" [154, 153]. It includes experiments and design, validation metrics, performance and safety [12]. This chapter presents "a verifiable simulation model for microscopic traffic simulations that is suitable for a variety of application domains and that can be used in several contexts" [30].

As now a complete specification for a verifiable simulation model exists that, moreover, has been validated successfully, a significant progress in the modelling and the development cycle could be achieved.

4.1.2 Programming languages

The real implementation phase of the OLSIMv4 simulation system started in 2009 after the conception completed and several implementation experiments succeeded. Mainly Habel and the author implemented the most relevant experimental version of a small subset of the functionality intended for OLSIMv4 during the work on Habel's Master Thesis [70]. As a part of the thesis, Habel examined the usefulness of the asynchronous lane changing extensions for the microscopic traffic model as described in Section 3.4.2. Habel also validated the extended microscopic traffic model for a part of the German A44 highway. The experimental implementation used the second interpreter Scheme programming language dialect (SISC) [143] mainly due to its comfortable Java foreign function interface. Several pieces of code had been developed in the years before 2009 that, for example, transferred data from the data distribution software DAV [177] to the internal traffic database.

In the years before 2009, several programming languages were considered to be chosen for the use in the OLSIMv4 development. To that time and probably up to now, "two languages dominate scientific computing: Fortran and C++" [178, Preface]. The already introduced case study [36] which preceded the conclusion about the dominating programming languages by four years, unveils that the scientific software development projects use mainly C++, C, Fortran, and very little Java only for some libraries. The study additionally states in 2007 that developers of scientific and engineering software generally tend to avoid higher-level programming languages [36, Sec. 5.3], i.e. languages that provide high-level abstractions such as object-oriented programming constructs or functions as first-class values. These kind of languages compound not only the object-oriented and the functional or declarative programming languages. They also compound simulation languages such as Matlab, Mathematica, and Simulink, as well as theorem-provers such as Coq [25, 21] and the Maude LTL model checker [56] and Real-Time Maude [158] which have experienced substantial progress in the last years [73, Tab. 1]. Even though the study identified the performance criterion as one that competes with correctness and code maintainability, the study substantiates the conclusion that the projects chose the programming language depending on the potential of machine code level optimisations. The study mentions further motivations for the choice of a specific language. According to them, a programming language has to be "easy to learn, offer reasonable high performance, exhibit stability, and give developers confidence in the validity of the resulting machine code instructions" [36, Sec. 5.3]. However, these personal reasons for preferring the low-level abstractions may also be part of a premature optimisation in the process of language choice and, additionally, the following considerations may also prove them as false.

According to [128, Sec. 4], many "computer scientists spend much of their time to discover efficient algorithms. Clearly, this effort is important as the difference for an application of, say, an $O(n \log n)$ versus an $O(n^2)$ algorithm may mean the difference between success and failure. This kind of argument has often been used to justify the low-level programming which typically takes place to implement efficient algorithms. However, there is another cost of programming which can easily be ignored by computer scientists and that is the cost of programmer time and the cost of maintaining and upgrading existing code. Often having the most efficient algorithm isn't so important; often a programmer would be very happy with a programming system which only required the problem be specified in some way and the system itself find a reasonable efficient algorithm." The aspect of how intensive it is to maintain and upgrade existing code is accompanied by the question how difficult the learning of the language is and how difficult it is to express a specific algorithm in it. The questions are examined in the following.

Concluding from the citation of the aforementioned study and according to [103, Preface to the first edition], the C programming language is "easy to learn." However, in his Turing Award lecture "Can programming be liberated from the von Neumann style?" from 1978, John Backus stated already that "conventional programming languages" are "fat and flabby" [14]. To that time, the sizes of some imperative programming language manuals exceeded more than 500 pages which is not that much compared to the volumes of today's most programming language standards. Apart from the pure volume, the orthogonality can help significantly in reducing the difficulty to learn a particular language. "Orthogonality in a programming language means that a relatively small set of primitive constructs can be combined in a relatively small number of ways to build the control and data structures of the language" [187, Sec. 1.3.1.2] (5th Edition) in [170, Sec. 4]. A language can ease the difficulty by providing a small set of orthogonal constructs. In addition to the language's volume and the orthogonality, "in an imperative language, the logic and control are mixed up together, programmers have no choice but to be concerned about a lot of low-level detail" [128, Sec. 4]. Even though modern object-oriented languages such as C++ and Java provide high-level abstractions, they still require the programmer to concern about low-level details such as memory (de-)allocation, assignments, evaluation strategies, exception handling, deadlock and starvation. Consequently, in that sense, they remain basically imperative programming languages. However, as the "limits of my language mean the limits of my world" [215, p. 5.6], the aspects of the language size, the language challenges, and the mixed logic and control seems worth it investigating with respect to the point of time where the development of OLSIMv4 started. It is, thus, sketched in the following.

In many cases, programming languages are not only difficult to learn. They may be also difficult to teach. Several studies discuss the pros and cons from their perspective [132, 140]. Both variants may result in a misunderstanding of certain concepts of the programming language. As part of an empirical study [144], first and second year students learned the C and C++ programming language and assessed after each year how difficult to learn certain concepts of the programming language were. The students assessed the degree of how hard it is to learn the language the hardest concepts on average with 3.923 to 4.6 on a scale from 7 ("hardest") to 1 ("easiest"). According to their opinion, the hardest concepts are copy constructors, operator overloading, templates, dynamic memory allocation using `malloc`, pointers, recursion, casting, function overloading, and virtual functions. The group that contained teachers for the majority of it, assessed the degree of how hard it is to learn the programming language the hardest concepts on average with 6.054 to 4.000. According to their assessment, the hardest concepts are pointers, virtual functions, dynamic allocation of memory using `malloc` and `new`, polymorphism, recursion, copy constructors, and operator overloading. These programming language concepts obviously further restrict the limits of the students' languages. For the most part of it, these difficult concepts deal with control, provide only technical benefit, and do not provide for the logic.

The C99 language standard [89] that takes 554 pages may serve as another example for a valid standard of the time of the OLSIMv4 development. Meanwhile with the C11 standard [90] that has 550 pages the situation did not change. Furthermore, although not required directly, the knowledge of the C programming language is generally fundamental for a deep understanding of the C++ programming language. The size of the language standards is 776 pages for the C++98 version [87] while the C++11 standard [88] compounds 1338 pages. The Java language specifications 7 and 8 have a comparable size with about 670 pages. However, in general programming in Java requires additional knowledge of Java Development Kits. Even though distinguishing between basic and advanced features of a programming language may help in reducing the learning effort, the C and C++ languages require an enormous effort to learn, to understand, and to apply properly. In projects with teams that have stronger and weaker software developers, the team skill spectrum questions which kind of language subset to choose for a certain project. To answer the question in another software development domain, the motor industry software reliability association recommends only a limited subset of the imperative programming languages C and C++ for the use in safety critical systems "because there are aspects of the [C and] C++ language that are hard to understand, compiler writers have been known to misinterpret the standard and implement it incorrectly" [9, 8]. However, the aspects difficult to understand are not limited to the particular language only. Some aspects such as the fragile base class problem [6] and subclassing vs. subtyping also apply for the object-oriented programming and design concepts [104]. In contrast to the considerably large sizes of the C and C++ object-oriented programming languages, Fortran has relatively small language standards with about 300 pages until 1995 but exceeds the 500 pages already with the 2003 standard. However, the fifth revised report (R5RS) and the sixth revised report (R6RS) on the Algorithmic Language Scheme contrast the size aspect of the standards of conventional programming languages even better as they require only 50 pages and 90 pages, respectively [1, 97]. The latter has additionally 71 pages for the standard library [98] as well as 140 non-normative pages. The volume of a programming language impacts also the time effort that is necessary to spent for teaching the language and has proven as a main obstacle in the case of teaching the Java programming language from the study in [192]. Therein, the pure volume of the language prevented the students not only from learning the entire language but also prevented them from learning to design programs systematically which is clearly the more important task over learning a particular programming language.

In addition to the complexity of the programming languages, the aspect of their affinity to the machine model is worth looking at and done in the following. Although the name "von Neumann architecture" may be too much honour as Burk and Goldstine also developed the architecture [33], [160, Sec. 1.10], according to Backus [14], the von Neumann machine model consists of a "central processing unit (CPU), a store, and a connecting tube" and has a bottleneck, namely the (single) connecting tube that Backus identifies as the "von Neumann bottleneck." On the von Neumann architecture, programs operate by changing "the contents of the store in some major way." However, "before a word can be sent through the tube its address must be in the CPU; hence it must either be sent through the tube from the store or be generated by some CPU operation. If the address is sent from the store, then its address must either have been sent from the store or generated in the CPU, and so on. If, on the other hand, the address is generated in the CPU, it must be generated either by a fixed rule (e.g., "add l to the program counter") or by an instruction that was sent through the tube, in which case its address must have been sent ... and so on." It is general wisdom that exploiting multi-processor parallelism will contribute to overcome that bottleneck [32, 152, 157, 141]. However, the architectural innovation that contributes to the performance growth has focused solely on the performance growth of uniprocessors at least until the first few years of this millenium [157]. Even though the processor performance has increased faster than Moore's law [157], the possibility of a slow down of the architectural innovation has been foreseen and discussed long in advance [160, Sec. 8.1] to its manifestations. So has the increase in the importance of multiprocessors in case of a slow down in the pace of progress in uniprocessors [160, Sec. 8.1]. Meanwhile, in the last decade before the year 2014, the computer architecture has advanced significantly and now provides concurrent and parallel computing possibilities for the most wide-spread computer architectures available in workstations and notebooks. "For example, the Intel Xeon E7-8870 processor, released in April 2011, contains 10 cores and supports 20 threads. On October 2012, AMD released the FX-8300 processor that contains 8 cores and supports 8 threads. Such architectures present significant challenges to traditional "von Neumann trained" software developers: Specifically, how does one best write von Neumann-style programs capable of utilizing the computational resources of such multi-core (soon to be massively multi-core) platforms?" [213].

As part of the reaction to the new challenges, the majority of conventional programming languages offer some adaptions of the most well-known and wide-spread functional concepts such as lambda expressions or closures, type inference, and higher-order programming [22]. "The most significant features of Java 8, the latest release of Java, are lambda expressions and method references. Such function values enjoy a pseudo-first class status, the most important of which being that they can be passed as parameters to other methods. This computational shift enables a profound change in how iterations over collections can be performed. Specifically, a collection API can now control the parallelization of iterations over the data it stores" [213]. Even though some of these language features may have been usable in prior versions of any of these languages, their indirect use would have required additional syntax and a lot of effort to convince other team members to use such features. In addition to the Java 8 standard, the C++11 standard [88] also introduced lambda expressions explicitly. However, the adaptions do not make them to functional programming languages as they still lack several fundamental functional principles such as type polymorphism, algebraic data types, and proper tail recursion. Additionally, the presence of functional concepts per se does not make the programming languages easier to learn. On the contrary, the programmer still has to reason about control, storage, and evaluation order. The main hindrance with these kind of programming languages is that their semantics is "extremely messy" [128] and, thus, reasoning about programs in these languages is very difficult [128]. The C++ programming language, however, contains also a Turing-complete, purely functional sublanguage with referential transparency and no assignments [163], [195] in [104].

The point is here that conventional programming languages such as Fortran, C, C++, and Java – of which Backus at least some titled as von-Neumann programming languages[1] – "use variables to imitate the computer's storage cells; control statements elaborate its jump and test instructions; and assignment statements imitate its fetching, storing, and arithmetic" [14]. The reason for the postulated similarity between imperative and object-oriented programming languages relies on the similar behaviour in the control flow analysis [142]. A program in the object-oriented semantics describes mainly how to compute a result instead of the superior approach to focus on what to compute which is where declarative and also functional programming languages come in. When dealing with concurrent or parallel computing architectures, the imperative programming languages require the programmer to deal with (dead-)locking and starvation. "In contrast, functional languages are based on a computational model well-suited to a form of concurrency based on message passing. Unlike imperative languages, whose computational models are abstractly isomorphic to von Neumann architectures, functional programming languages, such as Erlang and SML, trace their origins to mathematics" [213]. As they focus on describing what to compute instead of detailing how to compute, such systems "have semantics loosely coupled to states–only one state transition occurs per major computation" [14]. "In the context of multi-core computing, the key attribute of computations expressed as functional programs is that

---

[1]Backus did not intend to "blame the great mathematician" for the complexity of the so-titled languages [14] by using the term "von-Neumann programming languages". Neither does the author of this thesis intend to blame von Neumann.

they are (essentially) stateless" [213]. Larger programs can then be built by combining the computed states. Declarative programming also "includes logic programming and functional programming, and intersects significantly with other research areas" [128, Sec. 8].

In addition to the difficulties in learning and teaching programming languages that rely on thick compendia [14], the advance in computer architecture from single-CPU systems to nowadays widely available multi-CPU systems have changed the challenges for language designers and compiler writers that resulted in the (pseudo-)functional additions to Java 8 and the C++11 standard [88]. The architecture advance has led to some degree to think over the educational concepts and the intellectual foundations in teaching of programming languages. The Carnegie Mellon University is probably the most prominent University that switched from teaching object-oriented programming to functional programming in 2011 [48, 213]. They eliminated object-oriented programming "entirely from the introductory curriculum, because it is both anti-modular and anti-parallel by its very nature, and hence unsuitable for a modern CS curriculum" [76] in [213]. Even though the teaching of a functional programming language at the introductory course clearly can be advantageous, simply switching the language, for example, to Scheme does not turn "a bad course into a good one" [192]. Instead, the combination of a flexible, i.e. a highly extensible language, and functional programming language (and in particular Scheme and DrRacket) and the teaching of how to design programs systematically seems to be a key for the learning success of teaching beginners how to construct programs [192].

When studying the implementation of programs with parallel operation or interpretation, another advantage of declarative programming languages becomes obvious. A short fact might be worth noting prior to the following considerations. With one exception, all projects in the introduced case study [36] developed code that had to exploit massively high CPU hardware parallelism as they run on parallel supercomputers. Due to their focus on what to compute, declarative programming languages are implicitly parallel. Furthermore, the "lesson here is clear: the more declarative we can make a programming language, the greater will be the amount of implicit parallelism that can be exploited" [128, Sec. 6]. Imperative programming languages, in contrast, have to take care about how and, in particular, in which order to assign the results of a computation to a variable that represents a storage cell. This includes dealing with all low-level details such as deadlocks, starvation, load balancing, concurrency, and synchronisation. However, even though the declarative programming languages have obvious advantages through implicit parallelism, as of 2005 only Haskell and a few not very widespread used implementations of the ML programming language provided concurrency across CPU thread-level parallelism. The famous ML dialect OCaML and the DrRacket Scheme implementation lack CPU thread-level parallelism entirely. The difficult implementation of the garbage collector is probably the reason for it. In contrast to this situation, as of 2005 nearly all imperative programming languages provide a kind of CPU thread-level parallelism. However, the quick integration into the imperative programming languages may have unveiled several shortcomings intrinsic to these kind of languages and may have motivated for integrating new concepts into the language. Even though Java 8 is obviously still not a functional programming language in the first place, it may be that the "main motivation for lambdas in Java has been to facilitate stream-based declarative APIs, and, therefore, easier parallelism" [22]. The stream-based declarative API demonstrates nicely–at least syntactically–the difference between in-memory data structures and anonymous ad-hoc data structures that are traversable only once. The API's syntax simply chains methods with lambda expressions as arguments sequentially and assigns the resulting object to the storage pointed to by a variable. Typical program fragments that may serve as an example involve the Java 8 equivalents for the well known functions `map`, `filter`, and `fold`.

Regarding the aforementioned criteria, namely the difficulty in learning the programming language, the simplicity of the semantics, the support of functional concepts in the programming language, and the support for implicit parallelism, in the context of choosing a programming language for the development of OLSIMv4, the author decided to choose the Bigloo Scheme programming language due to the following reasons. Firstly, the author had to decide the question in the time between 2007 to 2009. At that time, neither C++ nor Java had functional extensions that could have eased the use of functional concepts in the programming language significantly. The volume of the language standard equalled compendia while the R5RS language standard [1] had only 50 pages. Additionally, for a large subset of the language an executable formal semantics [205] defined in the declarative programming language Maude [44] exists. Second, the simple syntax, the small set of language constructs, and Scheme's semantics that eases reasoning about the programs promised to prove the language as "easy to learn" and to apply. Scheme's high abstraction level looked promising to result in readable and maintainable code. Lastly, Scheme's already available implicit parallelism promised seamless integration of explicit parallel constructs into the simulation. However, to that time, only a limited number of Scheme dialects provided real CPU thread-level parallelism for their concurrency implementation. As one of them, Bigloo supported the POSIX threads API and was chosen, therefore. The idea to use a concurrent programming model on top of parallelism goes back to the author's work [29]. To provide the easy to understand message passing programming model to OLSIMv4, the author ported the Scheme version [34] of Concurrent ML [174] and the POSIX threads back-end of Bigloo.

Even though the Scheme programming language eases reasoning about the programs and Scheme programs in general are much shorter than analogous C++ or Java programs, the Scheme programs become quite longish for real-world applications. As the programmer explicitly has to take care about operational details, a lot of thinking effort when programming in Scheme still deals with the details how to get the result to a given computation. As an example from the implementation of OLSIMv4, the vehicle container structure for the occupation tables has a huge impact on the performance of the implementation. To implement accessing and constructing a suitable container requires thinking about a lot of operational details. In general, an implementation for such a container will also provide several optimized variants of accessing and constructing a container. As a result, the code of the implementation will grow. Further reasons for the code growth include input and output operations, error and exception handling, software requirements such as configurations and command line handling, and type constructions and conversions. As another difficulty of the implementation in the Bigloo Scheme programming language, it turned out that both, the OLSIMv4 implementation and the Bigloo implementation, access and maintain a hidden state. It is not transparent to the programmer where and when exactly a module maintains such a state. The pseudo-random number generator state is one example to this but certainly not the only one present in Bigloo and the required libraries for OLSIMv4. Hidden states represent synchronisation points and, thus, prevent programs that aim to exploit CPU thread-level parallelism from running in parallel.

An implementation that focusses solely on the modelling aspect could contribute to overcome these shortcomings and it could contribute at the same time in applying formal methods. Such an implementation requires a language that supports expressing the mathematical models of the previous chapter in a straightforward way. And it requires a language that also provides a computational model to perform the computations described by the mathematical models. The declarative programming language Maude is such a language and it is especially well-suited to model systems [45, Sec. 1.4]. In the Maude language, a "program is a logical theory, and a Maude computation is logical deduction using the axioms specified in the theory/program" [45, Sec. 1.2]. Maude supports functional modules that specifiy a theory in membership equational logic and system modules that specify rewriting theories. The functional modules are included in the the system modules as a special case [45, Sec. 1.2]. The functional modules in Maude specify initial algebras in form of theories consisting of a pair $(\Sigma, E \cup A)$ where Σ is called the *signature* and specifies the type structure [45, Sec. 1.2]. The latter consists of sorts, subsorts, and overloaded operators [45, Sec. 1.2]. The E element of the pair represents the "collection of (possibly conditional) equations and memberships declared in the functional module" [45, Sec. 1.2]. The A element of the pair represents "the collection of equational attributes (`assoc`, `comm`, and so on) declared for the different operators" [45, Sec. 1.2]. Under certain assumptions such as proper simplification rules in the functional modules, the Maude interpreter can simplify an initial term [45, Sec. 1.2]. The system modules consist of a "4-tuple $\mathcal{R} = (\Sigma, E \cup A, \phi, R)$, where $(\Sigma, E \cup A)$ is the module's equational theory part, ϕ is the function specifying the frozen arguments of each operator in Σ, and R is a collection of (possibly conditional) rewrite rules" [45, Sec. 1.2]. However, the functional modules are sufficient for the scope of the OLSIMv4 enhancement effort.

The membership equational logic is very similar to the logic of the semantic algebras used for the introduction of the models of the previous chapter. Corresponding signatures in the Maude functional modules can represent the operations of the previous chapter. Corresponding simplification rules in form of equations can also represent the function specifications of the previous chapter. The similarity eases a straightforward implementation and also implementation verification. Due to the similarity between functional and declarative programming languages, an implementation in the Maude language is well-suited as a specification for an implementation in a functional programming language such as Scheme. As the Maude language is a declarative programming language in the strict sense, there are no hidden states in a program and the terms represent the states for the case of a state transition system such as the ones presented in Section 3.7. As a consequence, such a model implementation is well-suited to find out parallelism bottlenecks. Additionally, as the Maude modules can be used in three different ways [45, Sec. 1.2], namely as programs, as formal executable specifications, and as models that can be analysed and verified formally, an implementation as part of an enhancement effort provides several benefits. The first way of use grants that an implementation results in a valid program that can be evaluated. The second way of use facilitates an iterative development style of models until a satisfying level of detail is reached. The third way of use provides the means to prove certain properties of an implementation thereby increasing the trustworthiness of an implementation. Such an implementation in the Maude language can later be used as a specification for an implementation in another programming language. Maude's expressive power in combination without the need to care about control and operational details facilitates to implement a simulation model such as the ones presented in Section 3.7 in a few thousand lines of code. Such a modelling implementation can then serve as an object of study for an iterative improvement of the model. The latter can then, finally, serve as a formal specification for a productive implementation. Furthermore, as in the Maude language the evaluation of an initial term results in a corresponding canonical term, a state transition in a Maude program is characterized uniquely by the two terms. This property can be used to compare the terms of a Maude simulation with the results of a productive simulation system as part of a result verification. The Maude system additionally supports a broad variety of formal methods such as model checking with the Maude LTL model checker [56] and proofs of invariants.

For the declarative programming language Maude, the "project homepage [206] contains an excellent manual, a lot of tutorials, articles, and related projects of the Maude system. Additionally, the book "All about Maude" [45] serves not only as a very good introduction but also as detailed study book for the interested reader. In addition to introductory materials, it provides implementation examples for data types, games, Lambda calculus, and object-oriented programming. The publication [44] gives a brief introduction to Maude. The Maude system provides a read-evaluate-print loop (REPL). The Maude REPL supports executing of already loaded code. For the purpose of this paper, only the `reduce` command initiates the rewrite process. Executing code on the Maude REPL reduces an initial term into its ground terms via the loaded code. Therefore, every execution requires at least one initial term called *axiom* in the context of Maude. The initial term in the example `> reduce in NAT : 1 + 1 .` is `1 + 1` and Maude will print the resulting term `result NzNat: 2` along with the total number of rewrites" [30].

A last topic may also provoke a controversial discussion, namely the question which technique and language to use for managing, storing, and accessing the data. It is generally accepted that the relational model and SQL still are dominating among the solutions to manage and access data [11]. One reason for the dominance of the relational database model may be the productivity boost as promised by Codd in his ACM Turing Award lecture from 1982 [50]. However, the productivity boost relies on the constantness of the underlying data model. Changes in the data model will generally introduce larger downtimes in real-world productive systems [11, Sec. 3.1]. This is one of the major criticisms on the relational database model and it may have contributed to the development of the so-called NoSQL database systems. Publication [37] provides an up to date overview of those kind of data management systems. In general, they manage "semi-structured data, unstructured data, continuous data, sensor data, streaming data, uncertain data, graph data, and complexly structured data" [11]. The semantics of these kind of systems differs from the relational database management systems. In general, they do not guarantee the transactional paradigm to achieve atomicity, consistency, isolation, and durability (ACID) [72, 67]. Instead some grant only basic availability, soft state, and eventually consistency [37]. The clinical data management may serve as a recent and promising example where in the study the NoSQL database approach turned out as a "viable alternative to relational database design" [126]. The example proved as promising because "clinical data is dynamic, sporadic, and heterogeneous in nature" [210] in [126]. This is not the case for traffic data and topological data.

Probably the main reason for the dominance of the well-established relational database approach may be seen in the strong mathematical foundations, i.e. the relational database theory and the equivalence of relational algebra and relational calculus query languages [49, 107], upon which the relational database model is based [11]. In combination with the transactional ACID paradigm to achieve the four principles atomicity, consistency, isolation, and durability [72, 67], the relational database approach provides a really powerful set of operations on top of which data can be accessed by the declarative language SQL. Sacrificing these benefits too carelessly in favour of a potential scalability and flexibility gain may most likely end in a premature optimisation thereby loosing reliability and correctness. While the flexibility gain is relatively convincing, the scalability aspect remains undecided as it is unknown whether NoSQL systems really scale better. The most appropriate database management approach for a productive application such as OLSIMv4 must, thus, be chosen carefully. It depends on the correct categorisation and classification of the data to be managed. As for well structured data such as accounts, customers, and loans in the banking example the relational model is still well suited [11, Sec. 2.1], the question for the most appropriate approach reduces itself to whether topological and traffic data is well structured or not. Or, in case it is, whether it is only well structured or shows some of the properties that would support a NoSQL solution as to be more appropriate.

Admittedly, traffic data may not necessarily require the ACID principles as traffic data is generally appended but not manipulated [11]. That may also apply for the topological data. However, the topological data results from a non-trivial merging process that unites several sources of digital maps into a final representation. Equipping the merging process with a powerful declarative language manifests a significant enhancement. Additionally, for SQL based database management systems several geographic information system (GIS) extensions such as PostGIS and GeoSPARQL exist that may ease the merging process and with some even supporting uncertainty. As another point, the dynamic topology that adapts the topology of the running system in real-time to the real-world road traffic network conditions requires a transaction mechanism to ensure correctness. As this has been a design goal for OLSIMv4, the relational approach was already required somehow. Additionally, the development of the OLSIM database started in the year 2005 where several NoSQL systems were not available.

Moreover, the traffic data provided by loop detectors is hierarchical data that has a great normalisation potential. Loop detector measurements combine a timestamp and traffic observables and they refer to detectors which in the second line refer to single lanes on roadway segments. The author's publication [114] has, however, taken great benefit from the normalised structure of the data. When observing loop detector data over a period of several years, a certain combination of traffic observables occurs on average a thousand times each across all times and detectors. Additionally taking into account the diversity of the detector types, i.e. the local, global, and prognosis detectors, the thousand times are in fact several thousand times. A potential normalisation may save, thus, a lot of storage space. The normalisation relies, however, on the transactional ACID paradigm. As the dynamic topology requires also the transaction mechanism, the relational database management system seems to be adequate. However, in the year 2005 the author was unaware of the NoSQL approaches. Additionally, they were not used as widespread as nowadays and did not have the maturity of today. Facing these facts, the author decided to rely on the well-established relational model.

4.1.3 Implementation goals

The following quote from the Introduction of the author's publication [30, Sec. 1] express the three most relevant considerations for any implementation goals of OLSIMv4: "Microscopic traffic simulations participate in multiple contexts and for varying purposes – to improve traffic network performance and to assist the infrastructure planner, or to assist drivers in route planning, in context of urban environments, or to simulate highway traffic flow. Microscopic traffic models describe vehicle motion as the result of a vehicle's interaction with other vehicles and with road infrastructure interaction according to the traffic rules. In addition to microscopic traffic models, a simulation system also involves a number of other model families. Among them are the vehicle models, the network models, the detector models, the accumulator models, the tuning element models, the traffic light models, and the control models. Typically, microscopic traffic simulations have to support many vehicular traffic environments and they have to simulate on a lot of varying road geometries. They have to support a broad spectrum of data sources, run on large networks exploiting hardware parallelism, and finally should grant correctness of the results. Besides the complexity of the involved models, the difficulty of implementing a microscopic traffic simulation is beyond the complexity of the various involved models for the following reasons.

First, there is no microscopic traffic model that fits all needs for the various kinds of vehicular traffic flow. Highway traffic, urban traffic with lane discipline, traffic without lane discipline, or simply scientific traffic evaluation are kinds of vehicular traffic. Moreover, a microscopic traffic simulation may not be limited to vehicular traffic. It may include molecular transport, packet transport, rail, and pedestrian traffic. Thus, microscopic traffic simulation systems have to provide several traffic models among which an implementer of a simulation system can choose from for any set of roads. As the traffic model has to apply the traffic rules of the road environments, i.e. speed limits, traffic lights, variable message signs, overtaking restrictions, lane closings or merges, it has to be aware of some kind of topology. [. . .] These dependencies limit the traffic model's portability and make it hard to develop a simulation system model for a variety of potential application domains.

Second, information accuracy, increasing network size, network independency, and system scalability are conflicting goals. In general, [. . . ,] a simulation system can take over data from a lot of different data sources that have the potential to increase information accuracy. For example, the data sources may include loop detector data, camera detector data, traffic light states, variable message sign states, static signs, road restrictions, floating car data, weather data, and toll station systems data. On the one hand, while processing more data does not lead automatically to improved information accuracy, most data has the potential to do so. On the other hand, processing more data requires more computing power. To overcome this bottleneck, microscopic traffic simulations have to exploit hardware parallelism. As the computing cost of a microscopic traffic simulation increases with the number of vehicles in the network, the vehicle update procedures are a primary subject of parallelisation for fine-grained approaches. However, vehicles that have been updated in parallel require synchronisation thereafter. When choosing a coarse-grained approach, the parallelisation gain depends on the road geometries. Thus, increasing information accuracy, network size, and system scalability are conflicting goals.

Third, the number and the complexity of the part-taking models pose the question of how to implement a simulation system with all these entities and still grant robustness and correctness of the computation. Additionally, the following fact further complicates the situation. For real-world simulation scenarios and apart from free flow traffic, the results of a formal simulation model are still unknown. Thus, it is not possible to develop a set of formal criteria that simulation results have to meet. Admittedly, with the use of empirical data it is possible to validate real-world simulation systems as it has been done in the context of OLSIMv4 [31] using a section control based method and as described in the following. On two subsequent bridges across a highway, digital cameras have taken traffic photos in multi-shot mode. Subtracting associated time-stamps of the photos for a unique set of vehicle attributes such as parts of the license plate, colour, and manufacturer yielded to travel times that the simulation results had to match. This process is called a validation process. However, because it takes place at a very late phase of the implementation process, differences from expected to measured results are hard to trace and may be caused by a lot of potential reasons – e.g. by software bugs, artefacts or measurement errors, unforeseen interaction effects of the models, poor calibrations, weak models, or combinations of any of the aforementioned. Even worse "programming errors in the computer code, deficiencies in the numerical algorithms, or inaccuracies in the numerical solution, for example, may cancel one another in specific validation calculations and give the illusion of an accurate representation of the experimental measurements" [154, 153]. Instead of expecting to detect the differences from ex-

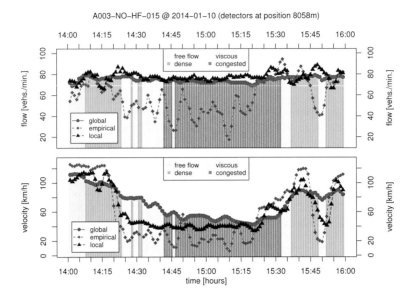

Figure 4.1: Spatio-temporal traffic information ("global" line) derived from stationary traffic data provided by loop detectors ("empirical" line) for the 8km long A3 section between Breitscheid and Duisburg-Wedau on Friday 2014-01-10. The "local" labelled line displays stationary data obtained from the simulation with a method that emulates the behaviour of the loop detectors (source [30, Fig. 1]).

pected to measured results posterior to the validation process model, code, and numerical verification in various degrees help to improve software quality during the design phase."

To summarise the just quoted considerations, the implementation goals for OLSIMv4 aimed at the following three orthogonal goals. Firstly, OLSIMv4 should provide a platform portable simulation system model that can use various provided or later extended microscopic traffic models, and, at the same time, is independent of a concrete road traffic network model. Secondly, take over or at least prepare the taking over of data from variable message signs, from static road restrictions such as speed limits, overtaking restrictions, lane merges or closings, and from dynamic road restrictions due to road works and thereby increase information accuracy. At the same time, any implementation for the needs of OLSIMv4 should guarantee platform independence by an appropriate road traffic network model. To process the increased amount of data, exploiting CPU thread-level parallelism forms another implementation goal for OLSIMv4. Thirdly and lastly, OLSIMv4 should provide robustness and correctness of the computation as well as ways to evaluate them which includes verification and validation activities.

However, at the time of the beginning of the OLSIMv4 implementation phase, a conceptual model such as the one described in Chapter 3 or as described in the next section was not available. Consequently, formal verification for the whole simulation model was not possible at all or strictly limited to certain parts of the simulation model such as the traffic model. Verification of the simulation model became first available with the Maude verifiable simulation model [30].

4.2   Simulation model implementation

The typical simulation scenario in the context of OLSIMv4 tries to adapt the stationary traffic data provided by loop detectors to the spatio-temporal traffic data as provided by the global detectors. Therefore, the tuning elements consume the loop detector values as their should-be values and compare them against the actual values that derive from the global detectors. The tuning elements insert then any vehicles into their associated occupation tables (or remove some from them) according to their actual should-be value comparison. As the detection area of a global detector could cover several loop detector measurement cross-sections, the measurement cross-sections in the scope of a global detector could potentially interfere with each other. To avoid interference effects, the tuning elements generally use only one of the measurement cross-sections in the scope of a global detector. The global detectors, finally, provide the spatio-temporal traffic information as part of the simulation result. Figure 4.1 depicts a typical spatio-temporal traffic information that results from the OLSIMv4 simulation of an 8 km long A3 highway section on Friday 10th January, 2014. The following lines describe briefly, how the simulation obtains the results. They apply not only to OLSIMv4 but also to the Maude implementation.

Section 3.3.3 introduced the roadways as having a rectangular structure. In the context of OLSIMv4, the tuning elements emulate the traffic behaviour of the on- and off-ramps (cf. Sections 2.1.3, 3.4, 3.6.2, and 3.7.3). The roadways, thus, are not connected laterally and do not rely on the switching areas. The rectangular structure simplifies the simulation model significantly as the roadways no longer depend on the state of each other. Section 3.7 could, thus, introduce the simplified simulation model as a set of functions that any simulation engine as an implementation of the simulation model can apply sequentially for one time step. The simulation model processes the various actions in each time step as follows. "As the first operation in each time step, the detectors collect measurements from the occupation tables of the simulation or from external resources. The occupation tables maintain all vehicles in the simulation that move on an associated roadway. In the step following the measure operation, the so-called accumulators group and aggregate the detector measurements for input to the tuning elements. In the next operation, the tuning elements insert or remove vehicles from the occupation tables based on the comparison of the actual and should-be accumulator values. As the last step, the simulation updates all vehicles of the occupation tables.

[...] Regardless of which kind of coupling between the roadway-based subsystems a simulation setup intends to use, the simulation model maintains and updates a state by iteratively applying a state transition function to it. Updating the subsystems in a parallel operation or a distributed environment, e.g. a cloud-computing environment, is possible and will result in a performance gain depending on the kind of coupling. The simulation stores the state in the network structure. It consists of a time stamp, an update interval, a list of roadway-based subsystems, and a possibly non-empty list of switching areas. Among them, the roadway-based subsystems consist of two vehicle update procedures, an occupation table, a list of detectors, and a list of tuning elements that belong to the occupation table. The switching areas specify a 'from' region of a roadway, a linking direction, a 'to' region of another roadway, and a pseudo random number generator state. As roadway-based subsystems do not store and maintain their associated switching areas, the simulation engine as presented in the following makes no use of the optimized implementation. However, to summarize the simulation model with respect to the state transition, the occupation tables form the common unit for all vehicle updates. As such, the state transition function of the presented simulation model takes a minimal form" [30].

The implementation issues and challenges of this kind of simulation model is the subject to the following subsections. This section continues with a discussion of the implementation issues that may arise when implementing the conceptual design of the previous chapter. It presents some parts of the OLSIMv4 implementation which uses the Bigloo Scheme dialect as well as the implementation of the enhancement investigation which uses the Maude declarative programming language. The discussion starts with the introduction of the general Scheme of a simulation engine and it introduces then the required libraries for each of the implementation variants in Section 4.2.2. Thereafter, it presents the interfaces for each of the involved model families in Sections 4.2.3 – 4.2.13.

4.2.1 Simulation engines

Although this section is quoted from the context of the Maude implementation [30], it applies also to OLSIMv4. As depicted in Figure 4.2, the simulation engine is a kind of a state transition system that relies on some libraries, provides transition functions and expects an initial state. Its implementation depends only on the libraries that Section 4.2.2 describes. It aims to demonstrate the fitness of the

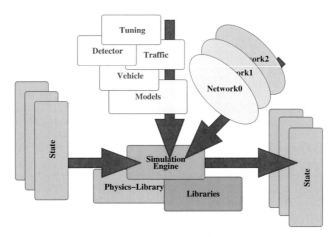

Figure 4.2: As a kind of a state transition system, the simulation engine supports loading of the model and the network definitions at runtime (source [30, Fig. 6]).

modelling approach. Therefore, linear algorithms such as `map` and `fold` scale well enough. A productive implementation requires, of course, more efficient algorithms and data structures. The transition functions of the simulation engine rely on abstract models and networks to process the update steps. As the engine does not depend on any concrete model or network, neither does a possible implementation. With the initial simulation network state given, the simulation engine then processes a finite number of discrete update steps on the simulation network. When completed, the simulation engine returns the final state of the simulation network. The following sections describe the simulation engine, the models, the networks, and the state of the simulation network.

The simulation engine as introduced in the following represents an implementation of the simulation model from the previous chapter. In addition to the libraries and the code provided by the Maude system, the engine relies on a physics library, a tuple library, and a self-implemented list library. For the purpose of this presentation, linear algorithms based on lists and tuples are sufficient. A productive implementation such as OLSIMv4 requires smarter algorithms. The engine implements each of the discussed kinds of couplings between the subsystems. The simulation engine performs the update operations measure, aggregate, tune, and vehicle update as introduced in the previous chapter. However, the simulation engine represents the subject-specific model families, namely the vehicle models, the detector models, the accumulator models, and the tuning models, only as abstract models. For example, the simulation engine knows the various model sorts and operations but, similar to virtual methods in object oriented programming, does not know how to further reduce their instantiations. The following subsections describe the simulation engine under various aspects. The simulation engine, the models,

```
(define-enumerated-type closed closed
  closed?
  closeds
  closed-name
  closed-index
  (cl:closed-for-cars cl:closed-for-trucks))
```

Listing 4.1: The `closed` enumerated type represents the closed property for cells per vehicle type and implements the closed cell property from [70, Sec. 2].

and the network specifications use the Maude declarative programming language. As the Maude system can execute the simulation engine, this approach provides iterative enhancement of the OLSIMv4 software via rapid remodelling and reverification of the various participating components. Figure 4.2 illustrates the input/output behaviour of the simulation engine as a state transition system. Figure 4.2 corresponds to the four `load` commands from Listing 4.52. The figure illustrates the engine's dependencies on the aforementioned libraries including the physics library. The first `load` command in Listing 4.52 ensures that the libraries are loaded into the Maude system. The second `load` command loads the code of the simulation engine itself. After the loading, the Maude system provides the simulation engine as executable code. The execution starts with an initial term. The horizontal arrows in Figure 4.2 indicate that the engine takes some initial state as its input (which corresponds to the initial term), processes the state transitions on it, and returns with the resulting state. Prior to processing the state transitions, the simulation engine requires loading of various model definitions for it. The figure from above illustrates the loading of the model definitions by the arrow from the top into the simulation engine. The model definitions implement the abstract models provided and used by the simulation engine. The engine uses the model definitions for the model families of the vehicle model, the microscopic traffic model, the detector model, the accumulator model, and the tuning element model. For example, each model definition in the detector model family defines a detector that provides a measurement method. However, the methods of the various detectors differ in the measuring details while they all result at the same time in an updated instance of the detector. The network specifications, by contrast, initially depend on the model definitions and then instantiate network objects as combinations of the elements of the various model classes. The simulation engine requires loading or defining of some network specifications prior to running the simulation. Figure 4.2 illustrates the loading by the arrow from the right top corner into the simulation engine. The network specifications complete the instantiation of the simulation engine. To initiate a simulation run, the simulation engine expects a simulation network as its initial state as well as the number of simulation runs, i.e. the number of time steps to simulate.

4.2.2 Libraries

This Subsection introduces the libraries required by the OLSIMv4 implementation as well as those required by the Maude implementation. Section 4.2.2.2 presents all Maude specific libraries. For the Bigloo implementation, the tuples and records requirements are complementary, as Bigloo provides a generic `struct` implementation that can serve as a tuples library. Bigloo implements, not surprisingly, the Scheme R5RS lists API which suits the needs of OLSIMv4 very well. Section 4.2.2.1 presents all Maude specific libraries.

4.2.2.1 Bigloo specific libraries

The Bigloo `struct`s implementation provides an accessor `(struct-ref::obj s::struct k::int)`[2] and a field modifier `(struct-set!::obj s::struct k::int o::obj)`, both close to and in the sense of the generic R5RS lists and vectors data types [1, Secs. 6.3.2, 6.3.6]. While both take a struct `s` and an field number `k` as their arguments, only the latter consumes the additional argument of an object that the particular struct element should be set to. The former returns the object stored in the location of the `k`-th field while the latter takes and sets the `k`-th field object to object `o`. As Bigloo maps the structs directly onto C structs they consume not any additional overhead. With the exception of the presented tuples, Bigloo in version 3.8c already provides libraries for pairs, lists, and records. The list library implements and extends the specification of the 1st "Scheme Request for Implementation" (SRFI-1) [190]. So does the 9th "Scheme Request for Implementation" (SRFI-9) [96] for "Defining Record Types". The SRFI-9 specification has the advantage of a partial portability across several Scheme implementations. The SRFI-9 Bigloo implementation uses the structs internally and provides syntactic sugar for it.

---

[2]The signature notation follows the Bigloo type annotation syntax [189, Sec. 25]. The double colon separates object identifiers from type annotations. Missing object identifiers indicate anonymous objects.

```
(define-finite-type v-limit v-limit
  (vl-km/h vl-m/s vl-c/s)
  v-limit?
  v-limits
  v-limit-name
  v-limit-index
  (vl-km/h v-limit-km/h)
  (vl-m/s v-limit-m/s)
  (vl-c/s v-limit-c/s set-v-limit-c/s!)
  ((vl:car-60    60 ((@ km/h->m/s fup-ca)   60) ((@ km/h->c/s fup-ca)   60))
   (vl:car-80    80 ((@ km/h->m/s fup-ca)   80) ((@ km/h->c/s fup-ca)   80))
   (vl:car-100  100 ((@ km/h->m/s fup-ca)  100) ((@ km/h->c/s fup-ca)  100))
   (vl:car-130  130 ((@ km/h->m/s fup-ca)  130) ((@ km/h->c/s fup-ca)  130))
   (vl:car-160  160 ((@ km/h->m/s fup-ca)  160) ((@ km/h->c/s fup-ca)  160))
   (vl:truck-60  60 ((@ km/h->m/s fup-ca)   60) ((@ km/h->c/s fup-ca)   60))
   (vl:truck-80  80 ((@ km/h->m/s fup-ca)   80) ((@ km/h->c/s fup-ca)   80))))
```

Listing 4.2: The `v-limit` finite type represents the maximum velocity cell property per vehicle type and implements the cell maximum velocity property from [70, Sec. 2].

```
(define-enum-set-type v-limit-set v-limit-set
                      v-limit-set?
                      make-v-limit-set
  v-limit
  (@ v-limit? zelle-finite-types)
  (@ v-limits zelle-finite-types)
  (@ v-limit-index zelle-finite-types))
```

Listing 4.3: The `v-limit-set` represents the set of maximum velocities that apply to a particular cell in the cell model from [70, Sec. 2].

However, unfortunately and in contrast to many other Scheme implementations such as Scheme 48 [99, Secs. 5.9–5.11] and R6RS [98] compliant implementations, Bigloo does not provide finite and enumerated types. As they are very elegant and practically for the needs of OLSIMv4, the author of this thesis implemented a more general variant for records that in addition to another SRFI-9 implementation provides the enumerated and the finite types. Listing 4.1 presents an OLSIMv4 example for enumerated types that represents the cell property `closed`. It describes whether a cell that employs the `closed` enumerated type is closed for a certain vehicle type or not. Habel introduced this cell model in the first position as part of his thesis [70, Sec. 2]. The cell property `closed` may participate in a rule of a certain cell configuration and as detailed in Section 3.3.1. The definition generates the enumerated type, the two constructor syntaxes `(closed cl:closed-for-cars)` and `(closed cl:closed-for-trucks)`, as well as a predicate function `(closed?::bool o::obj)`, the accessors `(closed-name::symbol ::struct)` and `(closed-index::int ::struct)`, and the vector `closeds` that contains all instances of the `closed` enumerated type. Thus, the for each enumerated type instance, a property such as `(eq? (vector-ref closeds (closed-index (closed cl:closed-for-cars))) (closed cl:closed-for-cars))` in the case of the `closed` enumerated type holds. The signatures of the generated procedures rely on the generic type `::struct` as the type for the enumerated type instances. The instantiated objects, however, are able to distinguish the various enumerated types and, thus, support optional runtime type checking.

The **define-finite-type** form extends the enumerated types by storing and providing constant field values per instance. Listing 4.2 presents a shortened variant of the OLSIMv4 velocity limit properties as an example for the finite types syntax form. They may also participate in a rule of a certain cell configuration that Section 3.3.1 has already described. The velocities were also part of the cell model by Habel. The **define-finite-type** expression makes use of the procedures `(km/h->m/s::long ::long)` and `(km/h->c/s::long ::long)` from the `fup-ca` OLSIMv4 module. They determine the corresponding velocity value in metres and cells per second of a velocity in kilometres per hour.

```
(enum-set-empty?::bbool ::struct)
(enum-set->list::pair-nil ::struct)
(enum-set-minimum::struct ::struct)
(enum-set-maximum::struct ::struct)
(enum-set-enumerand?::bbool ::struct ::struct)
(enum-set-member?::bbool ::struct ::struct)
(enum-set-adjoin::struct ::struct ::struct)
(enum-set-union::struct ::struct ::struct)
(enum-set-intersection::struct ::struct ::struct)
(enum-set-negation::struct ::struct)
(enum-set-difference::struct ::struct ::struct)
(enum-set?::bbool ::obj)
(enum-set=?::bbool ::struct ::struct)
```

Listing 4.4: The `enum-set` represents sets over enumerated and finite types.

```
(graph-node-ref::obj ::struct ::struct)
(graph-contains-key?::bbool ::struct ::struct)
(graph-has-neighbour?::bbool ::struct ::struct . ::obj)
(graph-add-node!::struct ::struct ::obj . ::obj)
(graph-del-node!::obj ::struct ::struct)
(graph-replace-node!::struct ::struct ::struct ::obj . ::obj)
(graph-add-neighbour!::struct ::struct ::struct ::obj)
(graph-del-neighbour!::struct ::struct ::struct ::struct)
(graph-replace-neighbour!::struct ::struct ::struct ::struct ::obj)
(graph-size::long ::struct)
(graph-map::pair-nil ::struct ::procedure)
(graph-for-each::unspecified ::struct ::procedure)
(graph-fold::obj ::struct ::procedure ::obj . ::obj)
(graph-traverse-pre-order::obj ::struct ::struct ::procedure ::obj ::procedure ::struct)
(graph-traverse-post-order::obj ::struct ::struct ::procedure ::obj ::procedure ::struct)
(graph-dfs::obj ::struct ::struct ::procedure . ::obj)   ; depth first search
```

Listing 4.5: Signatures of subset of the generic graph procedures.

The enumerated and the finite types in the presented form can only represent a single cell property but not sets of properties. They combine the vehicle type with the velocity limit implicitly and can, thus, implement a property of a single rule from Section 3.3.1. However, sets of rules require to represent sets over enumerated or finite types. The lane change and overtaking restrictions provide a typical example from the context of OLSIMv4 that require this functionality. These restrictions combine themselves disjunctively into sets of restrictions. The velocity limits, in contrast, combine themselves conjunctively into sets and, generally, the most restrictive element will dominate. The set theory with unions, intersections, member predicates, and join functions forms the common ground of both examples. However, the implementation of sets over finite types may look more trivial than it actually is. In the following, some reasons and properties are introduced.

Generally, the properties and requirements for any enumerated and finite type sets application programming interface (API) as well as a possible implementation consent likely into the following ones. The enumeration and finite types should be type safe and maybe even scopable as the enum classes that became available with C++11 [88, Sec. 7.2]. Any implementation should map simple enumerations onto bits and bitfields. But, at the same time, enumerations should have a generic API and use the same accessors and modifiers for instances of sets over varying enumeration and finite types. Additionally, any implementation should provide means to automatically choose or at least adjust manually the best underlying type for any given enumeration. Finally, any implementation should also provide excellent performance and minimal space. However, not all programming languages offered such types in 2008. Java 5 introduced in 2004 an implementation based on generics even though Java 5 may have had to pay a relatively high price with the introduction of the generics [170]. The Scheme standard R6RS [98, Sec. 14] provided a standardisation for them after several Scheme implementations such as [99,

```
traffic_nrw_1_0_0=# SELECT '(' || id || ' ' || node_sexpr || ')' AS sexpr
                    FROM topologie_2_4_9.matview_ov2_fsn_nodes LIMIT 10;
                sexpr
---------------------------------------
 (fs28720 (list "fs1" "100.0" "#f"))
 (fs28727 (list "fs1" "1950.0" "#f"))
 (fs28726 (list "hfs" "1950.0" "#f"))
 (fs28725 (list "fs2" "100.0" "#f"))
 (fs28724 (list "fs1" "100.0" "#f"))
 (fs28714 (list "fs1" "4100.0" "#f"))
 (fs28715 (list "hfs" "100.0" "#f"))
 (fs28712 (list "fs1" "100.0" "#f"))
 (fs28713 (list "hfs" "4100.0" "#f"))
 (fs28718 (list "fs1" "4300.0" "#f"))
```

Listing 4.6: Random nodes of the single lane graph that is in use for OLSIMv4.

```
(make-channel::struct)
(channel-send channel::struct message::obj)
(channel-receive channel::struct)
(choosing-rendezvous* choices::pair-nil)
(rv:synchronize rendezvous::struct)
```

Listing 4.7: The Scheme port [34] of the Concurrent ML offers also channels and rendezvous as part of a message passing abstraction.

Sec. 5.11] had already offered some variants. And, finally, C++ left the sets over finite types as an exercise for the interested programmer [162] in 2007.

For the Scheme programming language, the `enum-set`s provide the capability to form sets over the enumerated and finite types. Listing 4.4 gives a taste of how the supported operations look like. Most of them have rather self-explanatory names–e.g. the predicate function `enum-set-member?` tests a given element of an enumerated or finite type for membership in a certain enum set combination while the `enum-set-empty?` tests for empty enum sets. The `enum-set-complement` procedure calculates the complement of a given set using bit string inversion. The other procedures include forming unions, intersections, and differences of sets. The author of this thesis wrote an `enum-set` library in the year 2008 without having knowledge about the R6RS specification [98, Sec. 14] (even though it has already been present). The implementation of the `enum-set` library uses the "Integers as Bits" SRFI-60 library [91] that the author ported to Bigloo for this purpose. The `enum-set` implementation maps the enumerated and finite types onto binary digits of integer numbers or "two's-complement strings of bits" [91]. The set operations can then determine their results by using the arithmetic operations provided by the SRFI-60 library. The Bigloo port of the library provides not only the generic procedures from Listing 4.4 but also type and length specialised variants for each of them, e.g. the generic procedure `enum-set-maximum` dispatches any input enum set onto one of the following type specific variants `enum-set-maximumfx` for the integers, `enum-set-maximumllong` for the `long long` C type or machine dependent integer representations, and `enum-set-maximumbx` for the big numbers. As the Bigloo `enum-set` implementation maps the set operations onto integer arithmetic operations, it scales very well. Java uses also a bit length dependent dispatching method onto `RegularEnumSet` and `JumboEnumSet`.

An example for the definition of an enum set follows. The set type `v-limit-set` from Listing 4.3 defines a set over the `v-limit` finite type from Listing 4.2. The `enum-set` syntax creates the constructor `make-v-limit-set` as well as the constructor syntax `v-limit-set` that can be used as, for example, in `(v-limit-set vl:car-60 vl:truck-60)` to construct a set with the velocity limits `vl:car-60` and `vl:truck-60`. Furthermore, it creates the predicate function `v-limit-set?`.

However, the enum sets are not only useful in defining traffic properties for cells. In the further context of OLSIMv4, they also provide an elegant way to store and access the neighbour objects of a node which takes part as the basic element of a graph. The neighbours of a certain node of a graph are

```
(module plet
   (library fthread pthread srfi1 srfi19 srfi69 cml jobd)
   (main main))
(define-syntax plet
   (syntax-rules ()
      ((_ ((var expr) ...) body ...)
       (multiple-value-bind (var ...)
          (apply values (quasi-parallel (lambda () expr) ...))
          body ...))))
(define (main argv)
   (reset-quasi-parallel-workers! 4)
   (let ((n 100000))
      (plet ((x (map (lambda (x) (* 2 x x)) (iota n 1)))
             (y (map (lambda (y) (+ y y)) (iota n 1))))
         (let ((result (fold + 0 (map (lambda (x y) (/ x y)) x y)))
               (shouldbe (/ (* n (+ n 1)) 2)))
            (if (= result shouldbe)
                (print (format "plet: result ~a matched should-be value ~a." result shouldbe))
                (print (format "plet: result ~a did NOT match should-be value ~a." result shouldbe)))))))
; compile with bigloo -o plet plet.scm
; run with ./plet => plet: result 5000050000 matched should-be value 5000050000.
```

Listing 4.8: Example program for the use of a parallel `let` form (`plet`).

also known as arcs. The `enum-set`s provide the advantage of a name based access to the arc objects and they serve as a database of the node's neighbour relationships. The `enum-set-cells` component provides this functionality for OLSIMv4 as part of the graph library. They exploit the following characteristics of the bit sets. Enum sets as bit sets have a strict ordering of their elements from the lowest significant bit up to the highest significant bit. When a node wants to associate a neighbour object, the node provides the arc name and the neighbour object to associate with the arc name. The `enum-set-cells` structure looks up the underlying bit index of the associated arc name as an element of the `enum-set` structure. The position of the index in the set of all indexes currently in use serves then as an index in the vector that stores all neighbour objects. The graph library contains several generic procedures to access and modify graphs, nodes, and their neighbours in the same way as is the case with `struct`, `vector`, and `list`. Listing 4.5 gives a taste of how the generic graph procedures look like.

Most of the generic procedures have rather self explanatory names. The two graph traversing procedures and the `graph-dfs` enable graph traversing according to a given subset of the neighbours specified either by a single enumerated type argument or an enum set of the arc names. The graph library provides also automatic export into and import from a PostgreSQL database [68]. A specific PL/pg-SQL abstraction `SELECT graphs_2_1.init_schema('nrw_roadnet', '');` creates and initializes a fresh schema named `nrw_roadnet`. The schema can then be filled with the graphs instantiated from inside Bigloo. As the PostgreSQL implementation uses also the structuring into enumerated and finite types, enum sets, and graph types, navigating through the data inside the database is eased. Another special feature of the PostgreSQL graph library representation exists with the option to statically compile the graph data into Bigloo Scheme code. To choose this option, the implementer selects the graph node and neighbour data from the PostgreSQL database as Scheme list expression. The list expressions can then be typed and bound onto global variables. Listing 4.6 presents an example that relies on the single lane graph type similar to the one from Listing 4.9 as introduced below. In addition to the enum sets and enum set cells, the graph implementation relies on the SRFI-69 hashtables [93].

Listing 4.9 presents two examples to illustrate the neighbourship relations and demonstrate how to define a graph type and how to instantiate it thereafter. The examples include two graph type definitions, namely those of the `slg` and the `dng` graph type. The former represents sections of a single lane with their neighbour relationships such as left, right, predecessing, and successing sections. The latter represents a detector network where each detector may be neighboured to another detector by each of the relations left, right, next, and previous. The graph definitions are shortened versions of the OLSIMv4 implementation. The detectors' neighbour relationships support that detectors may depend on the measurements of neighboured detectors which may be essential for tuning algorithms.

The relationships `dn:in`, `dn:from`, and `dn:to` participate in bundling several detectors into a cross-section measurement detector which Section 3.6.1 treats as a special case of the accumulators.

The description of the examples from above did only cover graph types but left out their instantiation. Listing 4.9 covers also the instantiation by the demonstration for the single lane sections graph. The source code provides a complete example for the instantiation of the single lane sections graph `sl-net`. The single lane sections from Listing 4.9 implement in principle the sections from Section 3.3.2 with the addition constraint that lateral neighbour sections are required to have the same start and end point with respect to the longitudinal extent. The graph instantiation uses the constructor syntax created by the graph type definition. The graph library also provides generic constructor procedures such as `graph-add-node!`, `graph-replace-node!`, and `graph-replace-neighbour!`.

```
(define-graph-type slg sl-link                    ; graph with single lane sections
  (sl:prev sl:right sl:cell  sl:dets sl:up sl:left sl:next))
(define-graph-type dng dn-link                    ; detector network graph
  (dn:prev dn:right dn:in dn:pendant dn:von dn:bis dn:left dn:next))
(define (id->nr id . ignore)                      ; calculate the bucket for the hash table
  (let ((slen (string-length id)))
    (let ((num (string->integer (substring id 2 slen))))
      (if num num
          (error 'id->nr "id is not a number (list id n) : " (list id num)))))))
(define (make-id-gen id len)                      ; generate a unique id that wraps the bucket
  (let ((counter 0))
    (lambda ()
      (set! counter (+ counter 1))
      (let ((no (number->string counter)))
        (let ((slen (string-length no)))
          (if (> slen len) (error 'make-id-gen "counter out of bounds" no)
              (string-append id (make-string (- len slen) #\0) no)))))))
(define sl-id-gen (make-id-gen "sl" 5))           ; instantiate the id generator
(define-srfi9-record-type singlelane              ; the single lane nodes use a record type
  (really-make-singlelane id type length)
  singlelane?
  (id singlelane-id)
  (type singlelane-type)
  (length singlelane-length))
(define (make-singlelane type length)             ; wrap the constructor for the single lane nodes
  (really-make-singlelane (sl-id-gen) type length))
(define-enumerated-type slt slt slt? slts slt-name slt-index
  (sl0 sl1 sl2 sl3 sl4))                           ; single lane type: from right to left lane
(define sl-keys '())
(define sl-net
  (call-with-values (lambda () (slg singlelane-id id->nr ; slg is the constructor syntax for the graph
        ((hfb1/sl0 (make-singlelane (slt sl0) 1000)) ; instantiate nodes
         (hfb1/sl1 (make-singlelane (slt sl1) 1000))
         (hfb1/sl2 (make-singlelane (slt sl2) 1000))
         (hfb2/sl0 (make-singlelane (slt sl0) 1000))
         (hfb2/sl1 (make-singlelane (slt sl1) 1000))
         (hfb3/sl0 (make-singlelane (slt sl0) 1000))
         (hfb3/sl1 (make-singlelane (slt sl1) 1000))
         (hfb3/sl2 (make-singlelane (slt sl2) 1000))) ; now connect the neighbour relationships
        (hfb1/sl0 (sl:next hfb2/sl0)                    (sl:left hfb1/sl1)                         )
        (hfb1/sl1 (sl:next hfb2/sl1)                    (sl:left hfb1/sl2) (sl:right hfb1/sl0))
        (hfb1/sl2                                                          (sl:right hfb1/sl1))
        (hfb2/sl0 (sl:next hfb3/sl0) (sl:prev hfb1/sl0)                                         )
        (hfb2/sl1 (sl:next hfb3/sl1) (sl:prev hfb1/sl1) (sl:left hfb3/sl2) (sl:right hfb2/sl0))
        (hfb3/sl0                    (sl:prev hfb2/sl0)                                         )
        (hfb3/sl1                    (sl:prev hfb2/sl1) (sl:left hfb3/sl2) (sl:right hfb3/sl0))
        (hfb3/sl2                                                          (sl:right hfb3/sl1))))
      (lambda (sln keys) (set! sl-keys keys) sln)))
; 1:=> (graph-node-ref sl-net (car sl-keys))
; #{<singlelane> sl00001 #{<slt> sl0 0} 1000}          ; the first single lane section
; 1:=> (graph-node-ref sl-net (graph-node-neighbour-ref sl-net (car sl-keys) (sl-link sl:next)))
; #{<singlelane> sl00004 #{<slt> sl0 0} 1000}          ; 'next' neighbour of first single lane section
```

Listing 4.9: The graph type `slg` represents sections of a single lane with their neighbour relationships such as lateral and longitudinal neighbour sections. The graph type `dnet` represents a detector network where each detector may be neighboured to another detector by each of the neighbourship relations. The graph instantiation expression uses the constructor syntax generated by the graph type definition.

```
fmod PAIR{X :: TRIV,Y :: TRIV} is sort Pair{X,Y} .
  op <_;_> : X$Elt Y$Elt -> Pair{X,Y} . op 1st : Pair{X,Y} -> X$Elt . op 2nd : Pair{X,Y} -> Y$Elt .
  var A : X$Elt . var B : Y$Elt .
  eq 1st(< A ; B >) = A . eq 2nd(< A ; B >) = B .
endfm
```

Listing 4.10: Pairs for theory TRIV to illustrate the tuples library (source: [30, Lst. 11]).

```
view BoolEltSWO from STRICT-WEAK-ORDER to BOOL is
  sort Elt to Bool .
  vars X X' : Elt .
  op X < X' to term (not X and X') .
endv
```

Listing 4.11: View from sort Bool to theory STRICT-WEAK-ORDER (source: [30, Lst. 12]).

Apart from the graph library, OLSIMv4 relies on several other libraries that participate in providing CPU thread-level parallelism for Bigloo. Bigloo provides a POSIX thread application programming interface (API) that is widely compatible to the SRFI-18 [60] API. The API is quite low-level as the programmer has to deal with thread creation, explicit locking, and synchronisation. However, it supports the building of higher-level programming models with seamless language integration on top of it.

Section 4.1.2 has pointed to the parallelism theme and it thereby has discussed some of its aspects already. At the lowest level, the Scheme port [34] of the Concurrent ML [174, 175] provides message passing through synchronous and asynchronous channels. The library was designed for concurrent programming which is distinct to programming in the context of exploiting CPU thread-level parallelism. The Concurrent ML programming model is quite popular for use in concurrent programming tasks. It has, for example, been "adapted for use as a discrete event simulation language simply by equipping it with a mechanism for managing simulated time" [78]. Listing 4.7 presents some of the channels and rendezvous procedures that reflect at the same time the basic concepts of the Concurrent ML message passing abstractions. Even though Concurrent ML was not designed for parallel implementations in the first place [176], it is at hand to try implementing it in a parallel environment [176]. However, except of Haskell [38] and a few ML dialects such as ParallelML, MLton, the Manticore project, and the most famous or popular ML implementation, OCaML, do not offer concurrent programming models on top of parallel operating system threads. The number of functional language platforms shrinks further to Haskell [161] and (as of 2012) Manticore [20] when considering even more parallelism specific high-level concepts such as nested data parallelism. However, the Concurrent ML programming model significantly reduces the effort to implement similar concepts or exploit even implicit parallelism on top of the concurrency programming model.

As Bigloo provides already the SRFI-18 [60] API on top of POSIX Threads [193], the Concurrent ML API combined with CPU thread-level parallelism is tempting to implement it as a message passing API for the Bigloo Scheme programming language. While unaware whether the Scheme port [34] by Campbell of the Concurrent ML library already reflects the enhancements that provide support for parallelism [176], the author of this thesis ported the library to Bigloo. The explicit parallelism could then be used to offer alternative implementations of the standard forms such as let. The integration is a straightforward task and can be done as shown in Listing 4.8. The plet provides the syntax for a parallel let that uses the quasi-parallel procedure behind the scenes to compute the values for the right hand side expressions of the let.

The quasi-parallel abstraction expects a number of thunks[3] and returns a list containing the thunks' return values. The quasi-parallel procedure dispatches the thunks onto a fixed number of threads

---

[3]The word thunk is a mixture between thumb and function and means functions that are thumb in the sense of taking no arguments.

which in the case of Listing 4.8 has been set onto 4 by the use of the `(reset-quasi-parallel-workers!   4)` expression. In addition to the Concurrent ML Scheme port library, the `jobd` library relies on the `srfi19` library which is a port of an implementation of the SRFI-19 specification [63]. SRFI-19 specifies time and date operations for various time and date representations such as "Universal Coordinated Time (UTC), International Atomic Time (TAI), monotonic time (a monotonically increasing point in time from some epoch, which is implementation-dependent), CPU time in current thread (implementation dependent), CPU time in current process (implementation dependent), Time duration" [63].

The OLSIMv4 simulation engine makes also direct use of the SRFI-19 library, especially when accessing or storing data into the database. In these cases, OLSIMv4 has to convert timestamps from the simulation internal format which conforms to the monotonic time to the external time formats, currently mostly UTC.

As the last part of the libraries for OLSIMv4, a random number generator implementation is required. SRFI-27 [54] provides a good specification as well as some pointers to supplementary material such as the Diehard testsuite [134] for a random number sources interface. However, it does not provide any sources for the Bigloo Scheme dialect. Additionally, several factors influence the choice of a random number generator. In addition to the question which period a suitable random number generator should provide, there are several other characteristics of random number generators that might have an impact on the simulation result and its computation time. As random number generators require maintaining a state, the question arises, whether to share the state or not among several threads, i.e. using a thread-safe implementation. Instead of sharing the random number generator's state, splitting [122] and leap-frogging [123] provide additional options. In general, sharing the state of a random number generator that is used excessively (as is the case in OLSIMv4), is not a good idea as it requires a lot of synchronisation among the concurring threads. This turned out to initially massively limit the performance of the early OLSIMv4 implementation, as Bigloo's `random` implementation relies on the `rand` Standard C Library implementation. The `rand` function, however, is not thread-safe and it maintains a single state which will be updated on each call. Even in the case where an appropriate `rand` procedure would be thread-safe, multiple threads would update the state concurrently. The concurrent updates in turn would then require a lot of synchronisation in the case of OLSIMv4 where up to 150 000 vehicles can drive on the road network simultaneously. Updating them in the network consumes about 18 000 000 random integers in each simulation minute. After several test, the developers of OLSIMv4 decided to equip each occupation table with its own random number generator as this still leaves the option to keep some of them in sync. The other characteristics of the random number generator turned out as having not a big impact on the simulation results. As a consequence, the developers integrated several primitive pseudo-random number generators to choose among at runtime of the OLSIMv4 simulation. They all have signature `(random::long s::long)` for integers or even `(random::llong s::llong)` for long long integers and, consequently, maintain a hidden state.

4.2.2.2 Maude specific libraries

The implementation of the simulation engine requires libraries for lists, pairs, and tuples or records. The Maude language does not ship a tuples implementation but comes with a list implementation. The example from Listing 4.10 presents the tuples library for the special case of pairs. As most other Maude code snippets, too, the code of the example is taken from the author's publication [30]. The notation for triples and so on is analogous to the one from Listing 4.10. The tuples are available for theories `TRIV` and `STRICT-WEAK-ORDER` that come with Maude. Theories define interfaces in Maude. Theory `STRICT-WEAK-ORDER` requires a binary relation `op _<_ : Elt Elt -> Bool` over theory `TRIV`.

Because the list library of Maude treats a single element as the same domain kind as a list with a single element, the author decided to self-implement another list library. It is also available for both theories, `TRIV` and `STRICT-WEAK-ORDER`. In Maude, views define mappings between theories and sorts. To denote views to theory `STRICT-WEAK-ORDER` in this article the suffix `EltSWO` marks the name as being an element of the sort, e.g. as in `BoolEltSWO`. It is defined as shown in Listing 4.11. As an extension to the list library, a `map` and `fold` implementation for use in combination with lists and higher order functions is provided. The implementation idea is based on [55] and the functionality is a subset of the Scheme SRFI-1 list library [190] and inspired by it as well. The simulation engine further depends on a physics library that provides representation, access, combination, and modification of physics observables such as period, length, velocity, traffic flow, density, and level of service. Listing 4.13 presents the sort `Length` as an example for a subsort of `TrafficObservable`.

4.2.3 Vehicles

Section 3.2 has described the varying contexts where vehicle types may occur in. Even though certain aspects such as the formation of plugs [115] were known, the differences of the various contexts had not been considered thoroughly and the possible impacts on the traffic flow and the simulation results had not been studied carefully during the implementation phase of OLSIMv4. Consequently, OLSIMv4 lacks a clear concept for vehicle types or, to be more precise, sets of vehicle types. Such a concept could have prepared, for example, the modelling and implementation of vehicle type sensitive detectors that OLSIMv4 emulates [31, Sec. 3A] and presents already in the truck specific view as mentioned in Section 2.1. Listing 4.12 presents a part of the vehicle type definitions currently in use by OLSIMv4. The vehicle types are modelled by means of the vehicle constructor who stores and maintains a vehicle type instance per vehicle in a record. The record contains the vehicle type specific information maximum length, maximum velocity, speed limit tolerance, maximum acceleration and deceleration capability. Listing 4.12 presents some example constructor calls that wrap the vehicle type.

The vehicle types have not been modelled as presented using the finite types as shown in Listing 4.14 because during the startup and experimentation phase of the OLSIMv4 implementation the finite types library was not fully available. Additionally, there was some uncertainty wether to use a hyperrealistic approach or not with a kind of a randomisation of vehicle type properties. To preserve the highest flexibility for the implementation, the implementers decided to use the more loose approach from Listing 4.12 and not to store the vehicle type specific information in a fixed size database in form of

```
(define-enumerated-type typ typ typ? typen typ-name typ-index
  (t:dummy t:truck t:car))
;                 pos vel lane       vtype length vmax vmul amax dmax
; (make-truck     pos vel lane            10   18 1.15    1    2)
; (make-car       pos vel lane             5   22 1.15    1    2)
; their definitions wrap the vehicle constructor with different arguments:
; (make-vehicle   pos vel lane (typ t:truck) 10 18 1.15  1    2)
; (make-vehicle   pos vel lane   (typ t:car)  5 22 1.15  1    2)
```

Listing 4.12: Some of the OLSIMv4 vehicle types as enumerated types. Unfortunately they lack vehicle type specific properties such as maximum velocity, maximum length and similar.

```
sorts Unit LengthUnit PeriodUnit VelocityUnit .
subsorts LengthUnit PeriodUnit VelocityUnit < Unit .
ops Cells Kilometer Meter : -> LengthUnit [ctor] .
ops C/s Km/h M/s : -> VelocityUnit [ctor] .
ops Seconds Minutes Hours : -> PeriodUnit [ctor] .
sorts Observable Length . subsort Length < Observable .
op Observable : Qid Rat Unit -> Observable .
ops _<_ _<=_ : Observable Observable -> Bool .
ops _+_ _-_ : Observable Observable -> Observable .
op Length : Int LengthUnit -> Length .
ops Kilometer Cells Meter : Length -> Int .
ops Seconds Minutes Hours : Period -> Int .
ops Km/h C/s M/s : Velocity -> Int .
```

Listing 4.13: `Observable` subsorts `Period`, `Velocity`, and `Length` as parts of the physics library (source: [30, Lst. 13]).

```
(define-finite-type vehtype vehtype
  (len vmax vmul amax dmax)
  vehtype? vehtypes vehtype-name vehtype-index
  (len  vehtype-len)
  (vmax vehtype-vmax)
  (vmul vehtype-vmul)
  (amax vehtype-amax)
  (dmax vehtype-dmax)
;      name length vmax  vmul amax dmax
  ((vt:dummy       0   25 1.00    1    2)
  (  vt:car        5   22 1.15    1    2)
  (vt:truck       10   18 1.15    1    2)))
```

Listing 4.14: Implementation of the vehicle types from Section 3.2.1 using the finite types library.

```
sorts VehicleType DummyType . subsort DummyType < VehicleType .
op VehicleType : Qid Velocity Length -> VehicleType .
op VehicleType : Qid Nat Nat -> VehicleType .
op DummyType : Nat Nat -> DummyType .
vars Q : Qid . vars V L : Nat .
eq VehicleType(Q, V, L) = VehicleType(Q, Velocity(V, C/s), Length(L, Cells)) .
eq DummyType(V, L) = VehicleType('DummyType, V, L) .
eq DummyType = DummyType(0, 1) .
eq DummyType?(T) = IsVehicleType?(T, 'DummyType) .
```

Listing 4.15: Sort VehicleType combines a vehicle type name as well as its maximum velocity and length (source: [30, Lst. 14]).

the finite types from Listing 4.14. Theoretically, the conceptual design requires the vehicle types only to support or implement the accessors for the vehicle type fields. The vehicle types implementation in Listing 4.14 is, thus, a proper implementation of the conceptual design from Section 3.2.1—even though not the only one. However, the generic accessors for the vehicle types are defined together with the instantiation of the vehicle type elements.

Another implementation example of the conceptual design which is presented in Listing 4.15 has been taken from the author's publication [30]. It supports polymorphic vehicle types through subtyping sort VehicleType. Vehicles aggregate a specific vehicle type using the VehicleType constructor from Listing 4.15. It combines the vehicle type name of sort Qid, the vehicle type maximum velocity of sort Velocity and the vehicle type maximum length of sort Length into elements of sort VehicleType. The simulation engine uses and provides an additional vehicle type DummyType for passive vehicles. The Maude subtyping mechanism supports adding further vehicle types at runtime and, thus, provides a more flexible way which is the main difference over the finite types implementation from above. Section 4.3 introduces the vehicle type elements or subtypes of sort VehicleType.

The details of the conceptual design for the vehicles depend strongly on the implementation kind and purpose. For the reflective and rather minimal implementation of the author's publication [30], the following properties in accordance with Listing 4.16 are sufficient. A vehicle therein consists of a sort VehicleType element, a lane number of sort Int, a position of sort Length, a velocity of sort Velocity, and maybe also a velocity for lanes. Except for the last one, the simulation engine uses these attributes to arrange, order and maintain lists or sets of vehicles. Moreover, the simulation engine provides a kind of lane-position based access for vehicle lists. It also decides whether vehicles will be garbage collected or not. Vehicles with a lane or position attribute that refers to outside of the associated roadway may be–depending on the implementation–garbage collected or not with the following exception. In case of periodic boundary conditions, the vehicle update operation as described in Section 4.2.10 maps updated vehicles back into the roadway.

As has been the case with the vehicle types, the Maude sort Vehicle implementation supports also polymorphic vehicles through subtyping. Because subtyping is different to implementation inheritance

[104], the subsorts have to implement the operations that instantiate the vehicles with the given set of attributes lane, position, and velocity as shown in Listing 4.16. And they also have to implement accessors for each of the attributes. The appropriate operations to manipulate single attributes are then ChangeLane, SetPosition, and SetVelocity. The simulation engine uses ApplyOffsets and SwitchType as additional operations in combination with switching areas (see Section 4.2.11).

The presented Maude vehicle implementation supports only one set of coordinates at a time. The Maude vehicle code is not a complete implementation of the structures S_{veh}^4 and S_{coord}^4 from the conceptual design in Section 3.2.2. Instead it implements the rather minimal vehicle structure S_{minveh}^4 from Equation 3.6. The more general form of the conceptual design from Section 3.2.2 contained also vehicle trajectories of several time steps. Vehicle coordinates of several time steps are necessary to support debugging, to calculate the lane change velocity, and to provide alternative coordinates in scenarios where two vehicles would lane change onto the same position. With the additional attributes len, vmax, amax, dmax, and the multiple coordinates, the OLSIMv4 vehicle source code as presented by Listing 4.17 implements the S_{veh}^4 vehicle structure of the conceptual design from Section 3.2.2. For example, the (veh:pos::long k::vector) operation on a vehicle k, which represents its conceptual counterpart $\mathbf{k}_n^{(t)} \in S_{\text{veh}}^4$, implements the access on the position coordinate $(\mathbf{k}_n^{(t)} \# \mathbf{q}^{(t)}) \# x$. Similarly do the other procedures from Listing 4.17 implement the access and the modification of the other structure members even though the range and domain kinds do not always match completely. The vehicle position may serve as an example for the different domain kinds. The conceptual design specified $\mathfrak{X}_{\text{veh}}^4$ with $\{i \text{ cells} \mid i \in \mathfrak{X}_{\text{rway}}\}$ and $X_{\min}, X_{\max} \in \mathbb{Z}$ as the domain kind, whereas the implementation chose the long domain kind. As the longest roadway in NRW is 320 km long and the cells have length 1.5 m, the choice $X_{\min} = 0$ and $X_{\max} = 2^{18} - 1 = 262\,143$ would have been large enough for NRW and would have provided the additional benefit of combining the attributes position, velocity, lane, and others into a single size long element. However, at the beginning of the OLSIMv4 implementation phase things were not that clear and such a number format would then have been only space optimised but not performance optimised. The representation of $\mathfrak{X}_{\text{veh}}^4$ by long is, thus, not a bad choice.

In addition to the multiple coordinates, the vehicles for OLSIMv4 require more attributes. Two attributes, namely lcs-left and lcs-right, represent counters that store the number of successive and successful lane change initiations. The original symmetric multilane model by Pottmeier [164, Eq. 6.6] required this counter. However, Habel and the author replaced the counter by a more effective random number test that Equation 3.69 defines. The access times and the manipulation times of the vehicle attributes form another aspect which is of extreme significance for the simulation. The simulation engine has to process a lot of vehicle operations in each time step, therefore short delays are important. During the implementation of OLSIMv4, it turned out that determining the leading vehicles on the neighbour lanes forms a performance critical challenge. The author's friend's idea [130] of storing the indexes of the currently leading neighbour vehicles in the lcs-leader-left and lcs-leader-right attributes for each vehicle solved the case. Listing 4.17 summarises a subset of all supported operations by the

```
sort Vehicle .
op NonExistentVehicle : -> Vehicle [ctor] .
op ChangeLane : Vehicle Int -> Vehicle .
op SetPosition : Vehicle Length -> Vehicle .
op SetVelocity : Vehicle Velocity -> Vehicle .
op SwitchType : Vehicle VehicleType -> Vehicle .
op ApplyOffsets : Int Length Vehicle -> Vehicle .
ops Length Position BackPosition : Vehicle -> Length .
ops Velocity MaxVelocity : Vehicle -> Velocity .
op Lane : Vehicle -> Int .
ops VehiclesCrashed? _<_ _<=_ : Vehicle Vehicle -> Bool .
var K : Vehicle . var L : Int . var X : Length .
eq ApplyOffsets(L, X, K) = SetPosition(SetLane(K, - L + Lane(K)), Position(K) - X) .
```

Listing 4.16: Required operations for sort Vehicle (source for most of the signatures: [30, Lst. 15]).

```
(make-new-kfz::vector pos::long vel::long lane::long typ::struct len::long vmax amax dmax . args)
(veh:pos::long f::vector)
(veh:update-pos::vector f::vector pos::long . without-history?::obj)
(veh:vel::long f::vector)
(veh:update-vel::vector f::vector vel::long . without-history?::obj)
(veh:lane::long f::vector)
(veh:update-lane::vector f::vector lane::long . without-history?::obj)
(veh:vlane::long f::vector)
(veh:len::long f::vector)
(veh:vmax::long f::vector)
(veh:vmul::long f::vector)
(veh:amax::long f::vector)
(veh:dmax::long f::vector)
(veh:lcs-left::long f::vector)
(veh:lcs-right::long f::vector)
(veh:lcs-leader-right::long f::vector)
(veh:lcs-leader-left::long f::vector)
(veh:alt-pos::long f::vector)
(veh:alt-vel::long f::vector)
(veh:alt-lane::long f::vector)
(veh:alt-brake::bbool f::vector)
```

Listing 4.17: Operations supported by a vehicle implementation.

OLSIMv4 vehicle implementation. The `without-history?` arguments of the update procedures provide the optional operation of writing the updated position through into the trajectory.

For simulation scenarios where vehicles lack lane discipline such as the one described in [136], the modelling of the lane attribute as a discrete index value may be insufficient. In such scenarios, the sort `Length` for the case of the publication or sort `long` with unit 1 cm may be more appropriate for the lane attributes. More sophisticated implementations for vehicles could include lane change velocities, traces of previous lane and position coordinates, tolerance factors that model the strictness in adapting to speed limits, and counters for lane change probabilities. In any of the cases, the vehicle constructor has to reflect these additions.

For the purpose of the publication, it was sufficient to organise multiple vehicles in sorted lists that may be ordered by lane and position. For the productive implementation, lists were also used initially. However, after several performance tests, vectors turned out to significantly scale better due to the following reason. Apart from the concrete representation, a specific data structure has to provide at least a lane-position-wise and a lane-index-wise access for the looking up of vehicles with the certain properties. The data structures list and vectors have to support referencing, resorting, manipulation, and modification operations of individual elements. It turned out that vectors scale better under this specific aspect. The OLSIMv4 implementation adapted the concept of specifying the lane-index or lane-position attributes as arguments. However, this interface is not as generic as it could be. The data structure for the vehicles also has to provide access for bunches of vehicles on a specific area.

An area based access interface is more generic. And as such, it provides the ability, to specify the vehicle key attributes by means of an area structure such as the S_{area}^4 introduced in Equation 3.13 from the conceptual design in Section 3.3.2. The area based access of objects is not limited to the vehicles only. The roadways also provide area based access of individual cells and sections as will be seen later

```
op Area : Int Int Length Length -> Area [ctor] .
ops Insert Remove : List{VehicleEltSWO} Vehicle -> List{VehicleEltSWO} .
op VehicleLaneIndexRef : List{VehicleEltSWO} Int Nat -> Vehicle .
op VehicleLanePositionRef : List{VehicleEltSWO} Int Length -> Vehicle .
ops VehiclesOnArea VehiclesNotOnArea : List{VehicleEltSWO} Area -> List{VehicleEltSWO} .
```

Listing 4.18: Sort `Area`, list modifiers `Insert` and `Remove`, and accessors for individual vehicles and bunches of vehicles (source: [30, Lst. 16]).

```
(define-srfi9-record-type cell
  (really-make-cell id closed limit-v min-v limit-lc limit-ovt)
  cell?
  (id cell-id set-cell-id!)
  (closed cell-closed set-cell-closed!)
  (limit-v cell-limit-v set-cell-limit-v!)
  (min-v cell-min-v set-cell-min-v!)
  (limit-lc cell-limit-lc set-cell-limit-lc!)
  (limit-ovt cell-limit-ovt set-cell-limit-ovt!))
```

Listing 4.19: The OLSIMv4 cell record data type that implements the model from [70, Sec. 2].

in Listing 4.24. To illustrate the areas, Listing 4.18 presents the sort `Area` and some list accessors for individual vehicles and for bunches of vehicles. Moreover, it introduces two basic operations `Insert` and `Remove` that provide ways to manipulate the number of vehicles in a list.

4.2.4 Roadways

Roadways represent the real-world pendants of a road traffic network [30]. They provide the road dimensions and all topological constraints of the associated real-world roadway to the simulation. They consist of a collection of single-lane sections that contain a cell each. A section's cell stands prototypic for the whole section and it contains all the traffic rules on the section. In their role as a container for the cells, the roadways "serve as a database for topological information that is associated with the real-world locations. The simulation model encodes this topological information in form of cells that combine a list of rules into a bundle for each section of a single lane" [30]. For the purpose of the OLSIMv4 implementation, the simulation engine constructs the roadways dynamically on system startup from a set of roadway segments. It constructs them by traversing the graph based representation of the roadway segments and thereby appending contiguous segments into one resulting collection of single-lane sections. The single-lane sections in OLSIMv4 are structurally almost identical to the ones of the Maude implementation except that OLSIMv4 requires laterally neighboured single-lane sections to start and end at the same positions. The OLSIMv4 limitation over the structural properties of the Maude sections results from the graph-specific requirement that the single-lane sections must have unique lateral neighbour connections. This limitation may, however, not be necessary with an appropriate container for the single-lane sections. Such a container would then manage the sections' lane neighbourship property implicitly and provide the access by lane-index, lane-position, and area specification. Consequently, such a container makes explicit storage of the lateral section neighbours superfluous. Listing 4.20 introduces the sorts `Cell` and `Rule` for the Maude implementation.

Among the topological constraints for a microscopic traffic model that occur in real-world road traffic networks there are "the speed limits, overtaking restrictions, traffic light states, lane closings

```
sorts Cell Rule Change Direction Priority .
ops Left Ahead Right : -> Direction [ctor] .
ops Refuse Choose Encourage Force : -> Priority [ctor] .
op Change : Direction Priority -> Change [ctor] .
op Rule : VehicleType Change -> Rule .
op Rule : VehicleType Velocity -> Rule .
op Cell : List{RuleEltSWO} -> Cell [ctor] .
op Closed? : Cell VehicleType -> Bool .
op MaxVelocity : Cell VehicleType -> Velocity .
op LaneChange : Cell VehicleType Direction -> Change .
op LaneChange : Cell VehicleType -> Change .
```

Listing 4.20: Sorts `Rule` and `Cell` represent the rules and cells of the conceptual design from 3.3.1. They encode and combine the topological constraints into bundles (source: [30, Lst. 17]).

or merges, ascending or descending slopes of roads, arc radius, and visibility headways" [30]. For the Maude implementation, the sort `Rule` represents this kind of information and sort `Cell` combines a set of rules into a bundle. The rules combine all topological information with a vehicle type which, consequently, also serves as a parameter to query the information from a rule. As the rules for the Maude implementation were intended to represent a rather minimal set of rules, they do, in contrast to OLSIMv4, not implement the overtaking restrictions as defined by Equation 3.7(c) of the conceptual design from Section 3.3.1. The natural number subsort for the relative directions `Left`, `Right`, and `Ahead` form a query information unit of its own. The explicit direction helps to reduce the code complexity significantly as the code no longer has to inspect the data which is encoded in the cells and the enum sets from Listing 4.19 but can instead use procedures that consume the direction as a parameter. The implementation of the constructor `op Topology : Cell Cell Cell VehicleType -> Topology .` from Listing 4.21 demonstrates the use of the rule referencing. The constructor corresponds to the Scheme procedure `(ca:topo-3l::ulong cell-m::struct cell-l::struct cell-r::struct veh::vector ovt?::bbool)` from Listing 4.22 which depends on several other procedures among them the two-lane topological analysis procedures named `ca:topo-2l-left` and `ca:topo-2l-right`. The update procedures evaluate the topological analysis in each time step for each vehicle. They require, thus, very efficient operation. As a part of a size effective optimisation, the productive implementation packs the four resulting values from the topological analysis into a single unsigned long integer value.

For the topological model of the article [30] and for the purpose of OLSIMv4, considering only the lane closings, maximum velocities, lane change enforcements or restrictions, and overtaking restrictions as kinds of topological information is sufficient. The operations `Closed?`, `MaxVelocity`, and `LaneChange` provide per vehicle type access to the various information. In general the topological constraints have an impact on the vehicle's further motion as "the topological constraints on a vehicle's position on the current lane and on the neighbouring lanes limit the possible velocity of a vehicle as well as its lane changing behaviour. Respecting the constraints in the microscopic traffic model will result in an adapted velocity or in a different lane changing behaviour of the vehicle. To keep the microscopic traffic models minimal and widely portable across a lot of real-world road traffic networks, the microscopic traffic

```
sort Topology .
op Topology : Nat Velocity Velocity Velocity -> Topology .
op Topology : Cell Cell Cell VehicleType -> Topology .
op Topology : Change Change Velocity Velocity Velocity -> Topology .
op Priority : Topology -> Int .
op Velocity : Topology Direction -> Velocity .
var T : VehicleType . vars VL VM VR : Velocity . vars CL CM CR : Cell . vars L R : Change .
eq Topology(CL, CM, CR, T) = Topology(
    LaneChange(CM, T, Left),
    LaneChange(CM, T, Right),
    if Change(Left, Choose) < LaneChange(CM, T, Left)
        then Velocity(0, C/s) else MaxVelocity(CL, T) fi,
    MaxVelocity(CM, T),
    if Change(Right, Choose) < LaneChange(CM, T, Right)
        then Velocity(0, C/s) else MaxVelocity(CR, T) fi ) .
eq Topology(L, R, VL, VM, VR) =
    if (C/s(VL) == 0) and (C/s(VR) == 0)
    then Topology(min(PriorityAsNat(Change(Left, Refuse)),
                      PriorityAsNat(Change(Right, Refuse))), VL, VM, VR)
    else if Priority(L) == Priority(R)
        then Topology(PriorityAsNat(L), VL, VM, VR)
        else if Priority(L) < Priority(R)
            then if Velocity(0, C/s) < VL
                then Topology(PriorityAsNat(L), VL, VM, Velocity(0, C/s))
                else Topology(PriorityAsNat(R), Velocity(0, C/s), VM, VR) fi
            else if Velocity(0, C/s) < VR
                then Topology(PriorityAsNat(R), Velocity(0, C/s), VM, VR)
                else Topology(PriorityAsNat(L), VL, VM, Velocity(0, C/s)) fi fi fi fi .
```

Listing 4.21: Sort `Topology` represents per vehicle type topological constraints. The equations show the topological analysis (source: [30, Lst. 18]).

```
(define (ca:topo-31::ulong cell-m::struct cell-l::struct cell-r::struct veh::vector overtake?::bbool)
  (cond
    ((or overtake?
         (not (ca:is-open? cell-l veh))
         (not (ca:local-v-min<=veh:vmax? cell-l veh))
         (<= (ca:lc-right-priority cell-l veh) 2))
     (ca:topo-2l-right cell-m cell-r veh overtake?))
    ((or (not (ca:is-open? cell-r veh))
         (not (ca:local-v-min<=veh:vmax? cell-r veh))
         (<= (ca:lc-left-priority cell-r veh) 2))
     (ca:topo-2l-left cell-m cell-l veh overtake?))
    (else (let ((vl-r (ca:local-v-limit cell-r veh))
                (vl-l (ca:local-v-limit cell-l veh)))
            (let ((is-open? (ca:is-open? cell-m veh))
                  (not-too-slow? (ca:local-v-min<=veh:vmax? cell-m veh)))
              (let ((prio (ca:lc-priority cell-m veh overtake? is-open? not-too-slow?)))
                (let ((vl-m (if (and is-open? not-too-slow?) (ca:local-v-limit cell-m veh)
                                (if (< prio 2) 0 (ca:local-v-limit cell-m veh)))))
                  (topology-values prio vl-r vl-m vl-l))))))))
```

Listing 4.22: A part of the OLSIMv4 topological analysis (implemented by Habel and the author of this thesis).

```
sort Section .
op Section : Int Length Length Cell -> Section .
ops Length Start End : Section -> Length .
ops Lane Cells : Section -> Int .
op Cell : Section -> Cell .
ops NonClosedCells Cells : Section VehicleType -> Nat .
op Closed? : Section VehicleType -> Bool .
op MaxVelocity : Section VehicleType -> Velocity .
op LaneChange : Section VehicleType Direction -> Change .
op LaneChange : Section VehicleType -> Change .
```

Listing 4.23: Sort `Section` provides one cell as a prototype for a part of a single lane of a given roadway (source: [30, Lst. 19]).

```
op EmptySections : -> List{SectionEltSWO} .
op CompleteSections : List{SectionEltSWO} -> List{SectionEltSWO} .
op Cell : List{SectionEltSWO} Int Length -> Cell .
ops SectionsOnArea SectionsCoveredByArea : List{SectionEltSWO} Area -> List{SectionEltSWO} .
op ApplyOffsets : List{SectionEltSWO} Int Length -> List{SectionEltSWO} .
op Lanes : List{SectionEltSWO} -> List{IntEltSWO} .
op NonClosedCells : List{SectionEltSWO} Pair{AreaEltSWO,VehicleTypeEltSWO} -> Nat .
```

Listing 4.24: Lists of `Section`s provide area and vehicle type based access (source: [30, Lst. 20]).

models consume the topological constraints in form of a simplified and summarized set of information. The `Topology` operations from Listing 4.21 determine this minimal set of information from the real-world topological constraints with topological analysis per vehicle type. Sort `Topology` summarizes the four values lane closings or merges, speed limits, and overtaking restrictions. Thereby, the three sort `Velocity` values specify the maximum velocity of a vehicle's current lane and its neighbouring lanes. The fourth value describes the priority for a potential lane change to take place" [30]. Listing 4.21 presents sort `Topology` as a straightforward implementation of the S_{topo}^4 structure of the conceptual design from Section 3.3.4. The constructors `op Topology : Cell Cell Cell VehicleType -> Topology .` from Listing 4.21 and `(ca:topo-31::ulong cell-m::struct cell-l::struct cell-r::struct veh::vector overtake?::bbool)` from Listing 4.22 operate thereby on a subset of the domain kinds of function $\mathbf{F}_{\text{topo}} : S_{\text{roadway}}^4 \times S_{\text{veh}}^4 \longrightarrow S_{\text{topo}}^4$ from Equation 3.24. The vehicle update procedure for occupation tables derives the information subset to a given roadway and vehicle combination and wraps the `Topology` or `ca:topo-31` procedure appropriately.

```
traffic_nrw_1_0_0=# SELECT rwno, c0, m0, c1, m1, c2, m2
                    FROM topologie_2_4_9.view_roadway_sections_with_cell_properties
                   WHERE roadway = 'A003-NO-HF-034//A002-NO-HF-035'
                     AND track IN ('A003-NO-HF-019','A003-NO-HF-020')
                ORDER BY rwno;
 rwno |   c0   |       m0       |   c1   |         m1          |   c2   |        m2
------+--------+----------------+--------+---------------------+--------+--------------------
    1 | z11279 | lc:choose-left+| z11280 | lc:choose-left    +| z11281 | lc:choose-right   +
      |        | vl:car-160     |        | lc:choose-right   +|        | ov:truck-not-truck+
      |        |                |        | ov:truck-not-truck+|        | ov:truck-not-car  +
      |        |                |        | ov:truck-not-car  +|        | vl:car-160
      |        |                |        | vl:car-160         |        |
    1 | z11291 | lc:choose-left+| z11292 | lc:choose-left    +| z11293 | lc:choose-right   +
      |        | vl:car-160     |        | lc:choose-right   +|        | ov:truck-not-truck+
      |        |                |        | ov:truck-not-truck+|        | ov:truck-not-car  +
      |        |                |        | ov:truck-not-car  +|        | vl:car-160
      |        |                |        | vl:car-160         |        |
    2 | z11294 | lc:choose-left+| z11295 | lc:choose-left    +| z11296 | lc:choose-right   +
      |        | vl:car-160     |        | lc:choose-right   +|        | ov:truck-not-truck+
      |        |                |        | ov:truck-not-truck+|        | ov:truck-not-car  +
      |        |                |        | ov:truck-not-car  +|        | vl:car-160
      |        |                |        | vl:car-160         |        |
    2 | z26873 | lc:choose-left+| z26874 | lc:choose-left    +| z26875 | lc:choose-right   +
      |        | vl:car-160     |        | lc:choose-right   +|        | ov:truck-not-truck+
      |        |                |        | ov:truck-not-truck+|        | ov:truck-not-car  +
      |        |                |        | ov:truck-not-car  +|        | vl:car-160
      |        |                |        | vl:car-160         |        |
 ...
```

Listing 4.25: Cells together with their configured traffic rules for two tracks of the A3 roadway in NRW.

It determines the arguments to the constructors by referencing the cell on the position and the current lane as well as the cells on the neighbouring lanes to a given vehicle.

For the purpose of a comparison, Listings 4.19 and 4.20 presented parts of the OLSIMv4 and the Maude cells implementation. The OLSIMv4 cells consist of several enum sets that contain the lane change and overtaking restrictions as well as the velocity limits. Listing 4.3 has introduced the velocity limit enum set already. Listing 4.25 provides another example with the SQL query to list the cells of two subsequent multi-lane roadway segments together with their traffic rule enum set members. The example demonstrates also the advantage of having a simple yet powerful declarative language such as SQL [11, Sec. 2.2] to manage and access the topological data as structured and typed data. The advantage, however, relies on the well structuredness property of the topological data and the traffic data for which "the relational model and SQL are [still] well-suited" [11, Sec. 2.1]. For these kind of data, SQL provides powerful set-oriented operations [11, Sec. 2.2]. The example is, moreover, intended to provide a first impression for the simplicity in finding out, for instance, cells on the outer right lane with a lane change misconfiguration that would allow lane changing to the right.

While the conceptual design from Section 3.3.1 distinguishes between the three different rule types from Equation 3.7, it does not sort the rules by their rule type $t_{\text{name}}^{\text{rule}} \in \mathfrak{L}_{\text{rule}}^4$. The conceptual design offers, however, the filter function $\mathbf{F}_{\text{rule}} : S_{\text{cell}}^4 \times \mathfrak{L}_{\text{rule}}^4 \longrightarrow \mathcal{P}_{\text{cell}}^4(S_{\text{rule}}^4)$ from Equation 3.9 to extract all rules depending on a certain rule type. The cell structure of the OLSIMv4 implementation from Listing 4.19 reflects the distinction and offers direct access to each of the rule type dependent subsets thereby encoding the vehicle types and the directions implicitly. The encoding is less flexible and represents a bad compromise between design and scalability as the generic approach would have been available at almost no cost. However, at the time of writing the code, the solution was not that clear. As a consequence, the OLSIMv4 topological analysis consumes more lines of code because the `ca:topo-31` has to check several combinations of type of restriction, vehicle type, and direction. However, the OLSIMv4 cell implementation exploits a fixed order of the restriction types and, thus, probably a faster access to the topological information. As the topological analysis accesses the topological information for each vehicle update, the access time is important for a productive implementation.

Sections have a given extent on a part of a single lane of a roadway and they describe their extent

```
sort Roadway .
op Roadway : Qid Bool Area List{SectionEltSWO} -> Roadway .
op Name : Roadway -> Qid .
op Periodic? : Roadway -> Bool .
op Area : Roadway -> Area .
ops MinLane MaxLane : Roadway -> Int .
ops Length Start End : Roadway -> Length .
op Sections : Roadway -> List{SectionEltSWO} .
op Topology : Roadway VehicleType Nat Length -> Topology .
var R : Roadway . var T : VehicleType . var I : Int . var X : Length .
eq Topology(R, T, I, X) =
    Topology(Cell(Sections(R), I + DeltaLane(Left), X),
             Cell(Sections(R), I, X),
             Cell(Sections(R), I + DeltaLane(Right), X), T) .
```

Listing 4.26: Sort `Roadway` describes real-world roadways through its name, its area extent, and its list of sections (source: [30, Lst. 21]).

by lane number, starting and ending position. They combine the spatial extent with a prototype cell and represent an area with constant the topological information. Listing 4.23 quotes some signatures of the Maude implementation for the operations on sections while Listing 4.25 presented the prototype cells view of the database.

The operations `Closed?`, `MaxVelocity`, and `LaneChange` that have been introduced already in the context of the `Cell` domain, also apply for sort `Section` elements. They provide access to the various kinds of topological information again per vehicle type and per direction. In addition to the section elements, the Maude simulation engine requires some operations on lists of sections. Listing 4.24 displays each of the operations. The `SectionsOnArea` operation of them provides access to a subset of sections per area. It is probably the most important one, as it is –except for the domain kind– almost identical to the `VehiclesOnArea` procedure from Listing 4.18. For the Maude implementation procedures operate on lists. The Scheme implementation, however, operates on vectors of vehicles. A generic procedure might help simplifying the code here. However, the requirements for the two domain kinds may vary, depending on the implementation, significantly. While the `SectionsOnArea` procedure requires for most of the time fast access only of the specified sections, the `VehiclesOnArea` operation may operate on an underlying container structure that might require support for very frequent updates of the vehicles in the container when implemented not purely functionally using `vector-set!`, `list-set!`, or similar for the Scheme case. In the context of the OLSIMv4 implementation, hashtables turned out as not efficient enough as the vehicles container. However, in the next step of a productive environment it would be worth investigating for an efficient data structure as these two procedures are almost at the heart of the simulation engine. The `SectionsOnArea` accessor generalises the function $\mathbf{F}_{Y \times X}^{section} : S_{roadway}^4 \times \mathfrak{Y}_{veh}^4 \times \mathfrak{X}_{veh}^4 \longrightarrow S_{section}^4$ from Equation 3.20 by operating on the list of a roadway's $\mathbf{w}_i$ sections $\mathbf{w}_i \# \mathbf{q}$ only and instead of using a lane y and position x combination taking the more general area argument. For a given roadway $\mathbf{w}_i \in S_{roadway}^4$, lane $y \in \mathfrak{Y}_{veh}^4$, and position $x \in \mathfrak{X}_{veh}^4$ arguments, function $\mathbf{F}_{Y \times X}^{section}(\mathbf{w}_i, y, x)$ can be implemented by wrapping each function call as $\mathbf{F}_{Y \times X}^{section}(\mathbf{w}_i, y, x) := \mathbf{F}_{Y \times X}^{section}(\mathbf{w}_i \# \mathbf{q}, \langle y, y, x, x \rangle^{area})$.

For the Maude case in Listing 4.24, "the `Cell` operation provides access to the prototype cell that is associated with a certain lane and position." [30]. In combination with the wrapper method from above extended onto the cells, the `Cell` accessor implements, thus, function $\mathbf{F}_{Y \times X}^{cell} : S_{roadway}^4 \times \mathfrak{Y}_{veh}^4 \times \mathfrak{X}_{veh}^4 \longrightarrow S_{cell}^4$ from Equation 3.22. The remaining operations in Listing 4.24 are either variations, helper functions, or filter functions over the result sets. "The `NonClosedCells` operation calculates the number of cells that are available to a particular vehicle type on a given area. The `CompleteSections` operation completes a list of sections to a rectangular area of maximum extent using a default closed section to fill the potentially missing parts" [30]. Apart from the concrete implementation that relies on lists for the Maude case and on interval splitting procedures for the Scheme case, the Maude interface follows OLSIMv4 very closely.

Given a list of sections, the constructor of a sort `Roadway` element uses the operation `CompleteSections`

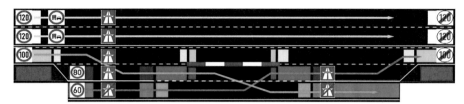

Figure 4.3: A complex yet possible roadway segment on a German highway. Longer sections start and end with the same color and may be interrupted by black when too long. Before and after the two merging areas, the second right lane consists of closed sections. The highway roadway can be encoded as a four lane roadway even though the second right single lane is only available in the case of exits or carriageways. The rightmost single lane is not part of the highway (source: [30, Fig. 4] with the traffic sign pictures taken from [194]).

to extend the list of sections in such a way that it covers the whole rectangular area as specified by the sort `Area` argument. To fill up missing sections, operation `CompleteSections` inserts a closed section for each unspecified section. The other properties of sort `Roadway` elements contain "a unique name and a flag that marks them as having periodic boundary conditions, i.e. having a ring structure as in the case of roundabouts" [30]. "Roadways do not depend on sort `Vehicle`. Instead they depend on the sort `VehicleType` only indirectly through sort `Cell`" [30]. Listing 4.26 introduces the sort `Roadway` for the Maude case.

Figures 3.2 and 4.3 show examples of complex yet possible but strictly hypothetical roadway segments as found on German highways. The highway segment from Figure 4.3 "consists of four lanes divided into several sections. The right outermost lane is not part of the highway. Before and after the two merging areas, the second right lane consists of closed sections. The merging areas lead onto a carriageway and back again onto the main roadway. At the second merging area, vehicles have to enter the main roadway, but are not allowed to switch back onto the carriageway. The two left outermost lanes limit the maximum velocity for vehicles to $120\,$km/h , prohibit overtaking for trucks, and restrict lane changing to take place only in between themselves. The third left lane limits the maximum speed to $100\,$km/h , permits lane changing to the left but restricts lane changing to the right. The sectioning of the lanes may differ for individual lanes in Figure 4.3. The introduced sort `Roadway` elements are able to represent the roadway of Figure 4.3" [30].

The presented roadway model is intended for one-way roadways only. Extending them for two-way traffic is possible in principle where this should be unavoidable. As a consequence of such environments, the traffic models also have to reflect the two-way traffic properties. The extension impacts not only the implementation of the traffic model but might also require a redesign of the traffic model interface even though the changes can be held minimal.

4.2.5 Occupation tables

The logical view as presented in the conceptual design combines a random number generator's (RNG) state, a periodic boundary condition flag, a roadway, and all the vehicles positioned on the roadway into an occupation table. For the Maude case, sort `Occupation` elements represent the occupation tables. The sort provides "per vehicle access to the neighbour vehicles and to the topological constraints that rule on a vehicle's lane and position" [30]. Listing 4.28 presents some signatures for the operations of the occupation tables. Prior to their introduction, the vehicle neighbours are introduced in the following.

For the Maude case, the vehicle update of an occupation table's vehicles relies on the sort `Topology` of the previous section and the sort `Neighbourhood` as introduced by Listing 4.27. "A vehicle's next two predecessors on the current lane and its successors on the neighbouring lanes and their next two predecessors form the vehicle's neighbourhood. The neighbourhood combines a vehicle's neighbouring

```
pr TRIPLE-SWO{VehicleEltSWO,VehicleEltSWO,VehicleEltSWO} .
pr TRIPLE-SWO{TripleVehicleEltSWO,TripleVehicleEltSWO,TripleVehicleEltSWO} * (
   sort Triple{TripleVehicleEltSWO,TripleVehicleEltSWO,TripleVehicleEltSWO} to Neighbourhood,
   sort Triple{VehicleEltSWO,VehicleEltSWO,VehicleEltSWO} to VehicleTriple,
   op <_;_;_> to <_|_|_> ) .
op VehicleTriple : Vehicle Vehicle Vehicle -> VehicleTriple .
op Neighbourhood : VehicleTriple VehicleTriple VehicleTriple -> Neighbourhood .
ops Left Ahead Right : Neighbourhood -> VehicleTriple .
op Lane : Neighbourhood Direction -> VehicleTriple .

vars V0 V1 V2 : Vehicle . vars L M R : VehicleTriple . var D : Direction .
eq VehicleTriple(V2, V1, V0) = < V2 | V1 | V0 > .
eq Neighbourhood(L, M, R) = < L | M | R > .
eq Left(< L | M | R >) = L .
eq Ahead(< L | M | R >) = M .
eq Right(< L | M | R >) = R .
```

Listing 4.27: Three triples of vehicles form a vehicle's neighbourhood which sort `Neighbourhood` represents (source: [30, Lst. 22]).

vehicles into a triple of vehicle triples. The general tuple operations as described in Section 4.2.2 apply. Listing 4.27 displays a constructor along with some accessors for sort `Neighbourhood`" [30]. The sort implements the neighbour structure S_{nbrs}^4 from Equation 3.32.

The OLSIMv4 implementation however, does not have an explicit neighbourhood structure. It references a vehicle's neighbouring vehicles on demand inline directly before invoking the vehicle update procedure. The referencing method to determine a vehicle's neighbour vehicles corresponds to the `op Neighbourhood : Occupation Vehicle -> Neighbourhood` . operation from Listing 4.28. The latter implements the $\mathbf{F}_{\text{neighbours}} : S_{\text{occ}}^4 \times S_{\text{veh}}^4 \longrightarrow S_{\text{nbrs}}^4$ function from Equation 3.31 while operation `op Topology : Occupation Vehicle -> Topology` . implements its counterpart, the topological analysis function $\mathbf{F}_{\text{topo}} : S_{\text{roadway}}^4 \times S_{\text{veh}}^4 \longrightarrow S_{\text{topo}}^4$ from Equation 3.24. Listing 4.28 presents also sort `Occupation`.

Operations `InsertDummyVehicles` and `RemoveDummyVehicles` insert dummy vehicles into (or remove from) occupation tables depending on the periodic boundary conditions property. They provide the neighbouring vehicles beyond the margins of the roadway for the traffic model update of the marginal vehicles. The occupation tables contain all information that microscopic traffic model updates require. Occupation tables without switching area connections do not require further data structures for their update.

The simulation model does not necessarily require the detectors and the tuning elements for proper functioning. But without the tuning elements, the occupation tables would contain no vehicles to process vehicle updates on. The occupation tables form, thus, the basic simulation unit of the simulation model. As the basic simulation unit, the occupation tables require special attention concerning the update process. Section 3.3.5 has already pointed out, that a vehicle update of any occupation table results in a sequence of vehicle updates where each vehicle update relies on the state of the previously updated vehicle. The dependency surprises as the vehicle update is a synchronous update. However, only the state of the random number generator causes sequential update character. The recursion formula $\mathbf{u}_i^{(t+\Delta t)} := (s_{\text{rand},i}, \mathbf{k}_i^{(t+\Delta t)}) := \mathbf{F}_{\text{drive}}(s_{\text{rand},i-1}, \mathbf{k}_i^{(t)}, \mathbf{b}_i, \mathbf{p}_i)$ with the neighbouring vehicles $\mathbf{b}_i := \mathbf{F}_{\text{nbrs}}(\mathbf{o}_i^{(t)}, \mathbf{k}_i^{(t)})$ according to Equation 3.31 and the topological information $\mathbf{p}_i := \mathbf{F}_{\text{topo}}(\mathbf{o}_r^{(t)} \# \mathbf{w}^{(t)}, \mathbf{k}_i^{(t)})$ according to Equation 3.24 manifests thereby the sequential character of the update. Listing 4.35 presents the operational semantics of the occupation table update and whitnesses also the sequential character by the use of the `fold` operation which is part of the implementation of operation `UpdateOccupation`. However, there are ways to circumvent the strict limitation of a sequential update or at least to break up the sequential update into pieces that are suitable for a concurrent or parallel executed update.

For the longitudinal update part of the vehicle motion, the state of the (pseudo-)random number generator is the only limiting factor. As the random number generator state consists, in general, only of a single integer with size 32 bit or size 64 bit, the state could be held in a CPU register as a first try to increase the parallelisation gain. Whenever the register solution is not applicable, thread local

```
sort Occupation .
op Occupation : Nat Roadway List{VehicleEltSWO} -> Occupation .
ops InsertDummyVehicles RemoveDummyVehicles : Occupation -> Occupation .
op ReplaceVehicles : Occupation List{VehicleEltSWO} -> Occupation .
op DeactivateVehiclesOnArea : Occupation Area -> Occupation .
op ReplaceRoadway : Occupation Roadway -> Occupation .
op ReplaceCounter : Occupation Nat -> Occupation .
op ReplaceVehicles : Occupation List{VehicleEltSWO} -> Occupation .
ops Vehicle VehicleLaneIndexRef : Occupation Int Nat -> Vehicle .
ops Vehicle VehicleLanePositionRef : Occupation Int Length -> Vehicle .
op Neighbourhood : Occupation Vehicle -> Neighbourhood .
op Topology : Occupation Vehicle -> Topology .

var V : Vehicle . var O : Occupation . var R : Roadway .
eq Topology(O, V) = Topology(Roadway(O), Type(V), Lane(V), Position(V)) .
eq RemoveDummyVehicles(O) =
   ReplaceVehicles(O, RemoveDummyVehicles(Vehicles(O), Area(Roadway(O)))) .
eq InsertDummyVehicles(O) =
   ReplaceVehicles(O,
     if Periodic?(Roadway(O))
     then InsertPeriodicDummyVehicles(Vehicles(O), Area(Roadway(O)))
     else InsertDummyVehicles(Vehicles(O), Area(Roadway(O))) fi ) .
```

Listing 4.28: Sort `Occupation` combines vehicles and roadways. Adding dummy vehicles depends on the periodic boundary condition property (source: [30, Lst. 23]).

storage may provide an alternative implementation. However both variants are relatively difficult to obtain in the Bigloo Scheme dialect and they attempt to improve the situation on the physical layer which is like curing symptoms. Additionally, they do not increase the parallelism for the updates of an occupation table's vehicles. As another approach that exploits the strict ordering of the vehicles, the update procedures could start the vehicle update in parallel for all vehicles with the same initial random number generator state as provided by the occupation table. By exploiting the leap-frogging technique which is frequently used in the context of random number generators [123], the vehicle update procedures calculate then the individual random number generator state for each vehicle update by successively applying the random number generator as much times as given by the total vehicle index in the occupation table. As a last approach, the occupation table update procedure could pre-calculate an array of random number generator states and equip each vehicle update with an individual random number generator state. These approaches are also useable for the lateral update part of the vehicle motion but would additionally require synchronisation of the updated vehicles thereafter. As this kind of vehicle update procedure depends only on the current states of the neighbouring vehicles as an additional input, the synchronisation restricts the performance gain not too much because the updated vehicles have already been copied and require to be stored in a container anyway. This kind of update procedure provides the most parallelism. It would be interesting to study and to verify the equivalence between the various approaches using the linear temporal logic model checker of Maude [56] recommended from the Maude Manual [46, Sec. 10.3].

However, these bottlenecks as well as the approaches to solve them were only seen with the knowledge and analysis provided by the conceptual design from Section 3.3.5. As one of its benefits, the conceptual design supports better reasoning about the simulation model and its components. During the implementation of OLSIMv4, the parallel update of the occupation tables worked not as perfect as expected and the causes were never tracked down exactly as the project's timeline was already stressed over the limit and the project's performance goals had already been reached. During updating all occupation tables in parallel the following observation was made. The CPU load of most of the threads never reached the full hundred percent. One explanation for this not unexpected behaviour may be the garbage collector. However, with the considerations from above, another explanation exists with the random number generator state and the sequence of vehicle updates as the random number generator's state had not been stored in the thread-local storage. Implementing the leap frogging technique into the vehicle updates of the occupation tables would free the dependence on the global state and,

```
sorts Detector TrafficObservable Measurement .
op TrafficObservable : VehicleType Observable -> TrafficObservable .
op Measurement : Period List{TrafficObservableEltSWO} -> Measurement .
op Reset : Detector -> Detector .
op Measure : Period Detector Occupation -> Detector .
op Accumulate : Detector Period -> Measurement .
```

Listing 4.29: Sort `Detector` represents polymorphic detector models. Each of them has to provide the operations `Reset`, `Measure` and `Accumulate` (source: [30, Lst. 24]).

consequently, might increase the CPU load per thread. With the new insights, a future implementation of OLSIMv4 would no longer depend on a shared state. Even though it is not very likely as there are always several threads that consume full hundred percent of CPU load, a last explanation remains, of course, with poor implementation or too much copying of objects and therefore invoking too much garbage collection. The author suspects, however, the random number generator state. As part of the lessons learned, this can now be avoided in a future implementation.

4.2.6 Detectors

Detectors refer to a detector type that the Maude implementation represents by sort `DetectorType` and the OLSIMv4 implementation models using subclassing. The detectors refer also "to a particular cross-section or area of a roadway. They evaluate their associated road occupation table partly or in full using a method that depends on their detector type. For instance, detectors can use an area based or a cross-section based measuring method to determine level-of-service, traffic flow, density, average velocity, or travel-time of a road occupation table. Sort `Area` elements specify a detector's associated part of an occupation table. Detectors do not modify road occupation tables. Instead, they update their internal list of measurements in each `Measure` operation. " [30].

The detector model from the conceptual design of Section 3.5.1 has introduced detectors as having a concrete detector type structure element $s_{\text{dettype}} \in S_{\text{dettype}}^4$ element from Equation 3.77. The structure defines a detector type through observable specifications that combine vehicle types and observable units. Except for the column names, this definition is similar to the data definition in the relational model, i.e. when creating a table in a database. However, the detector model from the conceptual design is the result of enhancing the existing OLSIMv4 and Maude models and is, consequently, implemented only partly by them. The models share, however, the measure operation from Equation 3.84 $\mathbf{F}_{\text{measurement}} : \mathfrak{T}_{\text{secs}} \times S_{\text{occ}}^4 \times S_{\text{det}}^4 \longrightarrow S_{\text{measure}}^4$ at least partly. For the Maude case, the operations from Listing 4.29 implement these functions partly. As the Maude implementation does not need to fetch measurements from external resources, it implements, apart from the argument order, the measure operation $\mathbf{F}_{\text{measure}}(t, \mathbf{o}_r^{(t)}, \mathbf{d}_i^{(t)}) := \mathbf{F}_{\text{det-update}}(t, \mathbf{o}_r^{(t)}, \mathbf{d}_i^{(t)}, \mathbf{F}_{\text{measurement}}(t, \mathbf{o}_r^{(t)}, \mathbf{d}_i^{(t)}))$ from Equation 3.85 by `op Measure : Period Detector Occupation -> Detector` . As the OLSIMv4 implementation uses an object-oriented approach to model the detectors, it mutates their internal state instead of instantiating an updated detector for the aggregate and the measure operations.

As the constructors for the particular detectors depend on the detector type and on the detector model, they are not introduced here. Section 4.5 will discuss the implementation of the various detector models in more detail. For the Maude implementation, "concrete detector models can be added through subsorts of sort `Detector`. Listing 4.29 introduces sorts `Detector` and `Measurement`. Operation `Reset` resets the detector with an empty list of measurements" [30]. Sort `Measurement` implements the measurement structure S_{measure}^4 from Equation 3.78. The detector implementations support at least partly the accessors and constructors of the conceptual design. For the Maude case, among them are "`Name`, `Roadway`, `Type`, and `Area` that provide access to the appropriate detector attributes. The simulation engine calls operation `Measure` directly when updating. In contrast to operation `Measure`, the tuning elements apply operation `Accumulate` indirectly through their accumulators" [30].

```
sort Accumulator .
ops Name Roadway : Accumulator -> Qid .
op Area : Accumulator -> Area .
op Type : Accumulator -> AccumulatorType .
op DetectorNames : Accumulator -> List{QidEltSWO} .
op Reset : Accumulator -> Accumulator .
op AddMeasurement : Period Accumulator Measurement -> Accumulator .
op Accumulate : Accumulator Period List{DetectorEltSWO} -> Accumulator .
op AggregatedMeasurement : Accumulator -> Measurement .
```

Listing 4.30: Sort `Accumulator` accumulates detector measurements (source: [30, Lst. 25]).

4.2.7 Accumulators

Accumulators implement their conceptual counterparts which are the elements of the accumulate structure S_{acc}^4 from Equation 3.122. However, the conceptual design represents already the result of an enhancement and, as a consequence, has a more general form than in the Maude and the OLSIMv4 cases. For the Maude case, "sort `Accumulator` elements provide the capability to accumulate measurements originating from several detectors in an `AccumulatorType` dependent way. For example, they can accumulate measurements of several single lane detectors into a cross-section measurement value. Or, they can calculate a balance between measurements originating from detectors that are located on different roadways. All submodels have to implement operation `Accumulate`. It provides the functionality to accumulate measurements. Listing 4.30 presents sort `Accumulator`" [30]. OLSIMv4 uses only the cross-sectional type of the accumulators and it models them as a subclass of the detector class.

The constructor for sort `Accumulator` takes a list of detector names as one of its arguments. OLSIMv4 implements the detector name references as neighbours of a node. In the Maude implementation, "accumulators maintain a list of all measurements that have been recorded since the last `Reset` operation. Operation `AddMeasurement` adds measurements to the list. The `AggregatedMeasurement` operation provides the aggregated value. The simulation engine calls operation `Accumulate` directly when updating" [30]. Operation `Accumulate` implements function $\mathbf{F}_{\text{accumulate}} : \mathfrak{T}_{\text{secs}} \times S_{\text{accu}}^4 \times \mathcal{P}_{\text{det}}(S_{\text{det}}^4) \longrightarrow S_{\text{accu}}^4$ from Section 3.6.1. In the OLSIMv4 implementation, it corresponds to the measure operation as the accumulators therein are subclasses of the detectors.

4.2.8 Tuning elements

For the Maude case, sort `TuningElement` implements the tuning elements of the conceptual design from Section 3.6.2. In contrast to the detectors that are only able to read the occupation table states, the tuning elements "are able to modify their associated road occupation table, i.e. they insert or remove vehicles into them. Their chosen action depends on an implementation dependent strategy and the comparisons of their actual to should-be accumulator measurements. Tuning elements contain the actual and should-be accumulators, a name attribute, a name reference to the associated roadway, and an area attribute that specifies their extent" [30]. It is possible to subsort sort `TuningElement` and provide a subsort dependent constructor for each. The signature of the constructor in Listing 4.31 demonstrates the minimum requirements for a constructor of the tuning elements structure S_{tun}^4 of the conceptual design from Equation 3.123.

```
sort TuningElement .
op TuningElement : Qid Qid Area Accumulator Accumulator -> TuningElement .
ops Name Roadway : TuningElement -> Qid .
op DetectorNames : TuningElement -> List{QidEltSWO} .
```

Listing 4.31: Sort `TuningElement` represents elements that modify road occupation tables (source: [30, Lst. 26]).

OLSIMv4 uses several types of tuning elements among them weather, travel time measurement, and the standard actual should-be tuning elements. As has been the case with the detectors, OLSIMv4 uses the object-oriented approach to implement them. The `Tune` operation represents the main purpose of the tuning elements which is the modification of the occupation tables. The result of the `Tune` operation is, however, a pair "as it intentionally modifies a given occupation table and additionally maintains an internal state". The `Tune` operation requires, thus, to introduce first "sort `TuningArea` which is done in the following" [30].

4.2.9 Tuning areas

As pairs of a tuning element and a road occupation table, the tuning areas represent the result of a `Tune` operation. They are only required for the Maude case, as OLSIMv4 uses the object-oriented approach and, thus, can mutate the state of the tuning elements and, at the same time, can return a new instance of the occupation table. In the Maude implementation, "the simulation engine instantiates the tuning areas only for the purpose of applying the `Tune` operation. The `Tune` operation can be thought of as returning multiple values. Thus, the `Tune` operation takes as argument (and provides the result as) a pair of a tuning element and an occupation table. Sort `TuningArea` is shown in Listing 4.32" [30].

```
sort TuningArea .
op TuningArea : TuningElement Occupation -> TuningArea .
op TuningElement : TuningArea -> TuningElement .
op Occupation : TuningArea -> Occupation .
op Tune : Period TuningArea -> TuningArea .
op Accumulate : Period TuningArea List{DetectorEltSWO} -> TuningElement .
```

Listing 4.32: Sort `TuningArea` pairs a tuning element and an occupation table (source: [30, Lst. 27]).

The `Tune` operation from Listing 4.32 implements function $\mathbf{F}_{\text{tune}} : \mathfrak{T}_{\text{secs}} \times S_{\text{occ}}^4 \times S_{\text{tun}}^4 \longrightarrow S_{\text{occ}}^4 \times S_{\text{tun}}^4$ from Equation 3.124 of Section 3.6.2. It "constructs the pair of the updated tuning element and the updated road occupation table as its function value. Tuning elements that operate on overlapping areas may interfere in consecutive `Tune` operations. Thus, appropriate tuning area models, non-overlapping tuning area networks or a special processing order for overlapping tuning areas may contribute to avoid interference or minimize possible impacts of interference" [30].

4.2.10 Vehicle update

Section 3.4 has introduced and discussed in great detail the vehicle update functions already and as part of the conceptual design. In general and according to [30], "vehicles can move longitudinally and laterally. Consequently, a vehicle's resulting motion is a superposition of both directed motions. Thus, most microscopic traffic models divide a vehicle's motion into these two parts. The drive update operation models the longitudinal motion. It changes a vehicle's position on the current lane. The lane change update operation models the lateral motion. It can change a vehicle's lane. The resulting vehicle motion can then be thought of as the superposition of both parts or as the composition of the two updates. Even though the vehicle motion is divided into two parts that are processed sequentially, the motion takes place in one atomic time step.

Microscopic traffic models describe the motion of vehicles (or assumed drivers, respectively) more or less in accordance with well-known traffic phenomena. In a microscopic traffic model, the headways to the preceding and even neighbouring vehicles as well as their velocities constitute a vehicle's velocity and its covered distance. Many traffic models take only one vehicle predecessor into account for the update procedure. The model as described in [124] uses two predecessors resulting in some kind of anticipation. Section 4.2.5 already introduced sort `Neighbourhood`. It has been chosen large enough to free the traffic model abstraction from the concepts of the occupation tables.

```
ops Drive LaneChange : -> VehicleUpdate [ctor] .}
op F : -> VehicleUpdate [ctor] .
var C : Nat . var T : Topology . var N : Neighbourhood .
eq F[V, < C ; T ; N >] = < C ; V > .
```

Listing 4.33: Signature for the vehicle update operation `VehicleUpdate`.

```
sort VehicleUpdate .
subsort VehicleUpdate < MapFunc{VehicleEltSWO,TripleIntTopologyNeighborhoodEltSWO,PairIntVehicleEltSWO} .
ops Drive LaneChange SafeInsert : -> VehicleUpdate [ctor] .
op InsertSafe? : VehicleUpdate Vehicle Triple{IntEltSWO,TopologyEltSWO,NeighbourhoodEltSWO} -> Bool .
var F : VehicleUpdate . var V : Vehicle . var C : Int . var T : Topology . var N : Neighbourhood .
eq InsertSafe?(F, V, < C ; T ; N >) = (NonExistentVehicle == 2nd(F[V, < C ; T ; N >])) .
```

Listing 4.34: Sort `VehicleUpdate` elements are `map` operations. They represent the microscopic traffic model update functions inside the simulation engine. `InsertSafe?` tests whether vehicles can be inserted safely into a neighbourhood constellation (source: [30, Lst. 28]).

Additionally, an implementation of a microscopic traffic model has to take topological constraints into account. However, it should not depend on any road traffic network but be independent of the complex roadway structure. Sort `Topology` from Section 4.2.5 has been designed for that purpose.

Moreover, stochastic microscopic traffic models may want to use a pseudo-random number generator (PRNG). Thus, the vehicle update procedure requires a counter that represents the state of such a PRNG. The vehicle update procedure returns the updated counter as part of the result."

The conceptual design modelled the vehicle update by the two functions $\mathbf{F}_{\mathrm{drive}} : \mathfrak{R}_{\mathrm{occ}}^4 \times S_{\mathrm{veh}}^4 \times S_{\mathrm{nbrs}}^4 \times S_{\mathrm{topo}}^4 \longrightarrow \mathfrak{R}_{\mathrm{occ}}^4 \times S_{\mathrm{veh}}^4$ and $\mathbf{F}_{\mathrm{chlane}} : \mathfrak{R}_{\mathrm{occ}}^4 \times S_{\mathrm{veh}}^4 \times S_{\mathrm{nbrs}}^4 \times S_{\mathrm{topo}}^4 \longrightarrow \mathfrak{R}_{\mathrm{occ}}^4 \times S_{\mathrm{veh}}^4$. As has been the case with the tuning areas, they also return a pair in case of the Maude implementation. The latter models the function type of both through sort `VehicleUpdate` as introduced by Listing 4.34. The appropriate functions have, thus, the signature from Listing 4.33 with `F` also from Listing 4.33. Thereby, the element constructor `F` could be a valid implementation for a given sort `VehicleUpdate`. Due to the limitations in the implementation of the Maude higher-order functions, the signatures differ slightly between the conceptual design and the Maude implementation. The OLSIMv4 implementation differs strongly from the conceptual design in the point that the random number generator maintains a hidden state as has already been mentioned at the end of Section 4.2.2.1. The hidden state turned out to be a hidden bottleneck and has been discussed during the vehicle update of the occupation tables in Section 4.2.5. A future implementation of OLSIMv4 can now remove the bottleneck. Even though the signatures for the conceptual design and for the Maude implementation differ, the latter represents, thus, the conceptual design better than the OLSIMv4 implementation. Listing 4.35 discusses the implementation of the traffic model in more detail.

The abstractions for the vehicle update use the two sorts `Neighbourhood` and `Topology` and a counter that represents the state of a PRNG. The "abstractions should be easy to implement for most of the known microscopic traffic models. They consist of the two procedures for the longitudinal and the lateral motion of a predicate procedure that allows testing of safe inserts of vehicles into a neighbourhood constellation. The `SafeInsert` sort `VehicleUpate` operation is another abstraction that a traffic model has to implement. It manipulates a vehicle in such a way that the safe insert test is passed. Sort `VehicleUpate` specifies a map function with the signature as presented in Listing 4.34. Operation `Drive` implements the longitudinal motion and operation `LaneChange` the lateral motion, both of sort `VehicleUpdate`. They update the vehicle and the PRNG state. Operation `InsertSafe?` tests whether a certain vehicle can be inserted safely into a given vehicle list. Operation `SafeInsert` constructs a vehicle as a one that it does not disturb the surrounding vehicles too much when inserted into the vehicle list, i.e. they keep the maximum acceptable distance defined by the safety rules of the traffic model. A possible implementa-

```
op UpdateOccupation : VehicleUpdate Occupation -> Occupation .
op ReplaceCounterAndVehicles : Occupation Pair{ListVehicleEltSWO,IntEltSWO} -> Occupation .
op InsertUpdatedVehicle :
   Vehicle Pair{IntEltSWO,VehicleEltSWO} List{VehicleEltSWO} Length Length
   -> Pair{ListVehicleEltSWO,IntEltSWO} .
op UpdateVehicle :
   -> FoldFunc{VehicleEltSWO,PairVehicleUpdateOccupationEltSWO,PairListVehicleIntEltSWO} .
vars C C' : Int . vars X X' : Length . vars V V' : Vehicle .
var L : List{VehicleEltSWO} . var O : Occupation . var F : VehicleUpdate .
eq ReplaceCounterAndVehicles(O, < L ; C >) = ReplaceVehicles(ReplaceCounter(O, C), L) .
eq InsertUpdatedVehicle(V, < C' ; V' >, L, X, X') =
   if NonExistentElement?(V') then < Insert(L, V) ; C' > ***** or may be even: < L ; C' >
   else if (X == X') or ((X < X') and (X <= Position(V')) and (Position(V') <= X'))
                    or ((X' < X) and (X' <= Position(V')) and (Position(V') <= X))
        then < Insert(L, V, V') ; C' >
        else < Insert(L, V, SetPosition(V', X, X')) ; C' > fi fi .
eq UpdateVehicle[V, < F ; O >, < L ; C >] =
   ***** sophisticated resolving of lane-change collisions for three lane roadways possible here
   if Periodic?(Roadway(O))
   then InsertUpdatedVehicle(V, F[V, < C ; Topology(O, V) ; Neighbourhood(O, V) >],
                             L, Start(Roadway(O)), End(Roadway(O)))
   else InsertUpdatedVehicle(V, F[V, < C ; Topology(O, V) ; Neighbourhood(O, V) >],
                             L, Length(O, Cells), Length(O, Cells)) fi .
eq UpdateOccupation(F, O) = ReplaceCounterAndVehicles(O,
   fold(UpdateVehicle, < F ; InsertDummyVehicles(O) >,
        < EmptyVehicleList ; Counter(O) >, ActivatedVehicles(Vehicles(O)))) .
```

Listing 4.35: Updating occupation tables. The vehicle update operation maps vehicles back in case of periodic boundary conditions (source: [30, Lst. 29]).

tion may use the traffic model specific optimal velocity function for the `SafeInsert` operation. The tuning elements may use this operation to construct and insert vehicles safely" [30].

With the signature for the vehicle update, the implementation of the occupation table becomes possible. Listing 4.35 presents the implementation for the update of the occupation tables. It implements the definition of function $\mathbf{F}_{drive} : \mathfrak{T}_{secs} \times S_{occ}^4 \longrightarrow S_{occ}^4$ of the conceptual design from Equation 3.33. It also whitnesses the sequential character of the occupation table update by the use of the `fold` operation in the implementation of the `UpdateOccupation` operation. The `fold` operation reflects the definition of the recursion formula which manifests the sequential character. The Maude implementation has not reflected the alternative approaches from Section 4.2.5 to circumvent the sequential character as they are a direct benefit of the conceptual design. In case of periodic boundary conditions, the simulation engine maps vehicles which have moved beyond the spatial extent of the occupation table back to the beginning of it. The `UpdateOccupation` operation does not update the deactivated vehicles of an occupation table in order to support the update of the switching areas. The switching areas are not part of the OLSIMv4 implementation. The merging back of the updated vehicles in the implementation of `InsertUpdatedVehicle` has been integrated in the `Insert` operation. Thus, the merging back limits a possible speed-up through parallelisation of the `UpdateOccupation`.

As discussed in [30] and demonstrated exemplary for the case of potential collisions, "operations `UpdateOccupation` and `InsertUpdatedVehicle` are good candidates for further formal investigations. As an example, comparing the number of vehicles in an occupation table before and after the vehicle update could ensure that vehicles do not get lost due to garbage collection. The fact that the vehicle update operation – denoted by `F` in Listing 4.35 – allows vehicles to drive onto colliding positions leads to the situation that operation `Insert` for lists of vehicles – depending on the implementation – might garbage collect overlapping vehicles. Therefore, it is possible that the number of vehicles differs before and after the update. As another example, a possible, optimizing implementation of the occupation tables might want to rely on the order reversing character of the fold operation. However, the vehicle update operations do not guarantee that the position order remains conserved. Possible reasons for the vehicle overlapping or disordering are not limited to implementation errors. They can be a model for characteristics as in [164, Sec. 5.5, 184] or intentional model design as is the case with the particle hopping models [39, 171, 83]. Consequently, updated vehicles would be inserted into the occupation table without maintaining the correct order. In the next simulation step, the vehicle update procedure may determine negative velocities. In real-world, large-scale simulations it is questionable whether implementation errors like this will be detected during the implementation phase. Model checking might help here and might proove the correctness of the model and an implementation of the simulation engine and the part-taking models."

```
(define (make-update-2step update-parallel* chlane-parallel* drive-parallel*
                           prepare update lc-update-1dir lc-update-2dir safe? cleanup)
  (let ((drive-update (make-drive-update drive-parallel* update))
        (chlane-update (make-chlane-update chlane-parallel* lc-update-1dir lc-update-2dir safe?))
        (simple-cleanup (lambda (bid rway occ-pasv occ-actv occ-updated)
                          (if (not (=fx (occ:num-vehs occ-actv) (occ:num-vehs occ-updated)))
                              (error 'simple-cleanup "wrong number of kfzs, diff="
                                     (-fx (occ:num-vehs occ-actv) (occ:num-vehs occ-updated)))
                              occ-updated))))
    (lambda (bid rway occ parts tail-k)
      (let ((min-pos (occ:min-pos occ))
            (max-pos (occ:max-pos occ)))
        (multiple-value-bind (rway occ-pasv occ-actvs)
            (prepare bid rway occ parts)
          (let* ((occ-actv (occ:append-occs* min-pos max-pos occ-actvs))
                 (thunks (map (lambda (occ-actv part)
                                (lambda ()
                                  (let ((k (lambda (occ-chlane)
                                             (simple-cleanup
                                              bid rway occ-pasv occ-actv occ-chlane))))
                                    (if (or (empty-occ? occ-actv)
                                            (<fx (occ:num-lanes occ-pasv) 2))
                                        (k occ-actv)
                                        (chlane-update bid rway occ-pasv occ-actv k)))))
                              occ-actvs parts))
                 (k (lambda (occ-lcs)
                      (let ((prev-bel occ)
                            (occ (occ:append-occs* min-pos max-pos occ-lcs)))
                        (multiple-value-bind (rway occ-pasv occ-actvs)
                            (prepare bid rway occ parts)
                          (let ((thunks (map (lambda (occ-actv part)
                                               (lambda ()
                                                 (let ((k (lambda (occ-drive)
                                                            (simple-cleanup bid rway
                                                             occ-pasv occ-actv occ-drive))))
                                                   (drive-update bid rway
                                                    occ-pasv occ-actv k))))
                                             occ-actvs parts))
                                (k (lambda (occ-drives)
                                     (let ((occ-updated (occ:append-occs*
                                                         min-pos max-pos occ-drives)))
                                       (tail-k (cleanup
                                                bid rway occ-pasv occ-actv occ-updated))))))
                            (update-parallel* k thunks)))))))
            (update-parallel* k thunks)))))))
```

Listing 4.36: The two step vehicle update algorithm of OLSIMv4.

The update algorithm in OLSIMv4 differs strongly from the presented one as the latter contains already the results of several enhancement efforts. Listing 4.36 presents the OLSIMv4 implementation. It contains the occupation table update for the longitudinal and the lateral motion. Procedure `make-update-2step` expects several procedures for applying an update in parallel, namely `update-parallel*`, `chlane-parallel*`, and `drive-parallel*`. While looking monstrous, the procedure works like a pipeline with producers and consumers. For the case of a lane change update, it distinguishes between roadways with two and three single-lanes. Before it returns the resulting update procedure, it combines the traffic model specific update procedures for the drive update, `update`, and for the lane change update, `lc-update-1dir` and `lc-update-2dir` with the parallel execution procedures, `chlane-parallel*` and `drive-parallel*` to the resulting drive and lane change update, `drive-update` and `chlane-update`, respectively. Each thread instantiates its own update procedure as the result of applying `make-update-2step` appropriately. The thread-specific resulting update procedure expects an occupation table `occ` and di-

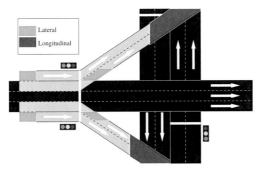

Figure 4.4: Switching areas connect roadways dynamically laterally and longitudinally (source: [30]).

vides it optionally into several partitions as specified by the `parts` argument. The remaining part of the update distinguishes then between active and passive parts of the occupation table. The update procedure updates only vehicles on the active parts. The vehicles on the passive parts are used to determine the neighbours of a vehicle. The update turned out as far too complicated. With the specification of the previous chapter, a future implementation can use a much simpler form for the update.

The optional runtime parallelism complicates the code as the scheduling procedures for sequential or parallel, drive, and lane change update have to be provided at instantiation time as arguments `update-parallel*`, `drive-parallel*`, and `chlane-parallel*`. The update algorithm supports partitioning of very large occupation tables such as the 320 km long `A003-NO-HF-034//A002-NO-HF-035` roadway of Figure 3.3 into with respect to the number of vehicles equally sized segments of the occupation table. The update procedure evaluates then the partitions in parallel and merges the updated partitions into a resulting updated occupation table. A future version could employ a combination of the multicomponent model and the network model instead of the merging process. The detailed actions of procedure `make-update-2step` are as follows. After the instantiation of the update procedures and the cleanup procedure, the `make-update-2step` procedure returns the procedure to update the occupation table `occ` with roadway `rway`. When evaluated, the update procedure partitions the occupation table `occ` into a list of partitioned occupation tables `occ-actvs` and instantiates a list of thunks for each of the partitions. In addition to the thunks, the parallel update algorithm requires a procedure `k` that consumes the results of the parallel evaluation and continues the computation. It is instantiated inside the same `let*` as the thunks and its code is very similar to the code discussed so far. It evaluates another parallel computation and forwards its results also to another procedure that performs the final cleanup. In contrast to the premature optimisation where the computation relies on the geographical segmentation of the roadways, the partitioning of the occupation tables takes their vehicle sizes into account and can, thus, distribute the computation onto several balanced threads.

4.2.11 Switching areas

Section 3.3.6 has introduced the switching areas as part of the conceptual design already. It has also sketched the switching areas in Figure 3.4 which is depicted also in Figure 4.4 for convenience reasons. OLSIMv4, however, has neither used nor implemented the switching areas, as OLSIMv4 implements the network specification simulation model as defined in Section 3.7.3. To support partitioning of occupation tables based on very long roadways, a future version of OLSIMv4 can implement the switching areas concept. Combining the multicomponent model with the network model approach from Section 3.7 provides then the partitioning approach. Their conceptual design from Section 3.3.6 is, thus, a result of the Maude implementation enhancement only. As occupation tables have already a defined update procedure and switching areas connect them pair-wise, the switching areas trace back their update procedure to the occupation table update procedure by simply instantiating a fresh occupation

```
sort SwitchingArea .
op SwitchingArea : Nat Qid Area Direction Qid Area -> SwitchingArea .
op SwitchingArea : Qid Area Direction Qid Area -> SwitchingArea .
ops LaneOffsetFrom LaneOffsetTo : SwitchingArea -> Int .
ops CellOffsetFrom CellOffsetTo : SwitchingArea -> Length .
ops JoinedName RoadwayNameFrom RoadwayNameTo : SwitchingArea -> Qid .
ops JoinedArea JoinedAreaFromPart JoinedAreaToPart AreaFrom AreaTo :
    SwitchingArea -> Area .
op PartitionArea : Area Int Int -> Area .
op PartitionArea : Area Length Length -> Area .
op JoinedRoadway : SwitchingArea Occupation Occupation Qid -> Roadway .
op JoinedOccupation : SwitchingArea Occupation Occupation -> Occupation .
op JoinedOccupation : SwitchingArea Occupation Occupation Qid -> Occupation .
op JoinedOccupation : SwitchingArea Occupation Occupation Roadway -> Occupation .
var S : SwitchingArea . vars O O' : Occupation . vars R : Roadway . vars Q Q' : Qid .
eq JoinedRoadway(S, O, O', Q) = Roadway(Q, false, JoinedArea(S), sort(
    append(ApplyOffsets(SectionsCoveredByArea(Sections(Roadway(O)), AreaFrom(S)),
                    - LaneOffsetFrom(S), Length(0, Cells) - CellOffsetFrom(S)),
        ApplyOffsets(SectionsCoveredByArea(Sections(Roadway(O')), AreaTo(S)),
                    - LaneOffsetTo(S), Length(0, Cells) - CellOffsetTo(S)))))  .
eq JoinedOccupation(S, O, O', R) = InsertDummyVehicles(Occupation(1, R,
    append(ApplyOffsets(VehiclesOnArea(Vehicles(O), AreaFrom(S)),
                    - LaneOffsetFrom(S), Length(0, Cells) - CellOffsetFrom(S)),
        ApplyOffsets(VehiclesOnArea(Vehicles(O'), AreaTo(S)),
                    - LaneOffsetTo(S), Length(0, Cells) - CellOffsetTo(S)))))  .
eq JoinedOccupation(S, O, O', Q) = JoinedOccupation(S, O, O', JoinedRoadway(S, O, O', Q))  .
eq JoinedOccupation(S, O, O') = JoinedOccupation(S, O, O', JoinedName(S))  .
```

Listing 4.37: SwitchingAreas define pair-wise links between road occupation tables (except for the first constructor for sort SwitchingArea taken from: [30, Lst. 30]).

table with the corresponding vehicle constellation as defined by the switching area. Depending on the linking direction of the switching area, the occupation table update procedure applies the vehicle update procedure to each vehicle on the freshly instantiated occupation table. Thereafter, the switching area update procedure assigns the updated vehicles back onto their corresponding occupation tables.

The first sort SwitchingArea constructor from Listing 4.37 implements the constructor of the switching area structure S_{swarea}^4 from Equation 3.34. The second one wraps the first one by initialising the random number generator state with a default seed. The remaining arguments to the constructor are the name references to the two road occupation tables, the two sort Area arguments that specify the particular areas on the road occupation tables, and the linking direction as well as a lane offset argument.

The JoinedOccupation operation constructs a fresh occupation table according to the vehicle constellation on the areas of the linked occupation tables. The operations JoinedAreaFromPart, JoinedAreaToPart, and PartitionArea transform the lane and position coordinates of the involved areas of the switching area. Operation VehiclesOnArea assists in selecting the corresponding vehicles on the joined area or on parts of it in order to place them back into their final destination occupation table. The VehiclesOnArea operations may have also been implemented as a variant of the aforementioned SectionsOnArea.

Switching areas may overlap. This is, for example, the case when lanes from various roadways lead onto the same destination lane in an intersection scenario. Figure 4.4 depicts such a scenario. In such a scenario, the results of the switching area update procedures $\mathbf{F}_{\mathrm{swarea-drive}}$ and $\mathbf{F}_{\mathrm{swarea-chlane}}$ of the multicomponent model from Equation 3.146 in Section 3.7.2 depend on the execution order. An implementation has, thus, to take special care about the case of overlapping switching areas.

The conceptual design from Section 3.7 binds the switching area concept to the multicomponent simulation model. This correspondence between the two concepts has not been seen at the time of writing of the Maude implementation. As a consequence, the Maude implementation mixes the concept of the switching areas with the simulation model instead of storing the switching areas component-wise. The unlucky mixture results in an implementation that is closer to the network simulation model from Section 3.7.3 and in having less procedure calls for the special case of having an empty set of switching areas. The Maude implementation should, however, be equipped with the corrected logic in a next

```
sort TrafficRoad .
op TrafficRoad :
   VehicleUpdate VehicleUpdate Occupation List{DetectorEltSWO} List{TuningElementEltSWO}
   -> TrafficRoad .
op TrafficRoad : TrafficRoad Pair{OccupationEltSWO,ListTuningElementEltSWO} -> TrafficRoad .
ops Tune Measure ChangeLane Drive : TrafficRoad Period -> TrafficRoad .
op Accumulate : Period TrafficRoad List{DetectorEltSWO} -> TrafficRoad .
var R : TrafficRoad . var P : Period .
eq Tune(R, P) = TrafficRoad(R, Tune(P, < Occupation(R) ; TuningElements(R) >)) .
```

Listing 4.38: Sort `TrafficRoad` combines all elements that are located on the same roadway in one element (source: [30, Lst. 31]).

```
op Tune : -> MapFunc{TrafficRoadEltSWO,PeriodEltSWO,TrafficRoadEltSWO} .
ops Measure Accumulate Tune : Period List{TrafficRoadEltSWO} -> List{TrafficRoadEltSWO} .
op UpdateTrafficRoads Drive ChangeLane :
   Period List{TrafficRoadEltSWO} List{SwitchingAreaEltSWO} -> List{TrafficRoadEltSWO} .
var P : Period . var L : List{TrafficRoadEltSWO} . var W : List{SwitchingAreaEltSWO} .
eq Tune[R, P] = Tune(R, P) .
eq Tune(P, L) = map(Tune, P, L) .
eq DriveTrafficRoads(P, L, W) = map(DriveTrafficRoad, < P ; W >, L) .
eq Drive(P, L, W) = DriveTrafficRoads(P, DriveSwitchingAreas(P, L, W), W) .
eq UpdateTrafficRoads(P, L, W) = Drive(P, ChangeLane(P, Tune(P, Accumulate(P, Measure(P, L)))), W), W) .
```

Listing 4.39: Some elementary operations for lists of sort `TrafficRoad` elements (source: [30, Lst. 32]).

version. A corrected logic would make the switching areas being a part of a sort `TrafficRoad` element that are introduced in the following section.

4.2.12 Traffic roads

The traffic road elements combine an occupation table with all other elements that are located on its roadway, i.e. the list of detectors and the list of tuning elements. Except for the switching areas and in some special circumstances the accumulators' detector name references, all name references to the roadways, detectors, accumulators, and tuning elements can be resolved in the traffic road elements itself. As part of the simulation model for the multicomponent and the network system specification in Sections 3.7.2 and 3.7.3, Equations 3.140 and 3.154 have introduced them as road structure elements, S_{road}^4 and S_{rdsys}^4, in the conceptual design section. Listing 4.38 presents sort `TrafficRoad` of the Maude implementation in Listing 4.38. It represents road structure S_{road}^4 elements for the special case $\mathbf{L}_{swarea,d} = \{\}$ in the multicomponent model from Section 3.7.2 and their corresponding structure elements S_{rdsys}^4 from Section 3.7.3 for the case of the network of systems specification. Section 3.7.3 has already discussed the correspondence between the two approaches at its beginning. For the Maude case, "a sort `TrafficRoad` element aggregates the two `VehicleUpdate` operations, `Drive` and `LaneChange`. [...] The operations for sort `TrafficRoad` as listed in Listing 4.38 wrap the appropriate operation of the contained sort, e.g. the `Tune` operation wraps the `Tune` operation for the appropriate tuning area, and constructs a new traffic road element that contains the updated entities.

The equation for operation `UpdateTrafficRoads` in Listing 4.39 describes that the Maude simulation engine rewrites the term of the left hand side into the five sequential operations of the right hand side operations `Measure`, `Accumulate`, `Tune`, `ChangeLane`, and `Drive`. The five operations are applied sequentially in one atomic simulation step. The first three of them do not depend on the switching areas. The remaining `VehicleUpdate` operations, `ChangeLane` and `Drive`, depend on the switching areas. As for an example, the first parts of the `Drive` operation are shown in Listing 4.39. The operation updates first the vehicles on the switching areas. Thereafter, it deactivates the already updated vehicles to avoid updating twice and then updates the other vehicles of the occupation table. Finally, it merges the results back again into one occupation table" [30].

The definition of the traffic road update function `UpdateTrafficRoads` from Listing 4.39 differs in the following parts from the one of the state transition function for the discrete time system specification of Equation 3.135. Firstly, the simulation time `P` is not part of the traffic road structure as it is the case for the two models. Secondly, the specifications for the traffic road elements `L` lack the switching areas. However, with the knowledge and the insights from the conceptual design, adjusting the Maude implementation for the case of the multicomponent model and for the case of the network model is now possible with only little effort.

In principle, OLSIMv4 employs the network model as its simulation model. However, OLSIMv4 additionally supports equidistant and dynamic partitioning of the occupation tables. Listing 4.36 presented the drive update and the lane change update as part of the implementation of the state transition function. The remaining parts evaluate the measure function and the tune functions. OLSIMv4 does not support accumulating the detectors measurements from detectors of neighbouring roadways. A future version of OLSIMv4 can, however, combine and implement the state transition function definitions of Equation 3.149 for the multicomponent model and of Equation 3.158 for the network model. The user can then choose between the two variants at run time.

4.2.13 Simulation networks

For the case of the multicomponent model from 3.7.2, the conceptual design specified the simulation system states as $\bigcup_{d \in D} Q_d = \bigcup_{d \in D} \{s_d^{(t)}\}$ with $s_d^{(t)} \in S_{\text{road}}^4$. For the Maude implementation, sort `SimulationNetwork` as presented in Listing 4.40 contains the simulation system state. This represents in the Maude implementation a state of a complete road traffic network and "combines a list of traffic roads with a list of switching areas in such a way that all name references of the switching areas refer to occupation tables of the contained traffic roads. Additionally, a simulation network contains the length of the time interval between two updates that the traffic models use in both iterations – called a time or update step. Listing 4.40 introduces sort `SimulationNetwork`. The constructor requires two sort `Period` arguments. One of them stores the time step of the simulation network, the other one stores the absolute time of the simulation" [30]. In contrast to the Maude implementation, the conceptual design stores the absolute time and the length of the time interval as part of the road structure S_{road}^4. As already mentioned in Section 4.2.12, the switching areas in the Maude implementation are not in accordance with the conceptual design but are part of the sort `SimulationNetwork`.

```
sort SimulationNetwork .
op SimulationNetwork : Period Period List{TrafficRoadEltSWO} List{SwitchingAreaEltSWO}
                        -> SimulationNetwork .
ops Period TimeScale : SimulationNetwork -> Period .
op UpdateTime : Period Period -> Period .
op Simulate : SimulationNetwork -> SimulationNetwork [iter] .
op Simulate : SimulationNetwork Period -> SimulationNetwork .
vars P P' : Period . var S : SimulationNetwork .
eq UpdateTime(P, P') = P + P' .
eq Simulate(S, P) = SimulationNetwork(P, TimeScale(S),
    UpdateTrafficRoads(P, TrafficRoads(S), SwitchingAreas(S))) .
eq Simulate(S) = Simulate(S, UpdateTime(Period(S), TimeScale(S))) .
```

Listing 4.40: `SimulationNetwork`s combine a list of `TrafficRoad`s and a list of `SwitchingArea`s in such a way that all references to roadways, detectors, accumulators, or tuning elements apply in itself (source: [30, Lst. 33]).

Operation `Simulate` performs a single simulation step on the simulation system state. It wraps the `UpdateTime` operation and instantiates a fresh simulation network using the constructor `SimulationNetwork`. The fresh simulation network contains the updated state of the simulation system. The updated state is the result of the `UpdateTrafficRoads` operation from Section 4.2.12. The `Simulate` operation can be applied iteratively with the number of time steps as an argument as in `Simulate^100(SimulationNetwork(...))`.

With sort `SimulationNetwork`, the simulation engine is complete. The source code of the engine takes only about 3000 lines of code. The implementation of the various models takes about 2000 additional lines of code. This sum is relatively small compared to OLSIMv4 which takes roughly estimated about 70 000 lines of Scheme code. Determining the exact number of lines is difficult as OLSIMv4 has a lot of libraries and not the clear distinction between simulation engine and models.

4.3 Vehicle models

Both implementations of the simulation engine require at least one vehicle model that represents a particular vehicle type such as passenger cars or trucks. The traffic model, detector model, tuning element model, and network model families rely on the vehicle models. Listing 4.41 presents the two vehicle types `CarType` and `TruckType` that are part of the Maude implementation. Apart from a few dummy vehicle types for the tuning elements, OLSIMv4 uses the same vehicle types. The vehicle types implement the vehicle type structure S^4_{vehtype} from conceptual design in Equation 3.2. The OLSIMv4 implementation has been presented in Listings 4.14 and 4.12 from Section 4.2.3.

```
sorts CarType TruckType .
subsorts CarType TruckType < VehicleType .
op CarType : -> CarType .
op CarType : Nat Nat -> CarType [ctor] .
op TruckType : -> TruckType .
op TruckType : Nat Nat -> TruckType [ctor] .
vars V L : Nat . var T : VehicleType .
eq CarType(V, L) = VehicleType('Car, V, L) .
eq CarType = CarType(25, 5) .
eq TruckType(V, L) = VehicleType('Truck, V, L) .
eq TruckType = TruckType(15, 10) .
```

Listing 4.41: Vehicle types `CarType` and `TruckType`. The constructors take the maximum velocity in cells per second and the maximum length in cells as arguments (source: [30, Lst. 34]).

For the vehicles, the number of attributes in addition to the vehicle type depends on the implementation requirements. Listing 4.42 presents the two vehicle models used by the Maude implementation and by OLSIMv4. The vehicle model implements the minimal vehicle structure S^4_{minveh} from Equation 3.6. The minimal vehicle structure results from the enhancement effort of the Maude implementation wherein "all vehicles in Listing 4.42 have either vehicle type `CarType` or `TruckType`, and they vary in length. The last argument to the constructor specifies a tolerance factor that models how strict vehicles, or assumed drivers respectively, will obey speed limits. Assuming that all vehicles in the simulation would follow speed limits strictly, the interaction between vehicles would come eventually to standstill, which in turn would break microscopic traffic model intrinsics [70, Chap. 5]. The magnitude of the standstill effect depends on the implementation. When dawdling takes place prior to the respecting of the topological constraints, the tolerance factor may even disable the dawdling" [30] .

In contrast to the minimal vehicle model, the OLSIMv4 implementation supported a rather blown vehicle structure that has evolved historically as a consequence of the bottom up development and where only a fraction of the available attributes where used in the end. The conceptual design from Equation 3.5 represented the resulting set of used attributes by the vehicle structure S^4_{veh}. Section 4.2.3 presented already some signatures for the accessors of the vehicle implementation in Listing 4.17. They also describe the attributes of the OLSIMv4 vehicle implementation. It also introduced the history which contains the fixed number of successive coordinate attributes that OLSIMv4 uses. A future implementation of OLSIMv4 can now rely on a vehicle model that provides all necessary details and at the same time contains no unnecessary information.

The classification into the passenger cars and trucks has been discussed already as part of the conceptual design in Section 3.2. The discussion pointed out that the two classes are already a rough simplification and that the classification in the reality varies and depends strongly on the context in which

```
sorts Car Truck . subsorts Car Truck < Vehicle .
op Car : Int Length Velocity Nat -> Car [ctor] .
op Truck : Int Length Velocity Nat -> Truck [ctor] .
```

Listing 4.42: Vehicles of sort `Car` or `Truck` (source: [30, Lst. 35]).

the vehicle models are used. The discussion also warned about the simulation risk of plug formation when using too diverse vehicle type and vehicle models. However, there may be simulation scenarios that require additional vehicle models. One of them has been introduced and described in [131] with the so-called privileged vehicles. Therein, emergency service vehicles such as ambulance, fire engine, or police vehicles can drive with higher privileges. Such a scenario may also require bidirectional roadways. The latter would also require adaptions in the microscopic traffic model.

4.4 Traffic models

Section 4.2.10 has introduced the vehicle update procedures by presenting the signatures of the Maude constructors of sort `VehicleUpdate` elements in Listing 4.34 for the drive and the lane change update part of the vehicle motion. However, the conceptual design from Chapter 3.4 presented the traffic models in large detail. Their implementation has not yet been discussed and, thus, will be done in the following.

Because the traffic models of the conceptual design need to implement only the interface as e.g. $\mathbf{F}_{\text{drive}} : \mathfrak{R}^4_{\text{occ}} \times S^4_{\text{veh}} \times S^4_{\text{nbrs}} \times S^4_{\text{topo}} \longrightarrow \mathfrak{R}^4_{\text{occ}} \times S^4_{\text{veh}}$ for the drive update, their implementation is, in general, a straightforward task. However, there are several severe pitfalls that have to be avoided. Rounding errors and improperly discretised values for the vehicles' position or velocity attributes occurred, for example, during the work on the Maude implementation [30]. Prior to the discussion of the real-world implementations, an example follows with the functional module `CONSTANT-TRAFFIC-MODEL` from Listing 4.43 that demonstrates how to implement a traffic model in the Maude case. The example implementation consists of the three operations `ConstantSafeInsert`, `ConstantDriveUpdate`, and `ConstantLaneChangeUpdate`. The constant traffic model does not change the vehicle attributes position and lane but returns instead the vehicle from the argument to each of the operations `ConstantDriveUpdate`, `ConstantLaneChangeUpdate`, and `ConstantSafeInsert`. The first two update the longitudinal and the lateral part of the vehicle motion, while the latter adjusts and returns a vehicle with appropriate velocity and position that is ready for safe insertion by the tuning elements, i.e. a vehicle with sufficient distance headway to the subsequent vehicles and without forcing the subsequent vehicles into extreme braking manoeuvres.

```
fmod CONSTANT-TRAFFIC-MODEL is
  pr NAGEL-SCHRECKENBERG .
  op IgnoreRandomNumber : Int Pair{IntEltSWO,VehicleEltSWO} -> Pair{IntEltSWO,VehicleEltSWO} .
  ops ConstantDriveUpdate ConstantLaneChangeUpdate ConstantSafeInsert : -> VehicleUpdate [ctor] .
  vars C C' : Nat . vars V V' : Vehicle . var T : Topology . var N : Neighbourhood .

  eq ConstantSafeInsert[V, < C ; T ; N >] = NaSchSafeInsert[V, < C ; T ; N >] .
  eq IgnoreRandomNumber(C', < C ; V >) = < C ; V > .
  eq ConstantDriveUpdate[V, < C ; T ; N >] = IgnoreRandomNumber(random(C), < C + 1 ; V >) .
  eq ConstantLaneChangeUpdate[V, < C ; T ; N >] = IgnoreRandomNumber(random(C), < C + 1 ; V >) .
endfm
```

Listing 4.43: Example about how to implement a traffic model (source: [30, Lst. 7]).

The constant traffic model from Listing 4.43 updates only the state of the random number generator, i.e. the number of random numbers generated so far represents the state. The model is a rather simplistic model as it does not update the vehicle. However, it is well suited as a test for the behaviour of the tuning elements and the robustness of the detectors. Each of the traffic model operations takes two arguments, the vehicle `V` and the triple `< C ; T ; N >`. The triple consists of the random number

```
op Drive : -> VehicleUpdate [ctor] .
ops NaSchSafeInsert NaSchDawdleWithoutReallyDrive
    NaSchDrive NaSchAccelerate NaSchDecelerate NaSchDawdle NaSchLimitSpeed :
    Pair{IntEltSWO,VehicleEltSWO} Pair{TopologyEltSWO,NeighbourhoodEltSWO} -> Pair{IntEltSWO,VehicleEltSWO
    } .
op vdiscrete : Velocity -> Velocity .
op xdiscrete : Length -> Length .
var C : Nat . var E : Velocity . var S : Length . var V : Vehicle . var T : Topology . var N :
    Neighbourhood .
eq vdiscrete(E) = 5 * Velocity(C/s(E) quo 5, C/s) .
eq xdiscrete(S) = 5 * Length(Cells(S) quo 5, Cells) .
eq Drive[V, < C ; T ; N >] = NaSchDrive(< C ; V >, < T ; N >) .
eq SafeInsert[V, < C ; T ; N >] = NaSchSafeInsert(< C ; V >, < T ; N >) .
eq NaSchDrive(< C ; V >, < T ; N >) = ( if V == 1st(Ahead(N))
    then NaSchDawdle(NaSchLimitSpeed(NaSchDecelerate(
        NaSchAccelerate(< C ; V >, < T ; N >), < T ; N >), < T ; N >), < T ; N >)
    else < C ; NonExistentVehicle > fi ) .
eq NaSchSafeInsert(< C ; V >, < T ; N >) =
    NaSchDawdleWithoutReallyDrive(NaSchLimitSpeed(NaSchDecelerate(
    NaSchAccelerate(< C ; V >, < T ; N >), < T ; N >), < T ; N >), < T ; N >) .
eq NaSchAccelerate(< C ; V >, < T ; N >) = < C ; Accelerate(V, Velocity(5, C/s)) > .
eq NaSchDecelerate(< C ; V >, < T ; N >) =
    if NonExistentElement?(2nd(2nd(N))) then < C ; V >
    else < C ; SetVelocity(V,
                    vdiscrete(min(Velocity(V), Headway(V, 2nd(2nd(N))) / Period(1, Seconds)))) > fi .
eq NaSchLimitSpeed(< C ; V >, < T ; N >) =
    < C ; SetVelocity(V, vdiscrete(min(MaxVelocity(V), Velocity(V)))) > .
eq NaSchDawdle(< C ; V >, < T ; N >) =
    < C + 1 ; Drive( if random(C) rem 3 =/= 0 then V
                    else SetVelocity(V, max(Velocity(0, C/s), Velocity(V) - Velocity(5, C/s))) fi ) > .
eq NaSchDawdleWithoutReallyDrive(< C ; V >, < T ; N >) =
    < C + 1 ; SetVelocity(V,
                max(Velocity(0, C/s),
                    vdiscrete(if random(C) rem 3 =/= 0 then Velocity(V)
                        else Velocity(V) - Velocity(5, C/s) fi ))) > .
```

Listing 4.44: The `Drive` part of the NaSch model as implementation for sort `VehicleUpdate`. The cell size is $7.5\,m$ in the classical NaSch model which yields in 5 cells of $1.5\,m$ used throughout this article. The `SafeInsert` operation is almost identical to the `Drive` part except that it does not move vehicles onto the next position (source: [30, Lst. 36]).

generator state `C`, the sort `Topology` element `T` as summarised by the topological analysis, and the sort `Neighbourhood` element `N`.

The Nagel-Schreckenberg (NaSch) model is a rather minimal real traffic model. It is well-suited to demonstrate the implementation of a real traffic model. Listing 4.44 presents the implementation of the longitudinal update part of the NaSch model [147]. The implementation of the lane change extensions [149] to the NaSch model is similar to the one in Listing 4.44. The implementation of the `NaSchDrive` operation is effectively a composition of the four operations `NaSchAccelerate`, `NaSchDecelerate`, `NaSchLimitSpeed`, and `NaSchDawdle`. While the space discretisation of the underlying simulation engine of Listing 4.44 assumes cells of length $1.5\,m$, the cells in the original NaSch model demand, however, length $7.5\,m$. To encode the $7.5\,m$ cells on top of the underlying $1.5\,m$ cells, one has to ensure that each position attribute and each velocity attribute also uses the discretisation of 5 cells with length $1.5\,m$. Otherwise, collisions will occur [30]. The operations `xdiscrete` and `vdiscrete` ensure the proper discretisation in Listing 4.44.

As the NaSch model does not reproduce synchronised traffic flow, it is not the best choice for the OLSIMv4. The conceptual design presented therefore, the more complex model by Lee, Barlovic, Schreckenberg and Kim and Pottmeier [124, 71, 164, 70] that reproduces the synchronised flow pattern. Listing 4.45 presents a part of the implementation of the longitudinal vehicle update. The implementation does not follow the Algorithm 1 completely as, at the time of implementing, the model was not that complete as it is in Section 3.4.1. At that time, several variants for the calculation of $\gamma_n^{(t)}$ were considered for OLSIMv4. $\gamma_n^{(t)}$ models the optimistic and the pessimistic driving behaviour. To achieve flexibility, the procedure for the calculation of the chosen $\gamma_n^{(t)}$ takes the argument `gamma`.

Procedure `pottmeier:make-c-next` from Listing 4.45 calculates the value of the "safe" maximum velocity for the next time step $c_n^{(t+\Delta t)}$ according to Equation 3.54. It already reflects the optimisation from Algorithm 1 of the conceptual design. It produces results that are in good accordance with empirical findings for German highway traffic. On evaluation, the procedure `pottmeier:make-c-next` instantiates a state that consists of several variables and functions, among them the predicate function `pessimistic?` and the method to calculate $\alpha_1^o = \alpha_0^p$ according to Equation 3.56. Thereafter it returns the function to calculate the "safe" maximum velocity $c_n^{(t+\Delta t)}$. Using the `cond` statement, the procedure distinguishes the cases as described in Algorithm 1. The rest of the expressions inside the `lambda` expression compute simple helper variables. As functions in Scheme are first class citizens, they can be returned just as normal values. With the simulation models and the system state definitions from Section 3.7, making several traffic models available for the update of occupation tables reduces to a technical task only.

For the further enhancement of the Maude implementation, there may be the need for "bidirectional" traffic inside the traffic model. In that case, the `Neighbourhood` argument to the `LaneChange` operation will still be sufficient. However, it may become necessary to add a per lane flag to sort `Roadway` that indicates the direction of travel. This flag would then be passed through to sort `Topology` elements.

In scenarios without lane discipline, at least the lane coordinate type for vehicles and for roadways would become subject of change from the discrete value presented herein to a length type such as centimeters. This kind of traffic flow can be found in developing countries such as India. A microscopic traffic model for that kind of traffic has been described in [136]. Additionally and still according to [136], it may become necessary to increase the vicinity of neighbouring vehicles compared to the presented ones. This may involve a change from the vehicle triples to a list type structure as well.

Another simulation scenario similar to [131] is the simulation of traffic flow with privileged vehicles. An implementation for that scenario requires only the above two alternative implementation ideas (i.e. no lane discipline and roadways with bidirectional traffic).

In addition to the vehicle neighbourhood argument, the traffic models also depend on the topology argument. The presented form tries to be quite minimal but still sufficient for most simulations. As an example in which this form may not be sufficient, exists with hyper-realistic simulations that want to regard ascending or descending slope, turning radius, or lane widths inside the traffic model. However, the principle of the microscopic traffic model remains the same" [30].

According to the discussion from Section 2.2.3, collisions do occur in the models by Lee, Barlovic, Schreckenberg and Kim [124] with symmetric multi-lane extensions by Pottmeier [164] or asymmetric lane-changing rules by Habel [70, 71]. However, the exact circumstances that lead to the collisions are unknown. Modelling the microscopic traffic model with Maude as demonstrated by Listing 4.44 and formalising the definition for the collisions as well as the situations that may not lead to collisions could help unveiling the circumstances under which the collisions occur. Under the proposition of a finite set of states, the Maude linear temporal logic model-checker [56] recommended from the Maude Manual [46, Sec. 10.3] can try to find counterexamples that violate the collision freeness property. The traffic model's complexity and the exact definition of the initial states that have to remain collision-free even after several steps of updates are probably the major difficulties with this approach.

4.5 Detector models

The conceptual design from Section 3.5 presented the various detector models in great detail. The presentation, however, resulted already from the enhancement effort over the Maude implementation. The OLSIMv4 and the Maude implementation, consequently, follow the conceptual design only partly. As an example for the Maude case that demonstrates how to implement the detector interface as introduced by Section 4.2.6, Listing 4.47 presents the operations for the rather minimal detector subsort `ConstantDetector`. The constructor consumes the following arguments. A symbol for the detector name, a symbol for the associated roadway name, a sort `Area` element, and a measurement value that defines the

```
(define (pottmeier:make-c-next t-safe g-add gamma delta tau-f tau-l limit verbose?)
  (let* ((pessimistic? (lambda (f0 f1 f2 vmax) (=fx 1 (gamma f0 f1 f2 vmax))))
         (to-exact (lambda (n) (let ((ex (flonum->fixnum (floor n)))) ex)))    ;; n::real, ex::bint
         (tsts+ts (+fx t-safe (*fx t-safe t-safe)))                            ;; ::bint
         (ts-l (-fx t-safe 1))                                                 ;; ::bint
         (ts+1 (+fx t-safe 1))                                                 ;; ::bint
         (g-add2 (+fx g-add g-add))                                            ;; ::bint
         (onehalf (/fl 1.0 2.0))                                               ;; ::real
         (onequarter (/fl 1.0 4.0))                                            ;; ::real
         (ninequarter (/fl 9.0 4.0))                                           ;; ::real
         (alpha (lambda (s*2/d)
                  (let ((basis (+fl onequarter s*2/d)))
                    (to-exact (-fl (if (>fl basis 0.0) (sqrtfl basis) 0.0) onehalf)))))
         (covhdw-pess (lambda (vn+1 dn+1)                           ; covered headway pessimistic case
                        (let ((a (fixnum->flonum (quotient vn+1 dn+1)))        ;; ::real
                              (b (fixnum->flonum (modulo vn+1 dn+1))))         ;; ::real
                          (*fl a (+fl b (*fl (*fl onehalf (fixnum->flonum dn+1)) (-fl a 1.0)))))))
         (covhdw-opti (lambda (vn+1 dn+1)                           ; covered headway optimistic case
                        (if (<=fx vn+1 dn+1)
                            0.0
                            (let ((dnts (fixnum->flonum (*fx dn+1 t-safe))))  ;; ::real
                              (if (<= vn+1 dnts)
                                  (covhdw-pess vn+1 dn+1)
                                  (-fl (fixnum->flonum (*fx vn+1 t-safe))
                                       (*fl (*fl onehalf dnts) (fixnum->flonum ts+1))))))))
         (c-next (lambda (sn dn tau)
                   (let ((tau+1 (fixnum->flonum (+fx tau 1)))                  ;; ::real
                         (tau (fixnum->flonum tau)))                           ;; ::real
                     (to-exact (/fl (+fl sn (*fl (*fl onehalf (fixnum->flonum dn)) (*fl tau tau+1)))
                                    tau+1))))))
    (lambda (f0::vector l0::vector l1::vector vmax::long) ;; f0 = 0. follower, l0 = 0. leader, ...
      (let ((vn (veh:vel f0))                                                  ;; ::bint
            (dn (veh:dmax f0))                                                 ;; ::bint
            (vn+1 (veh:vel l0))                                                ;; ::bint
            (dn+1 (veh:dmax l0))                                               ;; ::bint
            (hn (veh:headway f0 l0)))                                          ;; ::bint
        (let ((cn (if (pessimistic? f0 l0 l1 vmax)
                      (let* ((gn (fixnum->flonum (-fx hn (max2fx (min2fx g-add (-fx vn g-add)) 0))))
                             (bn+1 (covhdw-pess vn+1 dn+1))                     ;; ::real
                             (sn (+fl gn bn+1))                                 ;; ::real
                             (s*2/d (/fl (+fl sn sn) (fixnum->flonum dn))))     ;; ::real
                        (c-next sn dn (alpha s*2/d)))
                      (let* ((bn+1 (covhdw-opti vn+1 dn+1))                     ;; ::real
                             (sn (+fl (fixnum->flonum hn) bn+1))                ;; ::real
                             (s*2/d (/fl (+fl sn sn) (fixnum->flonum dn))))     ;; ::real
                        (cond
                          ((>= s*2/d tsts+ts) (c-next sn dn ts-l))
                          ((>= sn dn) (c-next sn dn (-fx (alpha s*2/d) 1)))
                          (else (to-exact sn)))))))
          (let ((c (max2fx (min2fx global:vmax cn) 0))) ;; bint
            (limit c f0 l0 l1 vmax)))))))
```

Listing 4.45: The core of the drive part OLSIMv4 implementation of the Lee-Barlovic and Pottmeier model [124, 164, 71, 70]. It calculates the value of the maximum safe velocity for the next time step according to Equation 3.54 and partly reflects the optimisation from Algorithm 1 in Section 3.4.1.2.

resulting traffic observables for each of the measuring cycles. The implementation of its main operation Accumulate simply constructs the resulting measurement value by combining the traffic observable values of the measurement value which has been provided to the constructor with the current timestamp of the simulation. The example aims to provide a first understanding of the Maude detector implementation. Besides their minimality, the constant detectors contribute, for example, to simulation scenarios with periodic boundary conditions that require a constant density. The simulation scenarios that participate in reproducing the fundamental diagrams as found in the scientific publications can be initialised that way.

As can be seen from the source code in Listing 4.49, the implementation of the global detector differs from the one of the constant detector in the arguments to the constructor and the operations Measure and Accumulate. Like the constant detector, the global detector takes the specification of its spatial extent

```
(define (sim:make-populate dtype measurement->csv conn tm tm-str)
  (let ((sql-partname (string-append
                "SELECT " (schema-olsim-prefix) "measurements_part_name('" tm-str "'::timestamptz)"))
        (tname (string-append "tmp_obsvs" dtype)))
    (let ((partname (car ((make-sql-select sql-partname (lambda (get-field) (get-field 1))) conn)))
          (create-tmp-obsvs-sql (string-append
                " CREATE TEMP TABLE " tname " ( detector_id varchar(50),               "
                "    j_any smallint, j_trk smallint, v_car smallint, v_trk smallint,    "
                "    p_occ smallint, density smallint, los smallint,                    "
                "    traveltime smallint, obsvs_req integer, obsvs_opt integer )    "))
          (texpr (string-append tname
                "(detector_id, j_any, j_trk, v_car, v_trk, p_occ, density, los, traveltime)"))
          (update-tmp-obsvs-sql (string-append " UPDATE " tname
                " SET obsvs_req = " (schema-olsim-prefix) "obsvs_req_pack(j_trk, v_car, v_trk, p_occ),"
                "     obsvs_opt = " (schema-olsim-prefix) "obsvs_opt_pack(density, los, traveltime)    "))
          (drop-tmp-obsvs-sql (string-append " DROP TABLE " tname)))
      (lambda (measurements)
        (let ((data (csv-strings->buffer (map measurement->csv measurements))))
          (insert-obsvs-sql (string-append
                " INSERT INTO " (schema-olsim-prefix) "obsvs(j_any, obsvs_req, mw_test)       "
                " SELECT DISTINCT w.j_any, w.obsvs_req, NULL::bit(19)                          "
                " FROM " tname " w                                                             "
                "       LEFT JOIN " (schema-olsim-prefix) "obsvs d                             "
                "         ON d.j_any = w.j_any AND d.obsvs_req = w.obsvs_req                    "
                " WHERE d.obsvs_req IS NULL                                                     "))
          (insert-measurements-sql (string-append
                " INSERT INTO " (schema-olsim-prefix) "partname "
                " SELECT '" dtype "'::char(1),                                                  "
                "        " (fixnum->string (sim:minute-of-the-day tm)) "::smallint,             "
                "        d.nr, w.nr, t.obsvs_opt                                                "
                " FROM " tname " t                                                             "
                "      INNER JOIN " (schema-olsim-prefix) "obsvs w USING (j_any, obsvs_req)     "
                "      INNER JOIN " (schema-olsim-prefix) "*sim:detektor-nodes* " d             "
                "        ON d.id = t.detector_id                                                "
                " WHERE schname = '" *pg:schema-topology* "'")))
          (let ((write-data-into-temp-table (write-table texpr data))
                (create-tmp-obsvs (make-sql-command create-tmp-obsvs-sql))
                (update-tmp-obsvs (make-sql-command update-tmp-obsvs-sql))
                (drop-tmp-obsvs (make-sql-command drop-tmp-obsvs-sql)))
            (create-tmp-obsvs conn)
            (with-handler (lambda (e) (pp e) (exception-notify e))
                          (write-data-into-temp-table conn))
            (update-tmp-obsvs conn)
            (let ((insert-obsvs (make-sql-command insert-obsvs-sql))
                  (insert-measurements (make-sql-command insert-measurements-sql)))
              (insert-obsvs conn)
              (insert-measurements conn)
              (drop-tmp-obsvs conn)))))))))
```

Listing 4.46: Populating the database with the output detector measurements from the OLSIMv4 simulation.

from the combination of the name of the associated roadway as well as the sort Area element that have been supplied to the constructor. During the Measure operation, the global detector determines the traffic flow, the occupation time, the average velocity, and the density that have shaped on the spatial extent of the detector's associated roadway segment. Listing 4.49 presents exemplary the implementation for the calculation of the density that reflects the specification as given by the formula $N_j^{(t)}/L_j^{(t)}$ of Equation 3.119 from the conceptual design in Section 3.5.6. Thereby, the evaluations of Function NonClosedCells calculate the total number of non closed cells $L_j^{(t)} = \mathbf{F}_{\text{non-closed}}(\mathbf{o}_r^{(t)}, \mathbf{d}_j^{(t)} \# \mathbf{r}_{\text{area}}, \mathbf{t}_{\text{name}}^{\text{vehtype}})$. The density per cell for each of the vehicle types results then from the fraction of the number of vehicles on the detector's spatial extent Area(G) and the number of non-closed cells. The first equation for the Measure operations in Listing 4.49 determines the vehicles on the detector's spatial extent with the evaluation of VehiclesOnArea(Vehicles(O), Area(G)).

The resulting density for each vehicle type follows then from the count after the vehicles have been filtered per vehicle type. The first equation for operation Densities counts and filters the vehicles. In contrast to the global detectors from the conceptual design, the Maude implementation does not

```
fmod CONSTANT-DETECTOR is
  pr DETECTOR . pr VEHICLES . pr CAR-AND-TRUCK-VEHICLES . pr CONSTANT-DETECTOR-TYPE . pr DETECTOR-COMMON .
  sort ConstantDetector . subsort ConstantDetector < Detector .
  op ConstantDetector{_,_,_,_} : Qid Qid Area Measurement -> ConstantDetector [ctor] .
  op ConstantDetector : Qid Qid Area Measurement -> ConstantDetector .
  op Measurement : ConstantDetector -> Measurement .
  vars Q Q' : Qid . vars A : Area . vars O : Occupation . vars P : Period .
  vars D D' : ConstantDetector . vars M : Measurement .
  eq ConstantDetector(Q, Q', A, M) = ConstantDetector{Q, Q', A, M} .
  eq Name(ConstantDetector{Q, Q', A, M}) = Q .
  eq Roadway(ConstantDetector{Q, Q', A, M}) = Q' .
  eq Type(D) = ConstantDetectorType .
  eq Area(ConstantDetector{Q, Q', A, M}) = A .
  eq Measurement(ConstantDetector{Q, Q', A, M}) = M .
  eq Measurements(ConstantDetector{Q, Q', A, M}) = List(M) .
  eq Measure(P, D, O) = D .
  eq Reset(D) = D .
  eq Accumulate(D, P) = Measurement(P, TrafficObservables(Measurement(D))) .
endfm
```

Listing 4.47: The `ConstantDetector` does not evaluate the occupation table. Instead it returns a constant measurement value.

```
sort LoopDetector . subsort LoopDetector < Detector .
op LoopDetector : Qid Qid Area List{MeasurementEltSWO} -> LoopDetector .
op HasTimestamp<=? : -> PredFunc{MeasurementEltSWO,PeriodEltSWO} .
var A : Area . var O : Occupation . var P : Period .
var L : List{MeasurementEltSWO} . var D : LoopDetector . vars Q Q' : Qid .
eq Measure(P, LoopDetector{Q, Q', A, L}, O) =
   LoopDetector(Q, Q', A, remove(HasTimestamp<=?, P, L)) .
eq Accumulate(D, P) = Aggregate(P, Type(D), Measurements(D)) .
```

Listing 4.48: The `LoopDetector` subsort provides empirical detector data to the simulation. Operation `Aggregate` is similar to the one in Listing 4.50 (source: [30, Lst. 38]).

calculate the level-of-service value, the travel time, and the occupation time. The global detectors of OLSIMv4, however, determine the level-of-service according to Figures 3.9(a) and 3.9(b). Apart from the order of the arguments, the `Measure` operation implements function $\mathbf{F}_{\text{measure}} : \mathfrak{T}_{\text{secs}} \times S_{\text{occ}}^4 \times S_{\text{det}}^4 \longrightarrow S_{\text{det}}^4$ from Equation 3.85.

Detector models either contribute in taking over traffic data into the simulation or in evaluating the simulation. To illustrate the implementation of the constant detectors and the global detectors, both kinds of detectors have been introduced. The loop detectors and the prognosis detectors represent the kind of detectors that contribute to the simulation by taking over data into it. Thus, their implementation follows the implementation of the constant detector very closely. Listing 4.48 displays the implementation of the operations. To keep the implementation of the loop detectors simple for the Maude case, the constructor instantiates detectors with a list of measurements that the detectors present to the simulation one at a time. "The `Measure` operation presented in Listing 4.48 removes old `Measurement` elements from the measurements list. It does not make use of the sort `Occupation` argument at all. In a real-world application, this operation could load measurement data for a particular time stamp from external resources" [30]. As the loop detectors, in general, do not derive the measurements from the simulation but instead use the data provided to the constructor or fetched from external resources, the loop detectors can represent several other detector types, too. "In addition to those already mentioned in Sections 2.2.2, 3.5, and 4.2.6, these can be weather data detectors, variable message sign detectors, vehicle tracking detectors, constant detectors, and traffic message jam detectors" [30].

In the context of OLSIMv4, the loop detectors fetch their measurements in each minute from the database. They store the measurements of the prognosis detectors as introduced by the conceptual design of Section 3.5.3. The prognosis detectors form a variant of the loop detectors as they provide the data to the simulation detectors through the same interface. They do, however, not present the raw data to the user, as the loop detectors do, but instead calculate the prognosticated data on demand. Except for the access time optimisation, the implementation of the calculation of the corresponding timestamps as well as of the prognosticated traffic data is a straightforward task. OLSIMv4 uses the procedural language feature of the PostgreSQL database. The OLSIMv4 7 days long-term prognosis provides measurements for the whole period of each minute within the next 7 days starting from some given timestamp. To speed up the calculation of the short-term prognoses, they reuse the 7 days long-term prognosticated measurements for their own prognoses. The implementation for the prognosis detectors relies, thus, on a service that computes the measurements for the 7 days long-term prognosis in advance. The common operation to all these kinds of detectors that provides the measurement value to the simulation is a more complex variant of the `Accumulate` operation from Listing 4.48.

The output detectors in the context of OLSIMv4 insert their measurements also into the PostgreSQL database. In contrast to the single select statement of the (input) loop detectors' measure operation, inserting of the output detector measurements as part of the detectors measure operation involves several atomic steps. Listing 4.46 presents the OLSIMv4 implementation of the steps. The first one creates a temporary table, the second one copies the data into it, and the remaining steps populate the database across the various tables with the measurements. The insert statement `insert-obsvs-sql` with the **LEFT JOIN** operation ensures a consistent state of the table that contains the traffic observables. As four simulation instances update this table concurrently in the case of OLSIMv4, this kind of data model requires the ACID properties. The database stores the measurements in per day partitioned tables. Even though the traffic data archive contains loop detector of the last 15 years, the database contains only 1321 table partitions that altogether consume roughly $3\,\text{TB}$ which occupies $75\,\%$ of the available disk space. At the time of writing, the partitions contain $44\,884\,899\,360$ measurements including those from the loop detectors and from the output detectors of the simulation. The insertion procedure `insert-obsvs-sql` places the traffic observables `j_any`, `j_trk`, `v_car`, `v_trk`, `p_occ` for the traffic flow, the velocity, and the occupation time into a separate table. These observables are common to the

```
sort GlobalDetector . subsort GlobalDetector < Detector .
op GlobalDetector : Qid Qid Area List{MeasurementEltSWO} -> GlobalDetector .
op AggregatedMeasurement : GlobalDetector -> Measurement .
op AddMeasurement : Period GlobalDetector Measurement -> GlobalDetector .
ops Densities Flows Periods Velocities :
    Period GlobalDetector Occupation List{VehicleEltSWO} -> List{MeasurementEltSWO} .
vars CA CC CT : Int . vars Q Q' : Qid . var P : Period .
var A : Area . var M : Measurement . var O : Occupation .
var G : GlobalDetector . vars L : List{MeasurementEltSWO} . var S : List{VehicleEltSWO} .
eq Measure(P, G, O) = Measure(P, G, O, VehiclesOnArea(Vehicles(O), Area(G))) .
eq Measure(P, G, O, S) = AddMeasurement(P, G, Measurement(P,
    remove(NonExistentElement?, append(
        append(Flows(P, G, O, S), Periods(P, G, O, S)),
        append(Velocities(P, G, O, S), Densities(P, G, O, S)))))) .
eq Densities(P, G, O, S, CC, CT, CA) = (
    if CC == 0 then TrafficObservable(CarType, Density(0, Vehs/Cell))
    else TrafficObservable(CarType, Density(size(filter(Car?, S)) / CC, Vehs/Cell)) fi
    if CT == 0 then TrafficObservable(TruckType, Density(0, Vehs/Cell))
    else TrafficObservable(TruckType, Density(size(filter(Truck?, S)) / CT, Vehs/Cell)) fi
    if CA == 0 then TrafficObservable(AllType, Density(0, Vehs/Cell))
    else TrafficObservable(AllType, Density(size(filter(CarOrTruckType?, S)) / CA,
                                    Vehs/Cell)) fi ) .
eq Densities(P, G, O, S) = Densities(P, G, O, S,
    NonClosedCells(SectionsOnArea(Sections(Roadway(O)), Area(G)), < Area(G) ; CarType >),
    NonClosedCells(SectionsOnArea(Sections(Roadway(O)), Area(G)), < Area(G) ; TruckType >),
    NonClosedCells(SectionsOnArea(Sections(Roadway(O)), Area(G)), < Area(G) ; AllType >)) .
eq AddMeasurement(P, G, M) = ( if Seconds(P) rem 60 == 0
    then GlobalDetector(Name(G), Roadway(G), Area(G),
        Aggregate(P, Type(G), M Measurements(G)), EmptyMeasurements)
    else GlobalDetector(Name(G), Roadway(G), Area(G),
        AggregatedMeasurement(G), M Measurements(G)) fi ) .
eq Accumulate(G, P) = Measurement(P, TrafficObservables(AggregatedMeasurement(D))) .
```

Listing 4.49: The `GlobalDetector` evaluates occupation tables with a density optimised method. Operation `Aggregate` is similar to the one in Listing 4.50 (source: [30, Lst. 37]).

loop detectors and the simulation detectors and correspond to the traffic observables j_{any}, j_{truck}, v_{car}, v_{truck}, and p_{occ} from the conceptual design in Section 3.5. The latter section emphasised the importance of having comparable input/output detector values. In the scope of the size discussion, they provide the additional benefit of being able to store, manage, and access the common traffic observables in the same data structure and by the same routines. Their values range over subsets of the natural numbers. The corresponding domain of the five tuples that combine all traffic observables would be $\mathbb{N}_0^{80} \times \mathbb{N}_0^{80} \times \mathbb{N}_0^{240} \times \mathbb{N}_0^{180} \times \mathbb{N}_0^{100}$. Under the aspect of a possible implementation of the conceptual design, maybe the most relevant question is, whether the representation of the combination in a database (or on any machine) fits into a single 32 bit integer or not. Concluding from the domain of the five tuples, the space of the five tuples does not fit into 32 bit number. It would instead require at least 36 bit[4] Even though nowadays 64 bit systems are widely available, the size of the primary key will impact the storage and access performance significantly.

Due to the functional dependence between the observables, the traffic observable tuples require normalisation, i.e. the traffic observables form a relation on its own and the measurements have to reference the traffic observables by primary key. Actions for further normalisation would also take into account that the velocity functionally depends on the traffic flow via the hydrodynamical relation $J = \rho \cdot v$ and viewed from the macroscopic traffic model perspective. Similarly does the occupation time in percent roughly depend on the product of the number of vehicles that passed a detector and the average vehicle length divided by the average velocity. As the exact functional dependence is apart from the mentioned generally unknown, further normalisation is not possible. However, it is quite obvious that not every theoretically possible combination will occur in the set of measurements. As an example from the study in [114], the high flow rates, i.e. traffic flow values above 50 vehs/min , occur with average velocities between 60 km/h and 120 km/h . The extremely high velocities above 120 km/h do not occur in combination with the high flow rates. The lack of many such combinations and the fact that the detectors measure some combinations more frequently than others results, consequently, in a traffic observables table that contains significantly less values than 2^{36}, namely $71\,659\,093$. The maximum number of combinations 2^{36} is roughly 1000 times larger than the number of values in the traffic observables table and the 32 bit integer space is 60 times larger. The table grows only very slowly, i.e. approximately $115\,000$ per week, and the growth decreases constantly. As a conclusion from the number of measurements contained in the 1321 table partitions, the detectors measure each combination of the traffic observables about 626 times.

Assuming enough disk space and a constant ratio for the frequency with which the detectors measure each combination of traffic observables, the traffic observable table could grow up to the volume that corresponds to roughly 200 years of measurements table partitions. As the database model manages the measurements in partitioned tables, they are not that space critical because it is possible to dump and restore individual partitions from or into a backup. An automated dump and restore mechanism could manage the import and export of the required set of measurements. The comfortable distance of traffic observables to the maximum possible number of rows in the tuples table (60 times) motivates to question whether it will be possible to add another traffic observable to the traffic observables relation. The density is the most obvious candidate as it functionally depends via the $J = \rho \cdot v$ on the traffic flow and the velocity. In principle, the density can grow up to a maximum value, even though the exact value varies between 133.33 vehs/km and 140 vehs/km across the publications [147] and, respectively, [150, Sec. 2.5.2] in [164, Sec. 2.1][5]. For the purpose of the traffic observable measurements, the range $\mathbb{N}_0^{150}$ should be sufficiently large. Extending the tuple definition by the density would be an interesting experiment. Too less is known about the variances between the traffic flow, the velocity, and the density. This situation is similar but not as worse as the initial situation of OLSIMv4 where also nothing was

---

[4]It might be possible to reduce the number of bits onto 34 bit by restricting the domain to smaller but still reasonable value sets for an optimised productive environment. As an example, the truck traffic flow may also be represented by a 6 bit field. Similarly, it might also be appropriate to truncate the truck velocities into to the range of values up to 127 km/h . The optimisation, however, still does not let the tuples fit into the desired size of 32 bit.

[5]The authors of [147, 150] did not define the maximum density explicitly. For the NaSch model, the maximum density results from the same length of the cells and the vehicles. Pottmeier derived the other density value from the plot in [164, Sec. 2.1].

```
var B : CrossSectionDetectorType .  var C : CrossSectionAccumulator .
var L : List{DetectorEltSWO} .  var P : Period .  var K : List{TrafficObservableEltSWO} .
eq Aggregate(P, B, K) = Measurement(P, remove(NonExistentElement?, append(
      append(map(Avg, < Flow(0, Vehs/Minute) ; Flows(K) >, AllType CarType TruckType),
        map(Sum, < Period(0, Seconds) ; Periods(K) >, AllType CarType TruckType)),
      append(map(Avg, < Velocity(0, C/s) ; Velocities(K) >, AllType CarType TruckType),
        map(Avg, < Density(0, Vehs/Cell) ; Densities(K) >, AllType CarType TruckType)))))  .
eq Accumulate(C, P, L) =
   AddMeasurement(P, C, Aggregate(P, DetectorType(C), Accumulate(P, Detectors(C, L))))  .
```

Listing 4.50: The `CrossSectionAccumulator` subsort accumulates detector measurements over a cross-section (source: [30, Lst. 39]).

known about how many traffic observable combinations will have place inside a single table. In fact, in addition to the unknown number of expected values their ranges were unknown also.

As the last aspect for the implementation issues of the detectors, the precision of the numerical calculation may also get some attention. The evaluations of operation `TrafficObservable` consume rational numbers as their arguments. In contrast to the vehicle attributes such as position and velocity in time and space discrete traffic models, the detector traffic observables do not necessarily need to be discrete. As OLSIMv4 uses a time discretisation of $1\,\mathrm{s}$ and a space discretisation of $1.5\,\mathrm{m}$ per cell, modelling the detector measurements as discrete values with $1\,\mathrm{km/h}$ as unit for the velocity is more than sufficient for the purpose. The loop detectors of the NRW highway network also use this unit. Therefore, the rounding to the nearest natural number should, thus, provide sufficient precision.

4.6 Tuning element models

OLSIMv3 did not support positioning of tuning elements at locations different from those of the checkpoints. Neither did it support choosing at runtime among several types of tuning methods. The only way to change the behaviour of the tuning elements existed with editing the source code, recompiling, and restarting the simulation. OLSIMv3 neither supported disabling of tuning elements at runtime which is important in cases where detectors do not function properly and submit unreliable measurements. It might also be necessary to enable other tuning elements in replace of the switched off ones. During the validation process, it is also unknown whether two subsequent tuning elements interfere with each other or not. Switching individual tuning elements on or off is, thus, essential for an online and real-time traffic information system such as OLSIMv4.

OLSIMv4 overcomes all these shortcomings. The OLSIMv4 tuning element models support different types of tuning elements by subsorting the tuning element type. The distinction into different types supports the implementation of the behavioural variations such as the actual should-be tuning elements and the traffic light tuning elements. However, as the various and the exact behaviour of the tuning elements are beyond the scope of this thesis, this section covers only the actual should-be tuning elements. Tuning elements keep track of the vehicles that they inserted or removed from the occupation tables. The OLSIMv4 tuning elements calculate balances with the traffic properties of these vehicles and the accumulator values. All the various kinds of tuning elements rely on the accumulators that also may vary in type and in functioning. In the contexts of OLSIMv4 and the Maude implementation, the most common accumulator is the cross-section accumulator. Thus, the following section presents it while the section thereafter presents the implementation of the actual should-be tuning elements.

4.6.1 Accumulator models

Accumulators sum and balance measurements provided by their associated detectors and provide the sums and balances to their tuning elements. The calculation of the sums and balances is a straightforward task. It involves looping through the measurements and applying a summation procedure. Listing 4.50 presents a piece of the code for it. The presented code sums and averages the traffic observables per vehicle type for the traffic flow, for the occupation times, for the velocities, and for the densities. In general, the calculation of the sums and averages does not require additional effort for the aspects of high precision and high performance.

In the contexts of OLSIMv4 and the Maude implementation, the simulation concept focuses only on the cross-section accumulators as represented by sort `CrossSectionAccumulator` in Listing 4.50. They solely accumulate measurements from detectors on the same roadway that stand on the same position but not on the same single-lane. Calculating balances of measurements from detectors of several roadways may be useful for intersection scenarios. When the detectors positioned on the on- and off-ramps provide reliable measurements, the accumulators can calculate a resulting traffic flow value for the main roadway. The advantage of this simulation scenario is that it does not require the multicomponent simulation model from Section 3.7.2. Instead, it can be used with the discrete time system network specification from Section 3.7.3 which provides significantly more parallelism as the network components states do not depend on the states of neighbouring component states. The components of the network system specification do only depend on other components' output. The measurements from detectors positioned on neighbouring roadways represent the components' output in this case. The difficulty, however, with this simulation model is where the detectors stand and where to position the accumulators and tuning elements to achieve a proper simulation result. As an example, whenever the main roadway contains detectors for the entire roadway cross-section that have a position in the middle of the on- and off-ramp but not downstream of the on-ramp, positioning a tuning element with a should-be accumulator that sums the measurements from on-ramp and the main roadway will probably provide promising results. In the contrary case, where the main roadway detectors are positioned downstream of the on-ramp, relying on the measurements of the main-roadway detectors will probably lead to better results.

4.6.2 Tuning element and tuning area models

The implementation of the tuning elements models the various tuning element types also via subsorting. Different types of tuning elements involve actual should-be tuning elements, traffic light, and weather tuning elements. The latter exploit weather data and tune their occupation tables with respect to the weather situation. The actual should-be tuning elements represent the most commonly used tuning elements in the context of OLSIMv4. They compare the measurements from the should-be accumulator to the ones from the actual accumulator and the tuned so far traffic values. Figure 5.3(d) presents an example for a purely flow adapting tuning element. As the result of the tuning elements' main operation, namely the `Tune` operation, is a pair, the Maude implementation of the tuning elements requires an additional data type, namely the tuning areas represented by the `TuneArea` type in Listing 4.51. They combine a tuning element with an occupation table.

Listing 4.51 introduces the Maude implementation of the actual should-be tuning elements. It presents parts of the `Tune` operation which implements function $\mathbf{F}_{\text{tune}*} : \mathfrak{T}_{\text{secs}} \times S_{\text{occ}}^4 \times S_{\text{tun}}^4 \longrightarrow S_{\text{occ}}^4 \times S_{\text{tun}}^4$ of Equation 3.124 from the conceptual design in Section 3.6.2. The implementation of operation `op`
`RecTuneDispatching : Period ActualShouldBeTuningArea Measurement ... -> ActualShouldBeTuningArea` contains the tuning elements' decision making of whether to insert or remove vehicles into the occupation tables. It corresponds to the first two cases in Equation 3.124. The implementation of operation `op`
`TuneDispatching : Period ActualShouldBeTuningArea ActualShouldBeTuningArea ... -> ActualShouldBeTuningArea`
covers the third case of the equation. The `Tune` operation wraps the aforementioned operations by calculating the appropriate balances between the actual and the should-be measurements. Its implementation is a straightforward task. The listing contains also parts of the implementation of the `TuneInserting` operation. The latter constructs a vehicle for insertion into the occupation table by using the safe insert operation of the traffic model. The expression to construct the pair of the vehicle and the updated random number generator state is `SafeInsert[V, < C ; Topology(Occupation(G), V) ; NeighbourhoodWithDummyVehicles(Occupation(G), V) >]`. For the case that no neighbouring vehicles exist, operation `NeighbourhoodWithDummyVehicles` constructs some. The implementation of the tuning elements requires numerical precision and robustness in the computational method. Due to rounding in combination with poor implementation it is possible that the tuning elements insert a vehicle and remove

one in the following time step. The steps to avoid this ping-pong effect involve the following ones. Calculate the traffic observables for the decision process only once in a minute. Decide the insertion or removing strategy including the quantity for the complete interval of one minute. Act depending on the strategy only and in equidistant time steps.

```
sorts ActualShouldBeTuningElement ActualShouldBeTuningArea .
subsort ActualShouldBeTuningElement < TuningElement .
subsort ActualShouldBeTuningArea < TuningArea .
op ActualShouldBe : Qid Qid Area Accumulator Accumulator List{MeasurementEltSWO}
    -> ActualShouldBeTuningElement .
op ActualShouldBe : ActualShouldBeTuningElement Occupation -> ActualShouldBeTuningArea .
op Tune : Period ActualShouldBeTuningArea -> ActualShouldBeTuningArea .

op TuneInserting : Period ActualShouldBeTuningArea Measurement Measurement Measurement VehicleType
    Flow Pair{IntEltSWO,VehicleEltSWO} -> ActualShouldBeTuningArea .
op TuneInserting : Period ActualShouldBeTuningArea Measurement Measurement Measurement VehicleType
    Flow Pair{IntEltSWO,VehicleEltSWO} Pair{IntEltSWO,VehicleEltSWO} -> ActualShouldBeTuningArea .
op RecTuneDispatching : Period ActualShouldBeTuningArea Measurement Measurement Measurement Flow
    Flow Flow Rat Rat Rat -> ActualShouldBeTuningArea .
op TuneDispatching : Period ActualShouldBeTuningArea ActualShouldBeTuningArea Measurement Measurement
    Measurement Flow Flow Flow Rat Rat Rat -> ActualShouldBeTuningArea .
vars V V' : Vehicle . var T : VehicleType . vars JA JC JT : Flow . vars NA NC NT C C' : Rat .
vars MA MS MD : Measurement . vars G G' : ActualShouldBeTuningArea . vars P : Period .
eq TuneInserting(P, G, MA, MS, MD, T, JC, < C ; V >, < C' ; V' >) =
    if NonExistentElement?(V') or NonExistentElement?(V)
    then TuningArea(TuningElement(G), ReplaceCounter(Occupation(G), C'))
    else TuningArea(AddTunedVehicle(P, TuningElement(G), V),
                    InsertVehicle(ReplaceCounter(Occupation(G), C'), V')) fi .
eq TuneInserting(P, G, MA, MS, MD, T, JC, < C ; V >) =
    if NonExistentElement?(V) then TuneInserting(P, G, MA, MS, MD, T, JC, < C ; V >, < C ; V >)
    else TuneInserting(P, G, MA, MS, MD, T, JC, < C ; V >,
            SafeInsert[V, < C ; Topology(Occupation(G), V) ;
                NeighbourhoodWithDummyVehicles(Occupation(G), V) >]) fi .
eq TuneDispatching(P, G, G', MA, MS, MD, JA, JC, JT, NA, NC, NT) =
    if NA == 0 then G' else RecTuneDispatching(P, G', MA, MS, MD, JA, JC, JT, NA, NC, NT) fi .
eq RecTuneDispatching(P, G, MA, MS, MD, JA, JC, JT, NA, NC, NT) = if (0 < NA)
    then TuneDispatching(P, G,
            TuneInserting(P, G, MA, MS, MD, JA, JC, JT, if NT < NC then CarType else TruckType fi ),
        MA, MS, MD, JA, JC, JT, -1 + NA,
                if NT < NC then -1 + NC else NC fi , if NT < NC then NT else -1 + NT fi )
    else if (NA < 0) then
    TuneDispatching(P, G, TuneRemoving(P, G, MA, MS, MD, JA, JC, JT, if NT < NC then CarType else
    TruckType fi ),
                MA, MS, MD, JA + Flow(1, Vehs/Minute),
                if NT < NC then JC else JC  + Flow(1, Vehs/Minute) fi ,
                if NT < NC then JT + Flow(1, Vehs/Minute) else JT fi ,
                NA + 1, if NT < NC then NC else NC + 1 fi , if NT < NC then NT else NT + 1 fi )
    else G fi fi .
```

Listing 4.51: The `ActualShouldBe` subsort of `TuningElement` (source for the lines up to the blank line: [30, Lst. 40]).

Tuning elements and their application areas are not limited to the ones mentioned above. They "can be used in a lot more fields of application. A control layer can use tuning elements that tag vehicles with route choices to implement explicit routing of vehicles. As another example, tuning elements can tag vehicles with time stamps to determine travel times in the simulation for particular road sections as it has been done in [80]. As a last example, to implement traffic lights, tuning elements can insert special dummy vehicles in front of selected vehicles in order to slow them down" [30]. The various tuning element types in combination with the detector types manifest "the type of simulation scenario. As an example, the standard simulation scenario in the context of OLSIMv4 [31] tunes the traffic flow and the velocity values from the measurements provided by the loop detectors to the values measured by the global detectors in the simulation. Another scenario exists with tuning elements that try to adapt the measurements from global detectors to the measurements from constant detectors. Using occupation tables with periodic boundary conditions, this scenario is an effective test case for the tuning elements. It is used later in Section 5.2.1 of this article for numerical solution verification of the simulation model

```
load libraries.maude
load engine.maude
load models.maude
load networks.maude

fmod OCCUPATION-EXAMPLE is
  pr CAR-AND-TRUCK-VEHICLE-TYPES . pr ROADWAYS . pr OCCUPATIONS .
  ops trp-11-10 tro-21-10 A057-NO-HF-030//A057-NO-HF-001 : -> Roadway [ctor] .
  ops occ-trp-11-10 occ-tro-21-10 : -> Occupation [ctor] .
  op occs-tr-10 : -> List{OccupationEltSWO} .
  eq trp-11-10 = Roadway('trp-11-10, true, Area(0, 0, Length(0, Cells), Length(9999, Cells)),
           Section(0, Length(0, Cells), Length(9999, Cells),
               Cell((Rule(CarType, Velocity(30, C/s)), Rule(TruckType, Velocity(15, C/s)))))) .
  eq tro-21-10 = Roadway('tro-21-10, false, Area(0, 1, Length(0, Cells), Length(9999, Cells)),
           Section(0, Length(0, Cells), Length(9999, Cells),
               Cell((Rule(CarType, Velocity(30, C/s)), Rule(CarType, Change(Left, Choose)),
                    Rule(TruckType, Velocity(15, C/s)), Rule(TruckType, Change(Left, Refuse)))))
           Section(1, Length(0, Cells), Length(9999, Cells),
               Cell((Rule(CarType, Velocity(30, C/s)), Rule(CarType, Change(Right, Choose)),
                    Rule(TruckType, Velocity(15, C/s)), Rule(TruckType, Change(Right, Force)))))) .
  eq A057-NO-HF-030//A057-NO-HF-001 = Roadway('A057-NO-HF-030//A057-NO-HF-001, false,
       Area(0, 2, Length(0, Meter), Length(117759, Meter)),
       Section(2, Length(1401, Meter), Length(1500, Meter), Cell((Rule(CarType, Velocity(160, Km/h)),
           Rule(CarType, Change(Right, Choose)), Rule(TruckType, Change(Right, Choose))))) ...
       Section(0, Length(1401, Meter), Length(1500, Meter), Cell((Rule(CarType, Velocity(160, Km/h)),
           Rule(CarType, Change(Left, Choose)), Rule(TruckType, Change(Left, Choose))))) ...) .
  eq occ-trp-11-10 = Occupation(trp-11-10) .
  eq occ-tro-21-10 = Occupation(tro-21-10) .
  eq occs-tr-10 = Occupations(trp-11-10 tro-21-10) .
endfm
```

Listing 4.52: Instantiation of the simulation engine with model definitions and network specifications. The network specifications contain definitions similar to the one of the functional module OCCUPATION-EXAMPLE provided after the load commands (source: [30, Lst. 1]).

implementation. More advanced scenarios may involve floating car detectors, travel time detectors, and similar. Whenever the detector has an appropriate implementation, the simulation does not require any changes. The following explanations introduce the various detector and tuning element types. They prepare the description of a complete simulation scenario setup that follows thereafter" [30].

4.7 Running the simulation

This section presents a complete example how to run the Maude simulation. Even though this section focuses on the Maude simulation, it also applies in parts to the OLSIMv4 simulation system as it demonstrates the complexity of a single instantiation of a simulation for a complete road traffic network. The Maude code for the instantiation differs, of course, from the code used for the instantiation of OLSIMv4. Most of the components' constructors are, however, very similar. Additionally, the example for the NRW highway network contains the same information as used in the productive variant OLSIMv4. The process of setting up the productive simulation system OLSIMv4 requires additional effort as the topological information is spread over several digital maps.

This section presents an example for the instantiation of the simulation engine and each of its components that participate in a simulation run. The examples for the components include the roadway elements, the occupation table elements, the detector elements, the accumulator elements, the tuning elements, the traffic road elements, and the simulation network, which represents the simulation state. Listing 4.52 presents the Maude code for an instantiation of the simulation engine. The four load commands in the listing provide the code of the libraries, the engine itself, the model definitions, and some network specifications to Maude's read-evaluate-print-loop interpreter. As an example for the instantiation of roadways, the code in the listing instantiates the three roadways trp-11-10, tro-21-10, and A057-NO-HF-030//A057-NO-HF-001. For two of the given roadways, the code in the listing instantiates the empty occupation tables, occ-trp-11-10 and occ-tro-21-10, by using the Occupation(...) constructor which

```
fmod DETECTOR-EXAMPLE is
  pr CONSTANT-DETECTOR . pr LOCAL-DETECTOR . pr GLOBAL-DETECTOR . pr LOOP-DETECTOR .
  pr OCCUPATIONS . pr LOOP-DETECTOR-MEASUREMENT .
  **** AggregatedValue : time j_any j_truck v_car v_truck p_occ -> Measurement
  op AggregatedValue : Period Nat Nat Nat Nat Nat -> Measurement .
  op Rect : Nat Nat Nat Nat -> Area .
  ops dn016680 dn016681 dn016682 : -> LoopDetector [ctor] .
  ops dc01p-10 dc02o-10 : -> ConstantDetector [ctor] .
  ops dl01p-11-10 dl01o-21-10 dl02o-21-10 : -> LocalDetector [ctor] .
  ops dg01p-11-10 dg01o-21-10 dg016680 : -> GlobalDetector [ctor] .
  vars Y0 Y1 X0 X1 : Nat .
  eq Rect(Y0, Y1, X0, X1) = Area(Y0, Y1, Length(X0, Cells), Length(X1, Cells)) .
  eq dn016680 = LoopDetector('dn016680, 'A057-NO-HF-030//A057-NO-HF-001,
      Area(0, 0, Length(2416, Meter), Length(2416, Meter)),
      AggregatedValue(Period(0, Seconds), 2, 0, 115, 255, 0) ...
      AggregatedValue(Period(25200, Seconds), 13, 3, 86, 77, 5)
      AggregatedValue(Period(25800, Seconds), 12, 4, 85, 81, 5) ...
      AggregatedValue(Period(86340, Seconds), 2, 0, 116, 255, 0)) .
  eq dg016680 = GlobalDetector('dg016680-31, 'A057-NO-HF-030//A057-NO-HF-001,
      Area(0, 2, Length(2000, Meter), Length(3999, Meter))) .               *** road-segment
  eq dc01p-10 = ConstantDetector('dc01p-10, 'trp-11-10, Rect(0, 0, 0, 9999),  *** single lane
      AggregatedValue(Period( 1, Minutes), 10, 0, 120, 0, 255)) .
  eq dc02o-10 = ConstantDetector('dc02o-10, 'tro-21-10, Rect(0, 1, 0, 9999),  *** double lane
      AggregatedValue(Period( 1, Minutes), 20, 0, 120, 0, 255)) .
  eq dl01p-11-10 = LocalDetector('dl01p-11-10, 'trp-11-10, Rect(0, 0, 5000, 5000)) . *** single lane
  eq dl01o-21-10 = LocalDetector('dl01o-21-10, 'tro-21-10, Rect(0, 0, 5000, 5000)) . ***   right lane
  eq dl02o-21-10 = LocalDetector('dl02o-21-10, 'tro-21-10, Rect(1, 1, 5000, 5000)) . ***    left lane
  eq dg01p-11-10 = GlobalDetector('dg01p-11-10, 'trp-11-10, Rect(0, 0, 0, 9999)) .    *** single lane
  eq dg01o-21-10 = GlobalDetector('dg01o-21-10, 'tro-21-10, Rect(0, 1, 0, 9999)) .    *** double lane
endfm
```

Listing 4.53: Examples for detector instantiation of the various detector types constant, local, global, and loop detector (source: [30, Lst. 2]).

is provided by the OCCUPATIONS module. Additionally, the list constructor Occupations(...) also provided by the OCCUPATIONS module as well as a list that contains the two occupation tables. The right-hand side terms in the equations from Listing 4.52 wrap the roadway constructor which is exported by the ROADWAY functional module. As part of the determation of the arguments of the roadway constructor call, the terms define the spatial extents by the Area(...) constructor and several sections that the Section(...) constructor instantiates. The sections combine a spatial extent on a single lane and a prototype cell that is instantiated by the Cell(...) constructor. The first three arguments to the sections constructor define the section's spatial extent on the single lane. A cell consists of several rules each instantiated by the Rule(...) constructor. The specification for real world roadways such as the ones of the NRW highway network differs only in size and number of rules and sections from the examples in Listing 4.52.

The Maude simulation, OLSIMv4, and even a future version of OLSIMv4 can access the same set of network elements such as roadways, sections, and cells as nodes of graphs that are stored in the database. OLSIMv4 constructs the occupation tables similar to the Maude implementation by mapping the empty occupation table constructor over the list of roadway objects. In addition to the Maude implementation, OLSIMv4 supports also filtering of roadways based on regular expressions at system startup. The filtering makes it possible to run a set of roadways in an individual process. As the POSIX thread-level parallelism lacked adequate efficiency in the beginning of the implementation phase, the filtering provided domain decomposition possibilities to OLSIMv4. In a system that implements the discrete time system network specification simulation model, this kind of domain decomposition provides a reliable, reasonable, and flexible way of running the simulation under the aspect of load balancing the individual simulation processes.

As the next step in the instantiation and running of a complete simulation, "Listing 4.53 presents the functional module DETECTOR-EXAMPLE. It contains examples for the instantiation of the detectors with respect to their various detector types, namely the constant, the local, the global, and the loop detectors. The constant and the loop detector instances instantiate their measurement values using the constructor op AggregatedValue : Period Nat Nat Nat Nat Nat -> Measurement. The constructor takes a time

```
fmod ACCUMULATOR-EXAMPLE is
 pr CROSS-SECTION-ACCUMULATOR . pr DETECTOR-EXAMPLE .
 ops ac02o-10 ac01p-10 ag01p-11-10 ag01o-21-10
     ap016680 ag016680 al01p-11-10 al02o-21-10 : -> Accumulator [ctor] .      *** relies on:
 eq ap016680 = CrossSectionAccumulator('ap016680, 'A057-NO-HF-030//A057-NO-HF-001,   ***    loop detectors
     Area(0, 2, Length(2416, Meter), Length(2416, Meter)), 'dn016680 'dn016681 'dn016682) .
 eq ag016680 = CrossSectionAccumulator('ag016680, 'A057-NO-HF-030//A057-NO-HF-001,   *** relies on:
     Area(0, 2, Length(2000, Meter), Length(3999, Meter)), List('dg016680)) .       ***    global detector
 eq ac02o-10 = CrossSectionAccumulator(                                        *** relies on:
     'ac02o-10, 'tro-21-10, Rect(0, 0, 0, 9999), List('dc02o-10)) .            *** constant detector
 eq ac01p-10 = CrossSectionAccumulator(                                        *** relies on:
     'ac01p-10, 'trp-11-10, Rect(0, 0, 0, 9999), List('dc01p-10)) .            *** constant detector
 eq al01p-11-10 = CrossSectionAccumulator(                                     *** relies on:
     'al01p-11-10, 'trp-11-10, Rect(0, 0, 5000, 5999), List('dl01p-11-10)) .   ***    local detector
 eq al02o-21-10 = CrossSectionAccumulator(                                     *** relies on:
     'al02o-21-10, 'tro-21-10, Rect(0, 1, 5000, 5999), 'dl01o-21-10 'dl02o-21-10) . *** local detectors
 eq ag01p-11-10 = CrossSectionAccumulator(                                     *** relies on:
     'ag01p-11-10, 'trp-11-10, Rect(0, 0, 0, 9999), List('dg01p-11-10)) .      ***    global detector
 eq ag01o-21-10 = CrossSectionAccumulator(                                     *** relies on:
     'ag01o-21-10, 'tro-21-10, Rect(0, 1, 0, 9999), List('dg01o-21-10)) .      ***    global detector
endfm
```

Listing 4.54: Example instantiations of detector type dependent accumulators (source: [30, Lst. 3]).

value and five natural numbers as its arguments. The first one contains the traffic flow for any vehicle type. It matches the total number of vehicles that passed the detector in the last measurement interval of one minute. The second natural number argument contains the traffic flow for the trucks only. The third and the fourth natural number arguments indicate the average velocities for the cars and for the trucks, respectively. The last argument to the AggregatedValue operation provides the occupation time as a percentage of the measurement interval within the detector covered by a vehicle. The AggregatedValue operation converts each of the numbers into the typed value representation of the simulation engine. While the constant detectors provide in each time step the same measurement value, the loop detectors provide for each minute an individual measurement value. The constructor for the loop detectors consumes, thus, a list of measurement values. The loop detector dn016680 is a real example from the OLSIMv4 simulation for the NRW A57 roadway A057-NO-HF-030//A057-NO-HF-001" [30]. Section 4.2.6 described the detectors in more detail. As the OLSIMv4 implementation stores the traffic values in the database, its instantiation does not require all measurements for the whole simulation run. Apart from the measurements, the instantiation of the detectors in OLSIMv4 is analogously.

A part of the detectors form the basis and, thus, the first step in the instantiation chain of the tuning elements. The next step thereafter involves the instantiation of the accumulators. The accumulators rely on the detectors as they aggregate their measurements. Listing 4.54 presents some examples of how to instantiate the accumulators. The listing presents the instantiation of the CrossSectionAccumulator that "calculates a cross-section wise sum of the measurements originating from the accumulator's detectors. The constructor for the CrossSectionAccumulator consumes two constant strings that specify the accumulator's name as well as the name of the associated roadway. The area type argument defines the accumulator's spatial extent. The last arguments to the constructor of the accumulator are the names of the associated detectors" [30]. OLSIMv4 uses a special detector type to model the cross section accumulators. Their instantiation follows, thus, the one of the detectors.

The tuning elements represent the last components in the series of the topological elements. Listing 4.55 presents some examples for their instantiation. As tuning elements combine accumulators pairwise, they contain actual and a should-be accumulator name references. As has been the case with the other topological components, the Area(...) and its variant, namely the Rect(...) constructor, define the spatial extent of the tuning elements. In combination with the detector and accumulator types, the tuning elements manifest the type of the simulation scenario. Due to the differences in the measuring methods, some parts of the traffic data may be more or less reliable. As an example, the loop detectors and the local detectors are not able to record the density correctly, i.e. they derive this value. In contrast, the global detectors measure the density directly but average the traffic flow over their entire spatial

```
fmod TUNING-ELEMENT-EXAMPLE is
 pr ACCUMULATOR-EXAMPLE . pr TUNING-ELEMENT . pr ACTUAL-SHOULD-BE-TUNING-ELEMENT .
 ops tn01p-lc-10 tn02o-lc-10 tn01p-gc-10 tn02o-gc-10 tn016680-pg : -> ActualShouldBeTuningElement [ctor] .
 eq tn01p-gc-10 = ActualShouldBe('tn01p-gc-10, 'trp-11-10,               *** tunes measurements from:
     Rect(0, 0, 0, 9999), ag01p-11-10, ac01p-10) .                       ***    global  -> constant
 eq tn02o-gc-10 = ActualShouldBe('tn02o-gc-10, 'tro-21-10,               *** tunes measurements from:
     Rect(0, 1, 0, 9999), ag01o-21-10, ac02o-10) .                       ***    global  -> constant
 eq tn01p-lc-10 = ActualShouldBe('tn01p-lc-10, 'trp-11-10,               *** tunes measurements from:
     Rect(0, 0, 5000, 5999), al01p-11-10, ac01p-10) .                    ***    local   -> constant
 eq tn02o-lc-10 = ActualShouldBe('tn02o-lc-10, 'tro-21-10,               *** tunes measurements from:
     Rect(0, 1, 5000, 5999), al02o-21-10, ac02o-10) .                    ***    local   -> constant
 eq tn016680-pg = ActualShouldBe('tn016680-pg,'A057-NO-HF-030//A057-NO-HF-001, *** tunes measrments from:
     Area(0, 2, Length(2000, Meter), Length(3999, Meter)), ag016680, ap016680) . *** global ->loop
endfm
```

Listing 4.55: Examples for instantiation of the tuning elements (source: [30, Lst. 4]).

```
fmod TRAFFIC-ROAD-EXAMPLE is
 pr NAGEL-SCHRECKENBERG . pr ACTUAL-SHOULD-BE-TUNING-AREA . pr TRAFFIC-ROAD .
 pr OCCUPATION-EXAMPLE . pr ACCUMULATOR-EXAMPLE . pr TUNING-ELEMENT-EXAMPLE .
 ops rlo-21-10 rlp-11-10 rgo-21-10 rgp-11-10 TRD//A057-NO-HF-030//A057-NO-HF-001 :
     -> TrafficRoad [ctor] .
 *** double lane, open boundary conditions, tunes measurements from local -> constant detectors
 eq rlo-21-10 = TrafficRoadFromOccupation(NaSchDriveUpdate, NaSchLaneChangeUpdate,
     Occupation(tro-21-10), dc02o-10 dl01o-21-10 dl02o-21-10 dg01o-21-10, List(tn02o-lc-10)) .
 *** single lane, periodic boundary conditions, tunes measurements from local -> constant detectors
 eq rlp-11-10 = TrafficRoadFromOccupation(NaSchDriveUpdate, NaSchLaneChangeUpdate,
     Occupation(trp-11-10), dc01p-10 dl01p-11-10 dg01p-11-10, List(tn01p-lc-10)) .
 *** double lane, open boundary conditions, tunes measurements from global -> constant detectors
 eq rgo-21-10 = TrafficRoadFromOccupation(NaSchDriveUpdate, NaSchLaneChangeUpdate,
     Occupation(tro-21-10), dc02o-10 dl01o-21-10 dl02o-21-10 dg01o-21-10, List(tn02o-gc-10)) .
 *** single lane, periodic boundary conditions, tunes measurements from global -> constant detectors
 eq rgp-11-10 = TrafficRoadFromOccupation(NaSchDriveUpdate, NaSchLaneChangeUpdate,
     Occupation(trp-11-10), dc01p-10 dg01p-11-10, List(tn01p-gc-10)) .
 *** triple lane, open boundary conditions, tunes measurements from global -> loop detectors
 eq TRD//A057-NO-HF-030//A057-NO-HF-001 =
     TrafficRoadFromOccupation(NaSchDriveUpdate, NaSchLaneChangeUpdate,
     Occupation(A057-NO-HF-030//A057-NO-HF-001),
     dg016680 dn016680 dn016681 dn016682 ..., tn016680-pg ...) .
endfm
```

Listing 4.56: Examples for instantiation of the traffic road subsystems (source: [30, Lst. 5]).

extent. Thus, some detector and accumulator type combinations may complicate the tuning algorithms. `tn016680-pg` represents the kind of tuning element that is used in the context of OLSIMv4. While the letter p symbolizes the periodic boundary conditions, the letter g symbolizes the dependence to the global detectors. The remaining four tuning elements `tn01p-gc-10`, `tn02o-gc-10`, `tn01p-lc-10`, and `tn02o-lc-10` setup tuning between constant and local detectors and constant and global detectors, respectively. Section 4.2.8 describes the tuning elements in more detail.

The instantiation of the `TrafficRoad` elements requires the vehicle update functions of a traffic model, an occupation table, and the list of the detectors and tuning elements located on the corresponding roadway. The functional module `TRAFFIC-ROAD-EXAMPLE` from Listing 4.56 demonstrates how to instantiate the three traffic road elements `rgp-11-10`, `rdo-21-10`, and `TRD//A057-NO-HF-030//A057-NO-HF-001`. The latter represents the traffic road as used in a simulation scenario of the A57 NRW highway. Section 4.2.12 described the traffic road elements in more detail. All components are complete now for a simulation with the discrete time system network specification. As demonstrated in Equation 3.153 from the beginning of 3.7.3, this is also equivalent to the multicomponent specification with an empty list of switching areas and a component-wise set of detectors. A generic implementation could, thus, wrap the multicomponent update function specifically and would not need to implement a separate update function. It remains, however, a question of taste which implementation variant is more appropriate. As

```
fmod SIMULATION-NETWORK-EXAMPLE is
  pr SIMULATION-NETWORK . pr TRAFFIC-ROAD-EXAMPLE .
  ops SimNetExample1 SimNetExample2 SimNetExample3 : -> SimulationNetwork [ctor] .
  op sw-tr-10 : -> SwitchingArea [ctor] .
  eq sw-tr-10 =
    SwitchingArea('trp-11-10, Rect(0, 0, 7000, 7999), Left, 'tro-21-10, Rect(0, 0, 2000, 2999)) .
  eq SimNetExample1 = SimulationNetwork(Period(0, Seconds), rlo-21-10 rlp-11-10, List(sw-tr-10)) .
  eq SimNetExample2 = SimulationNetwork(Period(0, Seconds), rgo-21-10 rgp-11-10, EmptySwitchingAreas) .
  eq SimNetExample3 = SimulationNetwork(Period(0, Seconds), List(rgp-11-10), EmptySwitchingAreas) .
endfm

set trace on .
set trace select on .
trace select PrintNumberOfVehicles .

reduce in SIMULATION-NETWORK-EXAMPLE : Simulate^180(SimNetExample2) .
...
*********** equation
eq PrintNumberOfVehicles(Q:Qid, Q':Qid, N:Int, N':Int) = N:Int == N':Int .
Q:Qid --> 'DriveUpdateDone
Q':Qid --> 'tro-21-10
N:Int --> 27
N':Int --> 27
PrintNumberOfVehicles('DriveUpdateDone, 'tro-21-10, 27, 27)
--->
27 == 27
...
```

Listing 4.57: Examples for instantiation of a complete simulation network (source: [30, Lst. 6]).

the traffic road elements result from the Maude enhancement effort, OLSIMv4 makes no use of them. However, a future version of OLSIM can now make use of them.

The terms and expressions in Listing 4.57 form the last part of the steps in preparing a simulation run. They configure the initial state by the SimulationNetwork structure. The latter combines the start time, the size of the time step which is equal to 1 s by default, the list of traffic road elements, and the optionally non-empty list of switching areas into a simulation network element. The network elements hold the complete state for the simulation engine. Listing 4.57 presents examples for the instantiation of sort SimulationNetwork elements. The simulation engine can iterate the state transition function Simulation for each of them. The listing provides an example for 180 state transitions of the SimNetExample2. Section 4.2.13 described the simulation network in more detail. Chapter 5 provides some results of the simulation. When all topological objects are available, the OLSIMv4 simulation is also ready to run. The program startup requires a lot of command line arguments such as database specific details, filtering expressions for the roadways, and the number of parallel CPU threads.

The OLSIMv4 simulation engine resulted from a bottom-up style development. One of the development goals were to have a simulator that is platform independent, that can be used in a variety of traffic simulation contexts, and that provides a high degree of parallelism. While the rough divisioning into the major components such as vehicles, detectors, tuning elements, and the network elements was relatively clear from the beginning on, the form of update procedure that implements the state transition function was not. As the latter depends even on the details of the divisioning into the components and the details were developing, the bottom-up style development of OLSIMv4 resulted in a poor and lengthy form of the update procedure. Listing 4.36 partly exposed the poor and lengthy form of the update procedure. As in the Scheme programming language, the programmer still has to care about a lot of details concerning how to achieve a particular result of computation, the declarative programming language and term rewriting system Maude was chosen to transform the update procedure into a simpler and shorter form. The effectiveness of the Maude approach can also be seen from the total lines of code. The Maude implementation of the simulation engine took only 3000 lines of code [30]. Rearranging the Scheme code in the hope of retrieving a clearer view on the details of the update procedure would have been much more difficult. The resulting form of the update procedure would still have been very complex as Scheme is not a modelling language. The OLSIMv4 implementation requires to deal with all the details such as database input and output, mapping of real-time and simulation time, compromise agreements between performance and readability aspects in the Scheme code, handling of explicit parallelism, and the remapping of locations and related traffic information between the digital maps. The Maude model dropped all these details to crystallise the essential parts. The differences between the two implementations represent the lessons learned and have been discussed up to this point. However, especially the update procedure of the Maude model still lacks elegance, scalability, comfortable parallelism, and usability for a variety of contexts. Based on the Maude implementation, the author, therefore, carved out the simulation models of Section 3.7 from the conceptual design.

For the implementation of a future version of OLSIM, the Maude implementation should first be adapted to implement the conceptual design. The Maude models can then serve as a formal specification for the implementation of a future version of OLSIMv4. Due to the strong correspondence between the semantic algebras of the conceptual design and the Maude implementation, a specification verification is now possible and does not require a great effort. The implementation of the OLSIM future version can then rely on the verified specification. This is a great advantage due to the following reasons. Firstly, several deficiencies and issues of the previous implementations have already been eliminated. Secondly, the conceptual design has been machine checked which excludes human errors from the design and hardens it into a formal specification. And lastly, the formal specification has already been refactored which involved the enhancements as presented in this chapter. With a formally verified specification, the implementer of the OLSIM future version can always verify the implementation by examining, for example, several single state transitions. As a result, a verification chain is now available. When applied properly, the verification chain will result in a significant degree of reliability.

4.8 Summary

This chapter presented the challenges that arise from the implementation of the microscopic traffic simulation as presented in the conceptual design of the last chapter. It started with Section 4.1 that raised some general considerations. The lack of an appropriate specification and an adequate software development process made the initial implementation for OLSIMv4 an experiment. The experimental nature lead to a bottom-up style development as it is common to most scientific computing projects. Verification and validation efforts, thus, were of experimental nature, too. The experimental nature of most scientific computing projects put also a burden on the software quality, i.e. the maintainability and code readability.

Additionally, as the programming languages C++ and Fortran dominate scientific computing, the reasoning about the bottom-up developed programs is also an extremely difficult task. Even though functional and declarative programming languages improve the reasoning possibilities significantly, they failed to crystallize the two different kinds of system specifications, namely the multicomponent model and the network of systems specification. The verbosity of the programming languages complicates the reasoning. Overthinking of the computational concepts on a mathematical basis proved superior over remodelling on the programming language level only. For a team of unexperienced software developers, functional programming languages proved also to be the superior approach when compared to industrial languages as the use of the functional languages result in better readable and maintainable code. The implicit parallelism of the functional programming languages supports smooth integration of explicit parallelism into the language or a given program. New architectures and especially increasing hardware parallelism call for parallelism in the programs even though exploiting hardware parallelism still lacks widespread support. The choice of the Scheme programming language, in particular the Bigloo Scheme dialect that aims to provide a replacement of the C++ programming language, proved as a good choice as it is relatively easy to learn. Compared to the traditional SQL relational databases, NoSQL approaches sound promising, however SQL has proven a reliable way.

The implementation goals of OLSIMv4 were to remove the known bottlenecks. Platform portability, support of a variety of application domains, information accuracy, increasing network size, network independency, system scalability, complexity, robustness, and correctness are conflicting goals. The adequate conceptual design could eliminate the platform dependence by an appropriate vehicle update procedure. However, the increased accuracy requires additional traffic data which, in turn, require additional computing resources. Robustness and correctness charge the implementation goals additionally. The solution to the conflicting goals requires to exploit CPU thread-level parallelism.

Section 4.2 introduced the simulation engine that implements one of the simulation models of the conceptual design from Section 3.7. The simulation engine turned out to be a state transition machine where the kind of coupling between the roadway based subsystems manifests the simulation model. In addition to the explicit models, the network objects, and the initial state, the implementation of the simulation engine consist of some libraries that—in the case of a productive implementation such as OLSIMv4—require a careful exploration of the peripheral environment. For the OLSIMv4 specific case, the libraries according to Section 4.2.2.1 involve the enumerated types library, the graphs library, the Concurrent ML on top of the POSIX thread library, the PostgreSQL library, and a random number generator library. The Maude implementation requires only a pairs, tuples, lists, and physics library. Due to the reduced complexity, the Maude model implementation provided a more abstract view on the simulation which unveiled several shortcomings of OLSIMv4.

The simulation engine uses the vehicles and vehicle types as provided by the vehicle models. The vehicle types frequently do not encounter adequate care. Their implementation impacts, however, several dependent components, namely the vehicles, the topological rules of the cells, the detectors, and the user interface. The challenges in the implementation of the vehicle model range between the minimal model as presented by the Maude implementation and a productive implementation such as OLSIMv4. Common to all vehicle implementations is the provision of a vehicles on area procedure which is quite similar to a sections on area procedure. The topological model that distinguishes between cells, sections, and roadways has proven stable as there has not been a change since the implementation of OLSIMv4. In addition to the vehicles on area and sections on area procedures, the topological analysis requires also high efficiency. The network and the topological model improve the OLSIMv3 models significantly as they provide more fine-grained modelling of the road traffic network and, at the same time, require only a minimal implementation. The impact of the simulation model on the efficiency has proven to reduce the synchronisation effort and thereby improve the parallelism. The occupation tables provide the neighbouring vehicles and the topology by two operations that require high efficiency. Additionally, the random number generator state requires special care in more parallel implementations, i.e. where each vehicle in an occupation table can be updated in parallel. As part of the simulation engine, the polymorphic detectors provide comparable input and output pairs by using a generic measurement procedure. The normalised measurements provide the basis for the use of the detectors in a variety of scenarios and application domains such as, for example, those modelled by the accumulators. The latter participate in the tuning elements and provide a generic tune procedure. These generic procedures grant the portability across a lot of platforms and, thus, represent a key concept of the simulation engine. Similarly does the unified vehicle update procedure support multiple traffic models as well as the development of test cases. The switching areas require the multicomponent approach as the simulation model. OLSIMv4 does not use them as they result in a significant drop of potential parallelism and the loop detector traffic data already provide their information. The traffic roads are also not present in OLSIMv4 although they are the essential and basic unit for the discrete time system network specification simulation. The last component forms the state of a simulation engine and is represented by the simulation networks. The state transition function has to iterate over this system state.

Section 4.3 discussed the implementation of several vehicle models. The challenge is to find the correct compromise between a minimal and a productive implementation. The traffic models from Section 4.4 presented a constant traffic model that strongly supports testing of the occupation table models. Further traffic models can be implemented in a day. Section 4.5 discussed the implementation of the detector models. The constant detectors turned out to provide testing capabilities for tuning elements. All detector models provided a significant improvement over OLSIMv3 under the aspects of several measurement methods as well as polymorphism with constant, local, global, and loop detectors. The loop detectors raised the question of efficient storage of measurements and their extremely space intensive character. As every single byte counts, normalisation of the data model proved to be a very important question. Although the implementation of the accumulator is a straightforward task, that does not encourage the implementation of the tuning element and the tuning area models as introduced by Section 4.6.2. The precision of their calculation counts as they risk to end in a non-terminating state. To avoid too complex logic, they first determine a strategy and then insert or remove vehicles into their occupation table depending upon that strategy. To avoid interference between several tuning elements, the tune operation has to be applied sequentially.

As Section 4.7 pointed out, the instantiation of the topological components requires an immense effort for a whole real-world road traffic network. Depending on the network size, the lines of code required for the instantiation exceed the lines of code required for the simulation engine by an order of a magnitude. As the network model remained mainly unchanged since OLSIMv4, a future version of OLSIMv4 can still use the network components. For the other components, the Maude implementation enhanced the OLSIMv4 models by reducing the necessary details to a minimum. With an adapted Maude implementation, the presented approach now supports specification verification of the conceptual design of the previous chapter. The so-retrieved machine checked conceptual design could then serve as a formal specification for the implementation of a future version of OLSIMv4. It can also serve to develop reliable tests for the implementation. Both the tests and the future version can then be verified formally, for example, by examining several single state transitions or by proving certain axioms in the model and the implementation. Based upon the semantic algebras of the previous chapter, the presented approach results in a complete verification chain. The retrieved lessons learned that result from the OLSIMv4 implementation, from the Maude enhancement effort, and from the conceptual design of the previous chapter motivate to develop a future version of OLSIMv4 that reflects the new insights.

Benefits of the presented work

This chapter describes the benefits of the presented work. These include the lessons learned of the OLSIMv4 implementation, enhancements of the Maude model implementation, and the reflections as manifested in the conceptual design from Chapter 3. As a first benefit, with OLSIMv4 a productive implementation exists now that provides spatio-temporal traffic information based on stationary traffic data such as the data provided by loop detectors. As a second benefit, a computer based model of the simulation exists now with the Maude implementation. Adjusting the model to contain the updates of the conceptual design can now be done and will not require a huge effort. As part of the adjustment, the conceptual design can undergo a specification verification. After a successful verification, the model can serve as a formal specification for the implementation of a future version of OLSIMv4. With the conceptual design as a semi-formal specification and the model as a formal specification, a verification chain with formal methods for future implementations exists. A future implementation can, thus, rely on the verification chain. This also contributes significantly to a reliable and trustworthy implementation. Such an implementation is then ready for validation and calibration. To summarise these benefits, this thesis provides a mature simulation model for a real-world microscopic traffic simulation and facilitates verification and validation possibilities for it. As a last benefit, the mature simulation model in form of the verified formal specification and in combination with the discussed implementation issues reduces the effort for a future implementation significantly. At the same time, the maturity of the simulation model increases code quality.

Section 5.1 describes the various kinds of traffic information. The road users, the road operators, and the scientists gain information from OLSIMv4 under different aspects. Road users are primarily interested in choosing the best route. When considering solely a highway network, the best route is the one with the shortest travel time. When taking several road networks into account, the best route depends on the different service qualities such as vehicle type specific service, safety, distance, fuel consumption, and touristic highlights. Road users collect their traffic informations from on-board devices such as radio and navigation devices but also from traffic information platforms on the internet in advance of a trip. In contrast to the road users, the road operators focus more on the proper functioning of the road. The functioning includes information about the current road state as well as information for a future road. It includes the proper functioning of loop detector devices, clean roads, safety of roads near to merging areas, service level, travel times, capacity, potential accidents, and traffic routes for emergency services. The road operators obtain their traffic information in planning centres and in operator control centres via specific traffic telematics solutions. As the last user group, the scientists focus on traffic information in more detail and concerning all aspects that covers the subject of traffic. The physicist and the traffic engineers are interested in the more fine-grained traffic information such as time-headway distributions, lane-change distributions, travel-times, just to name a few. The section closes with a description of further traffic information that includes historical traffic information and value added information.

Section 5.2 explains the details of the verification chain that is available now for future implementations of OLSIMv4 and it presents several verification and validation activities that have been applied in the past for the implementation of OLSIMv4 and the Maude enhancement effort. As part of the latter, the numerical solution verification for a simplified simulation scenario with periodic boundary has been demonstrated and the shortcomings have been discussed. However, being able to verify the correctness of a particular simulation model and its implementation for the limited scenario of periodic boundary conditions plays an important role in the overall verification process. Section 2.2.6 mentioned already that, for verification purposes, the simulation models are required to trace the simulation scenario for the entire network back to the single road case. The simulation models from the conceptual design support the tracing back. The successful numerical verification of the periodic boundary case of a single road forms, thus, the key for a successful verification process. The lack of collisions forms another key in the verification process and retrieves special attention next in the section. The section also describes further verification possibilities as part of the benefit. The possibilities include proofs on the computer code, considerations about the random number generator state, sensitivity analysis, and examinations about the maximum traffic flow handled by the traffic model. The section closes with a presentation of a travel-time and a performance validation. The travel-time validation has been done in an early phase of the OLSIMv4 implementation.

As the last part of the benefits of this thesis, Section 5.3 discusses the design decisions and the issues for a potential implementation of the presented and updated models. Due to the maturity of the participating models, the increased insights from the implementation issues, and the existence of a semi-formal and a formal specification, the work of this thesis discharges as another benefit in a significantly reduced implementation effort for a future version of OLSIMv4. As Section 5.2 has already examined the effectivity of the model algorithms, the section focuses on some considerations about the design and efficiency.

Section 5.4 completes this chapter by summarising the most relevant benefits.

5.1   Obtaining and distributing traffic information

The ultimate goal of a traffic simulation system is to obtain higher valued traffic information. The various target audiences of the traffic information include road users, road operators, scientists, companies, courts, and further user groups. While all users may access the same traffic information, each user group has its own set of devices and tools to access the traffic information optimised for its specific context. Even though the various user groups do not access the traffic information via SQL queries, this section presents the traffic information frequently in the form of the results from accessing the database. The data interfaces as presented in Figure 2.3 refine and provide the data to the various subscribers.

Subsection 5.1.1 describes the traffic information for the road users. They access the traffic information either published via the internet platform as well as in form of traffic messages that are available on-board via navigation devices or via radio broadcasting. Road users are mainly interested in the current level-of-service, travel-time losses, road works messages, and police warning messages. However, to support better trip planning, the internet platform also provides the prognosticated traffic state for various time horizons in advance.

Subsection 5.1.2 presents the OLSIMv4 traffic information that is of main interest for the road operators and the radio broadcasting stations. They access the traffic information in operator control centres via specific traffic telematics solutions such as the DAV data distributor [177] and also include the internet access to the restricted access area. Their relevant traffic information with respect to OLSIMv4 and the internet platform contains statistics about the proper functioning of the loop detectors, road works messages, and police warning messages. In addition to the various user groups of the OLSIMv4 traffic information system, the section also enlightens the limitations of the approach by presenting traffic information of a rare event with extreme snow conditions. The limitations occur due to the analogy and the differences in the functioning of a simulation that Chapter 1 explained already.

The limitations of the simulation approach prepare also the presentation of the scientific traffic information that Subsection 5.1.3 introduces. The latter provides some traffic information for the traffic scientist as well as for the simulationist. The provision of the traffic information contrasts the road user traffic information with the loop detector traffic data. It presents exemplary an excerpt of the online OLSIMv4 simulation result for a road-segment of the A3 NRW highway.

As the last part of this section, Subsection 5.1.4 details further traffic information that can be derived from the OLSIMv4 simulation system.

5.1.1 Road user traffic information

Figures 2.1 and 5.1 render different views on the restricted access area of the NRW traffic information platform `Autobahn.NRW.De` [146]. The views present a superset of the traffic information that is available to the general public. While both, the public version and the restricted access version, make the traffic information of the simulation available, the public version hides the detailed loop detector statistics and the police warning messages from the internet user. The detailed loop detector statistics is too sensitive to publish because of the following reasons. The loop detector traffic information contains the traffic flow and the related average velocity for each minute. Publishing these information would facilitate all internet users to search for road segments where the vehicles drive much faster than the tolerated maximum velocity. The loop detector traffic information would also let the simulation results appear odd to untrained persons as the subject is rather complex. The traffic information of the loop detectors may appear as contradicting the information retrieved from the simulation. These reasons run contrary to the intent of equipping the road user with traffic information to achieve a better capacity utilisation of the NRW highway network. Among general warning messages about accidents and the like, the police warning messages contain also automatically and manually generated warning messages about traffic queue lengths in case of congested traffic. As a consequence that arises from the automatic generation, the warning messages have a fixed run of validity that does not fully take the dynamics of the traffic into account. As these warnings can contradict the traffic information that results from the simulation, a lot of users might be irritated by the contradicting content. Even worse, they might also loose their trust in the accuracy of the OLSIMv4 results. To avoid this, the generator process for the visualisation refines the data. The generator publishes only a subset of the police warning, namely lane or road closings and entry or exit blockings. However, due to the flexible model of OLSIMv4, a special detector could feed the information from the warnings about queue length into the simulation with the goal of increasing the information accuracy in a future version of OLSIMv4. Of course, striving to this goal requires also an corresponding detector model.

The web front end as depicted in Figure 5.1 for the restricted access area shows a part of the NRW highway network. The points between the two directional roadways represent the exits and split the roadways into the road segments. Unlike in the public version, the restricted access area depicts each road segment with a circle in the middle of it. The circles contain and display the information on mouse over and are analogous to the information depicted in the green box. The information contains the most recent and averaged loop detector data for the road segment that are only available in the restricted access area. The level-of-service and the travel-times are, however, available to all internet users. The web front end presents the total travel time for each of the chosen routes to the internet users by letting them choose origins and destinations of their trips and calculating the total on the client side. The web front end additionally provides the level-of-service in four various time horizons, namely the current traffic situation, the $30\,\text{min}$ and $60\,\text{min}$ short-term prognosis, and the $7\,\text{d}$ long-term prognosis. The road segment colours indicate the level-of-service to the users, i.e. a light green road-segment indicates free flow, a dark green one dense, a yellow one viscous, and a red one congested traffic. Two special modes for internet users with dichromacy provide the option to toggle between colour modes. In addition to the level-of-service, the unrestricted version provides special travel-time views for the internet users. The views distinguish between the set of vehicle types that they regard to calculate the travel times. Section 3.2 has discussed the difficulty of this approach. However, one of the views presents the travel times for the trucks only. The other view presents the travel times without respect to any vehicle type.

OLSIMv4 distributes its traffic information as part of the simulation results not only to the internet users. It also feeds the information into the data distribution software (DAV) [177]. The latter distributes the information to several subscribers among them broadcasting stations and regional traffic information providers. Listing 5.1 presents the structure and the kind of the traffic information for the DAV. OLSIMv4 feeds the information for each of the four time horizons. The traffic information for the four horizons are the prognosis for the current time, the $30\,\text{min}$ and $60\,\text{min}$ short-term prognosis, and the $7\,\text{d}$ long-term prognosis. The subscribers all focus on providing the traffic information to the user in advance of a trip and not as part of an on-board information. Section 1.1 has already discussed the

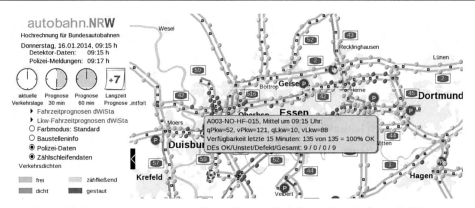

Figure 5.1: Restricted access area of the online traffic information platform [146].

superiority of this approach. As the internet platform equips the highway users route decision with more complete traffic information, the road users, thus, can make up a well-considered decision. On-Board navigation devices, in contrast, influence the highway user during the trip and tend to provoke spontaneous reactions including self-destroying prophecies. The latter may result from congestion warnings for road segments in the assumed case "when all drivers choose alternative routes to avoid the critical road segment, the latter may be free of congestion, whereas the alternative, formerly uncongested routes become congested" [110]. The facility to combine several traffic information of various sources in one view as demonstrated in Figures 2.1 and 5.1 results in more complex traffic information. The latter requires careful considerations in the context of route decisions which results in a plus in reasoning and prevents spontaneous mass reactions.

The web front end also presents traffic information from road works messages and from police warning messages about lane and exits closings to the internet users. The messages containing the information are ALERT-C [86] encoded. In addition to only decoding these messages, OLSIMv4 or a next version can use the same encoding to provide the spatio-temporal traffic information to broadcasting stations and/or on-board navigation devices. With the spatio-temporal traffic information, the topological data that is available in the database, and the event code and the location code list from the TMC forum, OLSIMv4 can also generate and format ALERT-C encoded traffic messages. Listing 5.3 demonstrates the generation of ALERT-C encoded message fragments for the A3 road segment subset that has either viscous or congested traffic state. Additionally, in the last years the smartphone technology has established a mixture of on-board navigation devices and on-board internet access. That is the reason why OLSIMv4 also provides an optimised version for mobile phone devices such as smartphones. The OLSIMv4 traffic information is, thus, also available on-board.

The OLSIMv4 simulation writes its simulation results into the database in the form of detector measurements depending on the detector type and similar to as presented in Listings 5.4 and 5.1. The web front end and the data distribution software (DAV) front end retrieve their traffic information from the database. The travel-time presentation of the web front end may serve as an example for the refinement of the traffic information. While OLSIMv4 calculates the travel-times for the individual road-segments only, the web front end displays the travel-times for complete routes to the internet users It is the web front end that adds the travel-times of the road-segments along a route to a total. As part of the travel-times refinement, the web front end lets the internet users choose origins and destinations of their trips and presents the total travel time for each of the chosen routes. Additionally, the web front end highlights the selected route and shows the level-of-service for each involved road-segment. The internet users can select several routes from an origin to a destination by simply selecting the intersections located on the way. The web front end presents also the road works along the chosen route. Additionally, the internet users may examine the future trend by the short-term prognosis functionality.

```
SELECT v.tmstmp, v.otdf_id, c.length, v.flow, v.density, v.velocity, v.traveltm, v.los
   FROM traffic_information('a'::bpchar)
       v(target, tmstmp, otdf_id, flow, density, velocity, traveltm, los)
   JOIN ( SELECT c.* FROM compat_otdf c WHERE c.otdf_id ~ '^A...-..-HF-...$'::text ) c
     ON v.otdf_id = c.otdf_id
  WHERE v.los > 5
ORDER BY 1, 2;
       tmstmp           |    otdf_id     | length | flow | density | velocity | traveltm | los
------------------------+----------------+--------+------+---------+----------+----------+-----
 2014-10-24 17:46:00+02 | A001-NO-HF-102 |   4500 |   22 |      29 |       44 |      343 |   6
 2014-10-24 17:46:00+02 | A001-SW-HF-085 |   2900 |   24 |      30 |       47 |      203 |   6
 2014-10-24 17:46:00+02 | A001-SW-HF-T84 |   7600 |   25 |      28 |       54 |      469 |   6
 2014-10-24 17:46:00+02 | A003-NO-HF-014 |   4600 |   25 |      29 |       51 |      302 |   6
 2014-10-24 17:46:00+02 | A003-NO-HF-015 |   8100 |   24 |      32 |       44 |      611 |   6
 2014-10-24 17:46:00+02 | A003-NO-HF-025 |   2200 |   25 |      25 |       66 |      126 |   6
 2014-10-24 17:46:00+02 | A003-NO-HF-026 |   1600 |   26 |      33 |       47 |      115 |   6
 2014-10-24 17:46:00+02 | A003-NO-HF-027 |   3600 |   25 |      23 |       64 |      188 |   6
 2014-10-24 17:46:00+02 | A004-NO-HF-015 |   2600 |   24 |      35 |       42 |      200 |   6
 2014-10-24 17:46:00+02 | A043-NO-HF-K17 |   2200 |   26 |      31 |       49 |      145 |   6
 2014-10-24 17:46:00+02 | A046-NO-HF-030 |   5400 |   23 |      28 |       49 |      355 |   6
 2014-10-24 17:46:00+02 | A046-NO-HF-032 |   2100 |   25 |      33 |       45 |      150 |   6
 2014-10-24 17:46:00+02 | A046-SW-HF-035 |   2000 |   25 |      36 |       43 |      152 |   6
 2014-10-24 17:46:00+02 | A052-NO-HF-025 |    700 |   27 |      26 |       83 |       37 |   6
```

Listing 5.1: Querying the traffic information from the database via the PL/pgSQL function `traffic_information(char(1))`. The function extends the global detector measurements of the main roadway segments by the travel time which equals the fraction of the roadway segment length and the average velocity on the roadway segment. The WHERE-clause limits the output to road segments with a level-of-service greater than 5, i.e. that have congested traffic.

The queries in Listings 5.1, 5.2, and 5.3 used the PL/pgSQL procedure `traffic_information('a'::bpchar)` to select the current traffic information provided by the global detectors. As OLSIMv4 runs 4 instances of the simulation in parallel, the real-time, online traffic information system of OLSIMv4 offers also traffic information in each minute prognosticated for a time horizon of 30 min, 60 min, and every additional full 30 min up to 7 d in advance. OLSIMv4 provides the information also as part of the procedure evaluation result. To obtain the 60 min and 7 d prognosis traffic information, the procedure must be evaluated for the `'s'::bpchar` and the `'l'::bpchar` argument, respectively. The truck specific traffic information follows similar for the upper case letter arguments. In earlier versions, OLSIM published the travel times for a predefined set of routes which is, of course, also supported by the database.

As detailed by Listing 4.46 from Section 4.5, the OLSIMv4 detectors write their measurements into the measurement table partitions of the database. Among the set of all OLSIMv4 detectors, the global detectors provide the spatio-temporal traffic information as is part of the query result from Listing 5.1. It compounds the traffic observables level-of-service, average velocity, density, and traffic flow on the particular roadway segment that is covered by the global detector. In contradiction to their definition in Equation 3.113 and their data definition of Equation 3.114 from the conceptual design in Section 3.5.6, the OLSIMv4 global detectors do, however, not provide the travel time as part of their measurements in the simulation. Instead, the travel time is generated by the database as a simple substitution that equals the fraction of the roadway segment length and the average velocity on the roadway segment. The reasons for the relatively simple calculation formulae of the travel times are that OLSIMv4 originally intended to use and, thus, introduced a specialised detector variant for the travel times. However, the travel time detectors were not validated and, thus, not built into the simulation.

A very promising approach to improve the information accuracy including the travel times significantly, exists at least theoretically with exploiting the traffic data from the toll stations on German highways. As the German government has serious plans in collecting toll from all German highway users in future and in contrast to the current situation where only truck drivers have to pay toll, a future version of the Toll Collect system in Germany might scan the number plates of all vehicles as part of the toll calculation process. The current laws allow the Toll Collect company and the German

```
SELECT v.otdf_id, v.velocity, b.direction || '\n' || b.message AS message
  FROM traffic_information('a'::bpchar)
       v(target, tmstmp, otdf_id, flow, density, velocity, traveltm, los)
    JOIN ( SELECT c.* FROM compat_otdf c WHERE c.otdf_id ~ '^A...-..-HF-...$'::text ) c
      ON v.otdf_id = c.otdf_id
    JOIN olsim_4_0_0.roadworks_current b ON v.otdf_id = b.otdf_id
  WHERE v.los > 5 AND v.tmstmp BETWEEN b.starts_at AND b.stops_at
ORDER BY 1, 3;
   otdf_id      | velocity |                          message
----------------+----------+-----------------------------------------------------------------
 A003-NO-HF-025 |       66 | Kölner Ring: Köln - Oberhausen                                 +
                |          | Zwischen AS Köln-Dellbrück und AS Köln-Mülheim                 +
                |          | in beiden Richtungen Dauerbaustelle,                           +
                |          | Höchstgeschwindigkeit: 60 km/h,                                +
                |          | Fahrbahn von 4 auf zwei Fahrstreifen verengt,                  +
                |          | Verkehrsbehinderung zu erwarten,                               +
                |          | Dauer: 24.10.2014 bis 27.10.2014 23:59 Uhr                     +
 A046-NO-HF-032 |       45 | Wuppertal - Düsseldorf                                         +
                |          | Zwischen AS Wuppertal-Katernberg und Sonnborner Kreuz          +
                |          | in beiden Richtungen Dauerbaustelle,                           +
                |          | Höchstgeschwindigkeit: 80 km/h,                                +
                |          | Verkehrsbehinderung zu erwarten,                               +
                |          | Dauer: 13.12.2013 bis 10.11.2014 23:59 Uhr                     +
```

Listing 5.2: Further investigation of the traffic situation on the road segments with congested traffic. The messages for the road segments describe the affected location and the direction of travel. They also contain the information about the reduced maximum velocity of $60 \, \text{km/h}$ and $80 \, \text{km/h}$, respectively, on the road segments. On the first road segment only 2 out of generally 4 single-lanes are available. Each message contains a guess whether to expect delays due to the roadworks or not. The messages end with the information about the time period within which the road segments are affected by the road works (output formatted/beautified manually).

administration for vehicle goods to scan and to store some vehicle unique information such as number plate. If one day, the data would become available to OLSIM, to the research, and to the industry for example in an anonymous form, there would be a great potential to improve the travel times and all the other traffic information published by OLSIM. Section 5.2.4 will present the results of a travel-time validation that is based on a section control approach. Using the section control method and the Toll Collect data, OLSIM would benefit from the data in the validation and the calibration process. As a result, OLSIM could significantly increase the information accuracy. A direct and unrefined use of the recorded travel times would probably not provide a great benefit. As the control stations check only whether a passing vehicle has paid the toll or not, the only 300 total number of control stations for all German highways seems sufficient. The small number of control stations leads to a very large average distance of two subsequent stations. The large distance begs the question of the usefulness in a direct and unrefined use.

Listing 5.1 demonstrates how to query the spatio-temporal traffic information from the traffic database. The query uses a PL/pgSQL function `traffic_information(char(1))`. The function searches for the global detectors and their measurements that cover the main roadways and outputs the names of the road segments under column `otdf_id` instead of the detector names. The function also extends the global detector measurements of the main roadway segments by the travel time as described above. The WHERE-clause of the query limits the output to road segments with a level-of-service greater than 5, i.e. that have congested traffic. The traffic observables in Listing 5.1 have the following units. The unit for the length values of column `length` is metre, the traffic flow of column `flow` is listed in vehicles per minute and single-lane, the density of column `density` has unit vehicles per kilometre and single-lane, the average velocity of column `velocity` refers to kilometres per hour, and the travel time of column `traveltm` is listed in seconds.

The demonstration also exhibits the approach of combining the topological data with the detector measurements and the time information. The database supports, of course, also querying the measurements for each of the remaining detector types. The real advantage of combining all the various traffic information into a single traffic database as discussed in Section 2.1.4.3 becomes even more obvious when further exploring the set of potential factors that may have contributed to the congested traffic on the mentioned road segments. Listing 5.2 demonstrates this by searching the road works messages for potential impacts on the affected road segments. The messages indicate road works for the two road segments A003-NO-HF-025 and A046-NO-HF-032. [1] In addition to the time period of the road works, the description of the affected road segments, the direction of travel, and the information about whether to expect delays due to the roadworks, the roadwork messages contain a velocity restriction to 60 km/h and 80 km/h , respectively, is one of the road works' consequences to the road users. Another one for the A003-NO-HF-025 road segment, reduces the number of available single lanes from 4 to 2 for the road users. Although the OLSIMv4's topology for the microscopic traffic simulation did not reflect the reduction in the number of available lanes, the average velocity of 66 km/h from Listing 5.1 corresponds in fact to the velocity limit restriction of 60 km/h from Listing 5.2. The detectors on the closed lanes of the real-world road-segment did not measure any vehicles. Consequently, the tuning elements on the two closed lanes did not insert any vehicles. The nonetheless existent match between the velocities without the knowledge of the reduced number of lanes may serve as an evidence for the effectiveness of the tuning approach to decrease the velocity by synchronously increasing the traffic flow [69].

The value for the measured velocity is even higher than the maximum tolerated velocity which raises the question whether the indicated level-of-service has a realistic value with the congested traffic. As the reduced maximum velocity due to the road works is part of the dynamic topology and the latter is not yet integrated to OLSIMv4, the simulation based the calculation on the default velocity value. As for this case the global detector applies the level-of-service classification scheme of Figure 3.9(a) with the mappings for maximum velocities above 100 km/h and with the value boundaries from Table 3.5.8, the level-of-service does not reflect the situation adequately. The other road segment that is affected by road works, in contrast, has a more tolerant velocity restriction of 80 km/h but with 45 km/h at the same time in fact a significantly lower average velocity on it. The indicated level-of-service reflects, thus, the traffic situation adequately.

5.1.2 Road operator traffic information

In addition to the traffic information for the road user, the transport ministry of NRW and the road operator retrieve more sensitive and more fine-grained information through a restricted access area of the publicly available online traffic information system. Figures 2.1 and 5.1 presented a snapshot of the restricted area. The area contains the summed travel times for a predefined set of road segments. The generator for the visualisation determines the travel times for the predefined set of routes analogously to the description in the previous subsection. The views for the travel times distinguish also between a truck only view and a view that regards any vehicle types. The area distinguishes the vehicle types into cars and trucks (see Section 3.2 for the difficulties of this approach) and sums their travel times. The road operator use the travel time estimations to indicate them along the roads with alternative route suggestions to the road users in high traffic flow conditions. The restricted access area contains more additional traffic information that will be described in the following. Additionally, the OLSIMv4 sends the simulation results to the road operator's traffic data centre for redistribution of the traffic information. Among the recipients are the broadcast station and several publisher of alternative traffic information platforms.

In case of several loop detector cross-sections for a particular road-segment, the average in Figure 2.1 expresses the mean value over all cross-sections. The green box displays also the number of available measurements during the last time period of 15 min . By dividing the number of measurements with 15, the total number of single-lane detectors for the particular roadway becomes available. The green box contains also the total number of single-lane detectors in the last field of the last row. The first field of the last row displays the number of single-lane detectors that work properly. The remaining two fields of the last line display the number of detectors that toggle between not working and functioning as well

---

[1] The road works message for road segment A003-NO-HF-025 came in fact as part of the police traffic warning messages. However, the way that the message came into the traffic database is irrelevant for the considerations.

```
traffic_nrw_1_0_0=# SELECT otdf_id, (lcd).d, (lcd).von AS sloc, (lcd).nach AS ploc,
  replace(array_to_string(ARRAY[descr, natdescr], E'\r\n'), 'Q', velocity::text || ' km/h') AS message
          array_to_string(ARRAY[descr, natdescr], E'\r\n') AS descr, velocity AS vel
    FROM (SELECT v.*, e.*, lcl_11_0.otdf_to_lcd(v.otdf_id) AS lcd
          FROM traffic_information('a'::bpchar)
               v(target, tmstmp, otdf_id, flow, density, velocity, traveltm, los)
          JOIN ( SELECT c.* FROM compat_otdf c WHERE c.otdf_id ~ '^A003-..-HF-...$'::text ) c
          ON v.otdf_id = c.otdf_id
          JOIN evl_4_01.eventlist e ON (CASE WHEN v.los = 6 THEN e.code = 101 ELSE e.code = 108 END)
          WHERE los > 4 ORDER BY 1, 2) t ORDER BY 1, 2;
    otdf_id      | d | sloc  | ploc  |                      message
-----------------+---+-------+-------+----------------------------------------------------------------
 A003-NO-HF-013  | + | 10924 | 10925 | queuing traffic (with average speeds 85 km/h)\r           +
                 |   |       |       | (L) stockender Verkehr (Durchschnittsgeschwindigkeit 85 km/h)
 A003-NO-HF-014  | + | 10923 | 10924 | stationary traffic\r                                     +
                 |   |       |       | (L) Stau
 A003-NO-HF-015  | + | 10922 | 10923 | stationary traffic\r                                     +
                 |   |       |       | (L) Stau
 A003-NO-HF-016  | + | 10921 | 10922 | queuing traffic (with average speeds 67 km/h)\r           +
                 |   |       |       | (L) stockender Verkehr (Durchschnittsgeschwindigkeit 67 km/h)
 A003-NO-HF-022  | + | 10913 | 10914 | queuing traffic (with average speeds 84 km/h)\r           +
                 |   |       |       | (L) stockender Verkehr (Durchschnittsgeschwindigkeit 84 km/h)
 A003-NO-HF-023  | + | 10912 | 10913 | queuing traffic (with average speeds 70 km/h)\r           +
                 |   |       |       | (L) stockender Verkehr (Durchschnittsgeschwindigkeit 70 km/h)
 A003-NO-HF-024  | + | 10911 | 10912 | queuing traffic (with average speeds 80 km/h)\r           +
                 |   |       |       | (L) stockender Verkehr (Durchschnittsgeschwindigkeit 80 km/h)
 A003-NO-HF-025  | + | 10910 | 10911 | stationary traffic\r                                     +
                 |   |       |       | (L) Stau
 A003-NO-HF-026  | + | 10909 | 10910 | stationary traffic\r                                     +
                 |   |       |       | (L) Stau
 A003-NO-HF-027  | + | 10908 | 10909 | stationary traffic\r                                     +
                 |   |       |       | (L) Stau
 A003-SW-HF-011  | - | 10928 | 10927 | queuing traffic (with average speeds 77 km/h)\r           +
                 |   |       |       | (L) stockender Verkehr (Durchschnittsgeschwindigkeit 77 km/h)
 A003-SW-HF-023  | - | 10914 | 10913 | queuing traffic (with average speeds 49 km/h)\r           +
                 |   |       |       | (L) stockender Verkehr (Durchschnittsgeschwindigkeit 49 km/h)
 A003-SW-HF-024  | - | 10913 | 10912 | queuing traffic (with average speeds 50 km/h)\r           +
                 |   |       |       | (L) stockender Verkehr (Durchschnittsgeschwindigkeit 50 km/h)
 A003-SW-HF-025  | - | 10912 | 10911 | queuing traffic (with average speeds 74 km/h)\r           +
                 |   |       |       | (L) stockender Verkehr (Durchschnittsgeschwindigkeit 74 km/h)
```

Listing 5.3: Demonstration how to automatically generate ALERT-C [86] encoded traffic messages as used by various broadcasting stations and navigation systems. The messages focus on the A3 subset of road segments that have viscous or congested traffic.

as the number of non functioning detectors. With these numbers, the total availability of the detectors is then expressed in percent of the measurements that were available and correct in the last $15\,\text{min}$. In cases where the availability value drops below some threshold value of about $90\,\%$, the colour of the circle changes like a chameleon from green into yellow. Below a second threshold value of $30\,\%$, the colour changes a second time into red. In case of a connection loss of the whole traffic data centre, the whole network toggles into red in a few minutes. In cases of partial loss of connectivity, the responsible sub-master station becomes also obvious. Such cases occur frequently in roadworks conditions where diggers unintentionally damage cables. Obtaining a rough overview about the proper functioning of the loop detectors is another kind of usage of the loop detector statistics. Of course, such a statistic can only give a first sketch and it will never replace a solid verification of individual loop detectors.

The approach of explaining the traffic situation results by means of the loop detector statistics has its limitations. An example that draws the limitations of the approach is given in the following. It also shows up the limitations of the microscopic traffic models. Section 1.1 has already discussed that the analogy and the differences between the real-world and the world of simulation exist only to a certain degree. Therefore, they must not be overstretched. The example became available as a result of the research facility provided by the OLSIMv4 database and the visualisation of the loop detector statistics. OLSIMv4 facilitates the road operators to compare the loop detector data against the simulation results. Having the averaged detector data on the same screen with the global detector simulation results, frequently helps in understanding the simulation results. As there are always situations that give riddles about what the simulation determined as value for the level-of-service, such a first indication often sheds light on the riddles. Such a riddle occurred during the development of OLSIMv4. On a part of the A45 highway and on a early evening of a normal working day, OLSIMv4 indicated free flow to the road users. However, at the next morning a lot of people were complaining about having been standing in a traffic jam for the whole night. Listing 5.4 shows ten minutes of a data provided by a loop detector cross-section. The measurements signalised also free flow as they showed a few vehicles with relatively slow velocities around 60 km/h . It turned out that, due to snow, the vehicles were driving not exactly in the single lanes but more in between of them. This vehicle driving behaviour is similar to the one described in [131] where privileged vehicles in emergency scenarios drive in between of two vehicle queues with higher velocities. Such scenarios require special microscopic traffic models. An appropriate microscopic traffic model might resemble a driving behaviour without lane discipline such as the one presented in [136].

Theoretically, the simulation should have behaved in the following way. When the loop detectors measure zero vehicles, the tuning elements do nothing [69]. In case of a too large velocity difference, the tuning elements would have inserted vehicles in order to increase the traffic flow and to adapt the velocity values against each other. As at all times at least one single-lane detector measured zero vehicles, the traffic flow has either been deactivated for the single lane and has been averaged across all single lanes. As a consequence, the traffic flow did not increase significantly. Additionally, the simulation lacks a pendant of the resistance that the snow causes to the vehicles in the real-world. Thus, vehicles inserted by the simulation would accelerate immediately after their insertion. The strategy to adapt the velocity by means of increasing the traffic flow would only be successful if the tuning elements could cause a sufficiently large congested area [69].

In the best case, the level-of-service classification scheme Figure 3.9(a) with the mappings for maximum velocities above 100 km/h and with the value boundaries from Table 3.5.8 would indicate viscous traffic for this situation. However, the classification scheme also permits the use of the previous level-of-service value which must have been the free or dense traffic in this case. The previous level-of-service value indicated free or dense traffic as the density had never reached high values. The traffic phase transition would, thus, match a phase transition from free flow to congested traffic which is a forbidden transition in the three phase traffic theory by Kerner [101]. However, Kerner most probably had not weather situations like this in mind when he developed the theory. From time to time some vehicles randomly passed the loop detector properly, the remaining gross of vehicles passed in between the middle of two neighbouring single lane detectors. Police warning messages can present another view on the traffic situation here. The restricted area access presents a subset of the data. However, as OLSIMv4 filters the police warning messages for a subset of event codes, only those with the appropriate event codes are displayed. In the case of the snow conditions, the presentation of a police warning message would have depended on the event code. In case of a snow warning message, it would have been filtered. However, the police warning messages are not available to the public and, thus, would have not helped here either.

Figure 2.3 depicted the way the road works messages are incorporated into the OLSIMv4 database. Like the other messages, the traffic data centre submits them in the ALERT-C [86] encoded format to OLSIMv4. With the spatio-temporal traffic information in combination with the topological data in the database and the event and location code list from the TMC forum, OLSIMv4 can also generate and format ALERT-C encoded traffic messages. Listing 5.3 demonstrates the generation of ALERT-C encoded message fragments for the subset of the A3 road segments that are in either viscous or congested traffic state.

The text atoms of the messages are part of the eventlist [85] that ships with the TMC location code list [86]. The query for message generation uses event code 101 for the stationary traffic and 108 for the queuing traffic. The event code list compounds almost 1500 text atoms for a lot of different traffic situations such as closed parking places and reasons for traffic restrictions. The event code list supports also the generation of more informational messages that may want to include the length of the queuing traffic and the traffic jams. Equipping OLSIMv4 with a special detector variant that measures the lengths of the traffic jams in the simulation could make these kind of messages available to the

```
       time          |     otdf_id      | detector | dpos | lane | j_any | j_trk | v_car | v_trk | p_occ
---------------------+------------------+----------+------+------+-------+-------+-------+-------+-------
2010-02-02 19:00:00+01 | A045-NO-HF-016 | dn011599 |  78  |  0   |   0   |   0   |   0   |   0   |   0
2010-02-02 19:00:00+01 | A045-NO-HF-016 | dn011600 |  78  |  1   |   0   |   0   |   0   |   0   |   0
2010-02-02 19:00:00+01 | A045-NO-HF-016 | dn011601 |  78  |  2   |   0   |   0   |   0   |   0   |   0
2010-02-02 19:01:00+01 | A045-NO-HF-016 | dn011599 |  78  |  0   |   0   |   0   |   0   |   0   |   0
2010-02-02 19:01:00+01 | A045-NO-HF-016 | dn011600 |  78  |  1   |   0   |   0   |   0   |   0   |   0
2010-02-02 19:01:00+01 | A045-NO-HF-016 | dn011601 |  78  |  2   |   0   |   0   |   0   |   0   |   0
2010-02-02 19:02:00+01 | A045-NO-HF-016 | dn011599 |  78  |  0   |   0   |   0   |   0   |   0   |   0
2010-02-02 19:02:00+01 | A045-NO-HF-016 | dn011600 |  78  |  1   |   0   |   0   |   0   |   0   |   0
2010-02-02 19:02:00+01 | A045-NO-HF-016 | dn011601 |  78  |  2   |   0   |   0   |   0   |   0   |   0
2010-02-02 19:03:00+01 | A045-NO-HF-016 | dn011599 |  78  |  0   |   2   |   2   |   0   |  54   |   4
2010-02-02 19:03:00+01 | A045-NO-HF-016 | dn011600 |  78  |  1   |   1   |   0   |  54   |   0   |   1
2010-02-02 19:03:00+01 | A045-NO-HF-016 | dn011601 |  78  |  2   |   0   |   0   |   0   |   0   |   0
2010-02-02 19:04:00+01 | A045-NO-HF-016 | dn011599 |  78  |  0   |   6   |   1   |  50   |  43   |   8
2010-02-02 19:04:00+01 | A045-NO-HF-016 | dn011600 |  78  |  1   |   4   |   0   |  48   |   0   |   5
2010-02-02 19:04:00+01 | A045-NO-HF-016 | dn011601 |  78  |  2   |   0   |   0   |   0   |   0   |   0
2010-02-02 19:05:00+01 | A045-NO-HF-016 | dn011599 |  78  |  0   |   5   |   4   |  48   |  48   |  11
2010-02-02 19:05:00+01 | A045-NO-HF-016 | dn011600 |  78  |  1   |   3   |   0   |  49   |   0   |   4
2010-02-02 19:05:00+01 | A045-NO-HF-016 | dn011601 |  78  |  2   |   0   |   0   |   0   |   0   |   0
2010-02-02 19:06:00+01 | A045-NO-HF-016 | dn011599 |  78  |  0   |   0   |   0   |   0   |   0   |   0
2010-02-02 19:06:00+01 | A045-NO-HF-016 | dn011600 |  78  |  1   |   1   |   0   |  41   |   0   |   1
2010-02-02 19:06:00+01 | A045-NO-HF-016 | dn011601 |  78  |  2   |   0   |   0   |   0   |   0   |   0
2010-02-02 19:07:00+01 | A045-NO-HF-016 | dn011599 |  78  |  0   |   2   |   1   |  38   |  35   |   5
2010-02-02 19:07:00+01 | A045-NO-HF-016 | dn011600 |  78  |  1   |   1   |   0   |  35   |   0   |   2
2010-02-02 19:07:00+01 | A045-NO-HF-016 | dn011601 |  78  |  2   |   0   |   0   |   0   |   0   |   0
2010-02-02 19:08:00+01 | A045-NO-HF-016 | dn011599 |  78  |  0   |   1   |   1   |   0   |  33   |   3
2010-02-02 19:08:00+01 | A045-NO-HF-016 | dn011600 |  78  |  1   |   1   |   1   |   0   |  39   |   3
2010-02-02 19:08:00+01 | A045-NO-HF-016 | dn011601 |  78  |  2   |   0   |   0   |   0   |   0   |   0
2010-02-02 19:09:00+01 | A045-NO-HF-016 | dn011599 |  78  |  0   |   0   |   0   |   0   |   0   |   0
2010-02-02 19:09:00+01 | A045-NO-HF-016 | dn011600 |  78  |  1   |   1   |   0   |  46   |   0   |   1
2010-02-02 19:09:00+01 | A045-NO-HF-016 | dn011601 |  78  |  2   |   0   |   0   |   0   |   0   |   0
```

Listing 5.4: Loop detector data of an extreme snow event for a complete cross-section.

public. The equipping requires, of course, careful validation of the data which becomes available with a verified model. In addition to the lack of the special detector variants and, consequently, the lack of the length quantification information in the traffic messages generated by OLSIMv4, there seems to be no market interest in this kind of information. Possible reasons may be the unknown quality of this information and the limited scope of the road network which covers NRW only. The web front end displays the road works messages to the general public.

The police warning messages contain information about the affected road segments in case of congested traffic as well as the lengths of the traffic jams. They frequently present another view on the simulation results and are most helpful in cases where no current loop detector data is available. In some cases, OLSIMv4 mixes their information with the simulation results for the visualisation only. However, as the traffic warning messages are either generated automatically or entered into the system manually, their validity period is frequently far too inexact to consider them for integration into OLSIMv4. The lane closings may be an exception. As they involve a concrete police measurement, they may be generally more accurate.

5.1.3   Scientific traffic information

One of the first things that one might want to do with a simulation system such as OLSIMv4 may be the evaluation of the system. Such a first evaluation is not the same as a validation. In contrast to verification and validation, it can only provide a first feeling for the soundness of the results. However, the first results also introduce a new era in the implementation phase of such a simulation. If the first results of OLSIMv4 would have been in the form of Figure 5.2, this would have been a great help. In general and in particular for OLSIMv4, this is not the case. The mentioned plots are important to verify whether the simulation results are reasonable or not. As mentioned above, when regarding at the simulation results of a complete highway network in the size of NRW, there are a lot of circumstances that may question the simulation results. Having a tool to produce such kind of plots is, thus, essential

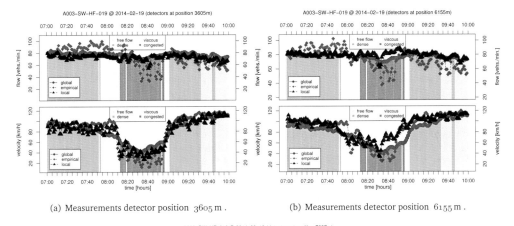

(a) Measurements detector position $3605\,\mathrm{m}$.   (b) Measurements detector position $6155\,\mathrm{m}$.

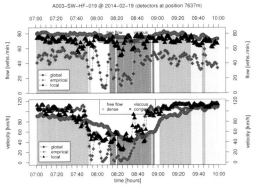

(c) Measurements at detector position $7637\,\mathrm{m}$.

Figure 5.2: The three plots 5.2(a), 5.2(b), and 5.2(c) contrast the spatio-temporal traffic information ("global" line) with the stationary traffic data ("empirical" line) as recorded on 2014-02-19 by the loop detectors at positions $3605\,\mathrm{m}$, $6155\,\mathrm{m}$, and $7637\,\mathrm{m}$ of the A3 in direction of travel south-west and before Hilden intersection. The "local" labelled line displays stationary data obtained from the simulation with a method that emulates the behaviour of the loop detectors.

for a productive environment with online traffic information. The production of such plots depends on having comparable input and output detector measurements. This is the case for OLSIMv4. The plots have been generated by a short R script. This evaluation mode has been integrated as part of the internal developer area. Starting the evaluation requires, therefore, only choosing an appropriate road segment and initiating the plotting by clicking on the appropriate button.

The three plots of Figure 5.2 contrast the spatio-temporal traffic information provided by the global detectors ("global" line) with the stationary traffic data provided by the loop detectors ("empirical" line) and the local detectors ("local" line) from the simulation. The traffic data has been recorded on Wednesday 19th February, 2014 between 7:00am and 10:00am. The data refers to the 4 lanes wide, $7700\,\mathrm{m}$ long, A3 highway segment from Mettmann onto Hilden intersection with direction of travel South-West. The stationary detectors measure cross-section wise and are located at three positions $3605\,\mathrm{m}$, $6155\,\mathrm{m}$, and $7637\,\mathrm{m}$. The tuning elements at the first and at the last detector position had been enabled. Even though the scale varies between the plots, the measurements by the global detector are the same in each of the three plots. While the most obvious difference between the global detector and the loop detector measurements appears to be the traffic flow at the last detector cross-section in Figure 5.2(c), the velocity of the global detector seems to reflect the loop detector at least qualitatively. The traffic flow differs significantly in this case due to the active tuning either at the last detector or thereafter. The section leads to the Düsseldorf intersection at which ending a lot of vehicles in general leave the main roadway on the right outer lane. The last local detector is, thus, not expected to produce comparable results. The traffic flow of the global detector can, of course, differ significantly as it refers to the complete road segment. Viewed from the last loop detector's traffic flow and velocity, the global detector's level-of-service transition from dense to viscous traffic at around 7:52am is justified. Admittedly, the presented assessment is purely qualitative. A quantitative assessment such as the L^2-norm would be more desirable, of course. As the tuning elements try to assimilate the measurements by the velocity values of the loop detectors by increasing the traffic flow values, a direct comparison is difficult and has not been done until today. The velocity drop in the loop detector data reaches the second detector at about 8:00am and the third one about 8:10am. The second local detector reflects the velocity drop qualitatively and the third one reflects it even quantitatively. While the velocity drop ends about 8:40am at the first detector, it ends at the second one about 8:50am and at the third about 8:57am. The global detector's velocity displays an adequate mean value to the three velocity values as it shows the velocity drop between 8:10am and 8:55am. Additionally, the plot from Figure 5.2(a) for the first detector displays excellent qualitative and quantitative agreement between the various kinds of measurement data. The transitions from dense to viscous and from viscous to congested traffic and back again reflect the traffic data also properly.

The discussed example demonstrated impressively the benefit of having comparable input and output detector data. The comparable data enable reasoning about the simulation system which would complicate things significantly and unnecessarily otherwise. Another important design principle is the support of possible combinations of all kinds of participating data, e.g. the topological data and the traffic data. Figure 2.2 presented the example for the generation of the origin-destination matrices. In addition to the global detector densities as part of the simulation results for a complete day, the generation requires also having pairs of subsequent road segments. The calculation simply sums the density differences between pairs of subsequent road segments over half a day. As the density on a road segment without any on- and off-ramp remains constant, the density difference between a pair of subsequent road segments can represent the sums of the in- and outflows at the on- and off-ramps between the two road segments on the daily scale. OLSIMv4 supports the generation of these matrices.

Of course and admittedly, the plots and the origin-destination matrices are only the very beginning of the interesting scientific traffic information. Further scientific traffic information is not limited to but include at least time and distance headway distributions, velocity distributions, and lane change distributions. OLSIMv4 supports their integration with the detector concept. In addition to the accumulated distributions, a productive implementation as well as scientific traffic information such as space-time plots (e.g. [70, Fig. 4.14]) require sometimes time step grained dumping of all individual vehicles on a road segment. OLSIMv4 supports this by an extended logging facility. A detector can switch its logging mode to extended logging at runtime when a certain property in the database toggles. As the dumping of all vehicles on a road segment requires an immense storage volume, it is not very elegant to store them into the relational database while still maintaining the ACID properties.

5.1.4 Further traffic information

OLSIMv4 supports exploring further traffic information. One aspect covers neither current traffic information nor prognosticated traffic information but looks instead in the past. In addition to an archive that provides the simulation results of the past years in visual form, OLSIMv4 supports also running the simulation for any time of the past and the future. However, such situations may eventually have to take inconsistencies in the data into account. As an example, the road network infrastructure may have changed significantly in the meantime. This turned out true during the estimation of travel times for an individual simulation scenario that looked back from 2013 into the year 2006. The number of lanes on a particular highway had increased significantly. The increased number of lanes is the minor problem of the possible cases. As OLSIMv4 supports closing lanes, a current topology can easily emulate this case by simply closing the affected lanes.

In Section 2.1.2 the idea of the hypothetical task of an optimisation for a prospective location of a hub regarding the travel times has been introduced. The section also expressed OLSIMv4's capability in supporting these kinds of optimisation problems. With the set of on-ramps where the employees enter into the highway networks, the archive of travel times, and the successor-predecessor relation between the road segments, the set of required data is complete for starting a search. As the number of off-ramps from the NRW highway network is finite and, fortunately, limited to only 703 exits, the exhaustive search should be possible and should not require too much time. For each of the starting points, the travel times to each of the possible highway exits can be determined directly by summing up the travel times of the particular road-segments. As there may exist several routes to the destination, these may require special attention, too. The destination that has the minimum sum of travel times for each set of routes and for each employee matches then the best location with respect to the travel times. Of course, there are smarter algorithms and even heuristics to compute these kinds of optimisation problems such as ant colony optimisation.

The variety of possible detector types support also the publication of further information that may come and go more or less into focus. The cities in NRW suffer from the vehicles CO_2 emission. Using a specific detector type for the calculation of the amount of CO_2 emission may provide another view on the traffic situation. The different views on the traffic may help the road users in choosing the best time and route for any intended journey. With growing trust in the system, the traffic informations may also contribute in the generation of traffic information statistics.

5.2 Verification and validation

Verification and validation co-operate in substantiating and assessing a system's proper functioning and its correctness. Verification, as the first activity domain of the two, ensures "that a model implementation accurately represents the developer's conceptual description of the model and the solution to the model" [154]. Taken strictly, this understanding could lead to implementations that pass the verification process but will fail in giving valuable results. On the other side, this understanding can also lead to implementations that give a mathematical or computer-based proof that the numerical algorithm converges to the right solution. According to that understanding and in contrast to verification, does validation determine "the degree to which a model is an accurate representation of the real world from the perspective of the intended uses of the model" [154]. In the understanding of this thesis, validation is also the process of applying the computational model to measured data. It is done with the tuning elements in each cycle. As result verification in the case of microscopic traffic simulations requires to validate the simulation, the given definitions for verification and validation are only of limited use. Verification compounds several aspects such as code verification, formal verification, and result or accuracy verification [12]. Sargent distinguishes additionally between specification verification and implementation verification [180]. Specification verification applies formal methods such as formal proof [73] and model checking [56, 30] to prove the correctness of a specification. Code verification ensures that the code does not contain errors. In general, it involves more than just compiling the program. Using theorem provers, for example, this activity can prove characteristic axioms in the computer code. Formal verification uses compiler like methods such as compilers or theorem provers to verify an implementation. And result verification compares the computed results against known analytic solutions [12]. Validation focuses on certain aspects of the result such as travel times in the traffic simulation world, compares them with measured results, and assesses the accuracy. Figure 5.2 may also serve as an example for validation. Verification and validation is not limited to implementations only. Models in the development phase and specifications require also verification and validation.

As validation activities "presume that the computational model result is an accurate solution of the mathematical model" [154, Sec. 2], they come first over validation activities—at least theoretically. In the general practice, verification takes–if at all–only place for the microscopic traffic model components as part of the microscopic traffic simulation. Even though the publications mostly affect the basic models such as the NaSch, a lot of publications deal with exact solutions for special cases or for simplifications of the models. However, the simulation process has never been examined and been verified. Apart from that the simulation used standard floating point and fixed point arithmetic, nothing was known about the simulation. With the lack of a specification, verification and validation are kind of superfluous and could not be achieved in a meaningful manner. The conceptual design of Chapter 3 has specified and defined the domains and ranges of each set of input and output parameters. Therefore, a potential next implementation of OLSIMv4 can advance to the next class in the following hierarchies.

The four-tier hierarchy developed in [13] and demonstrated in [12] distinguishes four classes of systems with increasing verification degrees. The lowest class C4 requires a system to use standardised arithmetic (like IEEE 754-2008) resp. standard data types for real numbers which applies to OLSIMv3 and OLSIMv4. The next class in the hierarchy requires a system to be divided into its constituent parts, use of at least a standardised arithmetic resp. standard data types for real numbers, and sensitivity analysis having been carried out for uncertain parameters. It also requires that "uncertainty is propagated through the subsystems, using e.g. Monte-Carlo methods" [12]. The remaining two classes have even stronger requirements. In the case when a next implementation of OLSIM should conform to the Chapter 3 specification, it also results at least one class higher as all data sets are specified and all constituent components are described by the functions that participate in it.

OLSIMv3 provided only little system knowledge. In particular, the exact update order of the interconnected road-segments was unknown. The functional relationship between input and output of the simulation was unknown. The conceptual design of Chapter 3 introduced simulation models that fill this gap. According to the hierarchy of systems specification [216, Sec. 1.3], a future implementation of OLSIMv4 can now claim to be a level 3 system whereas OLSIMv3 would have been at level 0.

Instead of directly implementing all the components in a suitable programming language, verifying them, and getting ready for production, the part taking models should first be concretised, analyzed, examined, and refined if necessary. With the conceptual design it is now possible, to examine the new simulation model variants, i.e. the multicomponent and the network specification approach, under the special aspect of a concurrent or a parallel implementation. The Maude system has proven well-suited for these sorts of tasks. The discrete event simulation approach as described in Section 3.7.5 may also be a reasonable alternative. The traffic model requires investigation under the aspect of potential collisions and, additionally, clarification about the undefined values for the configurations as described in Subsection 3.4.1.3. The tuning element models require first specification and verification of their effectivity especially under the aspect of proper termination in each state. Additionally, sensitivity analysis of scenarios such as the tuning of the should-be velocities by increasing the corresponding actual traffic flow values would substantiate their effectiveness significantly.

The advantage of first refining and verifying the models over a direct implementation and potential refinement thereafter results from the following facts. The implementation in a systems modelling language requires the formal model only to be effective, i.e. the computation terminates and gives the desired result. The efficiency aspect which affects in general the most part of the code remains thereby secondary. The refinement of the models is, thus, less work and much faster done. The insights of these verification activities will enhance a potential implementation almost automatically. The direct implementation and inline refinement approach, in contrast, deals with optimisation details, is frequently difficult, error-prone, and laborious to change, and the high amount of expectable code changes bear the risk of reducing the code quality. As a side effect of the verification activities and the models' proper functioning, an excellent specification will exist. With a proper specification and the advanced knowledge about the degree of parallelism, proper termination of all components, and correctness of the computation, an implementer can make reliable design decisions. The implementer can then also focus on the implementation requirements such as architectural constraints, efficiency, resource usage, and real-time requirements. Another advantage of an implementation in a term rewrite system exists with the automated proving of equivalence between the specification and the implementation. The formal specification may not be able to perform a complete simulation but may be able to process a single state transition in reasonable time. Assuming that the implementation could also process the same state transition, the results of both state transitions should be equivalent for any arbitrary state transition.

```
*** run the simulation with 1201 iterations
reduce in SIMULATION-NETWORK-EXAMPLE :
  NumberOfVehicles(Occupation(
    TrafficRoadRef(TrafficRoads(Simulate^1201(SimNetExample3), 'trp-11-10)))) .
rewrites: 16799107951 in 4554984ms cpu (4554700ms real) (3688071 rewrites/second)
result NzNat: 143
```

Listing 5.5: Simulation with Maude on a single lane roadway with $10\,000$ cells and periodic boundary conditions. The constant detector provides a value of 10 vehicles per minute for the traffic flow and is used as the should-be value by the tuning element. This is a part of the code that has been used to produce the results of Figure 5.3 (source: Listing 9 from [30]).

The emphasis on verifying the correctness for every single component of a microscopic traffic simulation, may look too pedantic in the first place. However, the components depend on the correct functioning of each other. The tuning elements depend on the detectors. The occupation tables depend on the tuning elements. The detectors depend on the vehicle update. Potential errors may include but not be limited to the following enumeration. Discretisation errors can occur when components use an inconsistent discretisation size and strategy. As an example, they may occur when rounding takes only place in the code to move a vehicle to its next position but forgets about the correct discretisation in the velocity update. Frequently, collision errors accompany these kinds of errors. In the case where they remain undetected, they can also produce negative velocities, relatively high densities, and bad lane changing behaviour, just to name a few. The microscopic traffic model rules are also quite complex and may have to deal with undefined situations, e.g. where vehicles stand on closed cells or neighbours have too small distance headways. They may require additional specification for, e.g., vehicles with negative distances. The implementation has to take care about such situations and behave in a clearly defined and well understood way. The implementation of local detectors requires also special care when vehicles cover the detector or change a lane at the detector position. As a local detector has to track vehicles sometimes over several time steps, the vehicles may appear to the detector as ghost vehicles. Otherwise, the detectors might measure half vehicles in such situations. As a consequence, the inadequate discretisation will eventually cause erroneous behaviour of the tuning elements.

Having a verified specification and implementation forms the basis of successful validation activities. Section 4.1.1 has already mentioned that OLSIMv4 has been validated "using a section control based method and as described in the following. On two subsequent bridges across a highway, digital cameras have taken traffic photos in multi-shot mode. Subtracting associated time-stamps of the photos for a unique set of vehicle attributes such as parts of the license plate, color, and manufacturer yielded to travel times that the simulation results had to match" [30]. The method as such was successful in obtaining an almost complete set of measured travel times and it has, thus, proven well-suited for the assessment of the accuracy of travel times determined by the simulation. However, certain aspects that will be discussed have not been considered in the aforementioned validation activity. Section 2.2.6 has explained already that a successful validation involves regarding a representative mixture of general, wide-spread field data. To find the representative mixture is, thus, one of the key tasks during the validation activity. Calibration activities, in contrast, rely on prior successful validation and include also the adjustment of the simulation results in form of traffic information quantitatively or the deformation of an existing model. With the conceptual design of this thesis, the verified specification, a future implementation of OLSIMv4, and the experiences from the OLSIMv4 validation and calibration, the calibration effort is reduced significantly. Scenarios such as already cited in Section 2.2.6 and as described by the case study for calibrating VISSIM in modelling the Zurich network [66] and by official "Guidelines for Applying Traffic Microsimulation Modeling Software" of the U.S. Department of Transportation from 2004 [169, Sec. 5.2] should never happen in a future implementation.

The list of verification issues is far from being complete. As a first addition Section 5.2.1 starts, thus, with a short demonstration of how to perform numerical solution verification for the NaSch microscopic traffic model. It also demonstrates a solution for the longitudinal update of the model by Lee,

Barlovic, Schreckenberg and Kim and with asymmetric lane change extensions by Habel and Schreckenberg. Subsequently the numerical solution verification subsection, Subsection 5.2.2 demonstrates how to verify collision freeness for the whole simulation model. Subsection 5.2.3 finalises by giving a more complete overview of verification needs. The verification of the components of the simulation model became possible with the conceptual design from Chapter 3. The Maude implementation and verification effort forms a starting point in the verification effort. The thesis tries to enhance the verification possibilities by pointing out what can and should be done for the next version. After the verification specific Subsections, this Subsection continues with Subsection 5.2.4 that explains and presents the section control method to validate OLSIMv4 under the aspect of travel times. Subsection 5.2.5 completes the verification and validation benefits by describing some of the performance issues that occurred during the performance validation of OLSIMv4.

5.2.1 Numerical solution verification

As only a fraction of the set of results are known and they refer to an even smaller set of special cases, numerical result verification is of only limited use for the domain of microscopic traffic simulations. However, it can give a first impression of the correctness of the results.

Looking at the simulation results is generally one of the first things an implementer might want to do after the implementation of a microscopic traffic model. The results for a traditional NaSch model may then look like Figure 5.3(b) or like 5.3(c). The simulation scenario of both plots is identical, runs on a basic single-road network with periodic boundary conditions, and is also identical to the one that has been used for example in [164, Sec. 5.1.1]. It consisted of a system length of 10000 cells and tuned the density on the periodic system to the values of a constant detector for the should-be densities from 5 vehs/km to 80 vehs/km . The Maude system executed the simulation model to reproduce the well known fundamental diagrams of the NaSch microscopic traffic model. The code from Listing 5.5 used the functional module SIMULATION-NETWORK-EXAMPLE from Listing 4.57 to run the simulation. The listing simulates for the should-be density of 10 vehs/km .

The plots of Figure 5.3 show the fundamental diagrams for "the velocity and the flow mean values of maximum 400 minutes of simulation steps per density. Figure 5.3(c) shows the flow density relation with collisions having occurred. The lines are estimated mean values for the continuous relation between the observables. However, the maximum velocity plateau in Figure 5.3(a) is given exactly by $\hat{v} = (1 - p_{\text{dawdle}}) * v_{\text{max}} + (v_{\text{max}} - 1) * p_{\text{dawdle}}$, where $v_{\text{max}} = 5\text{cell/s}$ and $p_{\text{dawdle}} = 1/3$. After the critical density $\rho_{\text{crit.}}$ in Figure 5.3(a), i.e. where the constant maximum velocity line ends, the density dependent velocity has been fitted with $v(\rho) = p_{\text{dawdle}}^2 \rho_{\text{max}}^2 (1/\rho - 1/\rho_{\text{max}})$, where $\rho_{\text{max}} = 1331/3\text{vehs/km}$. The flow density relationship which is depicted in Figure 5.3(b) follows from the equation $\hat{J}(\rho) = \rho\hat{v}$ and $J(\rho) = \rho v(\rho)$, where J denotes the traffic flow and ρ denotes the density. Under the assumption that the NaSch model follows this linear relationship between density and flow, the critical density would be at $\rho_{\text{crit.}} \approx 13.2\text{vehs/km}$. The fundamental diagrams follow this approximation roughly.

Figure 5.3(d) shows the saturation of the number of vehicles at a destination density in percent. The tuning elements adapt to the expected density values after several minutes. Thereafter, the tuning elements do no longer change the number of vehicles in the occupation tables and the density remains constant. The ideal saturation of 100% was not reached in the simulations due to the global detectors' rounding to the nearest integer with operation Vehs/Km(...)" [30]. The tuning element's saturation ratio in percent varies among the densities as the number of vehicles in the system and the system length are discrete values. The peak of the saturation ratio reaches more than 100 % because the detectors have a delay of one minute and the number of vehicles depicted in Figure 5.3(d) is not a detector value but instead has been taken directly from the tuning elements.

The Maude system is not the best tool for retrieving numerical solutions from a simulation system. For simple simulation scenarios, simulation tools such as SimuLink, Mathematica, and MatLab are a good starting point. A numerical solution verification can also be done for a simplified setup of Figure 5.4. Therein, the results of the deterministic variant of the vehicle update function $\mathbf{F}_{\text{drive}}$ from Section 3.4.1 have been depicted in relation to the velocity differences and the distance headways

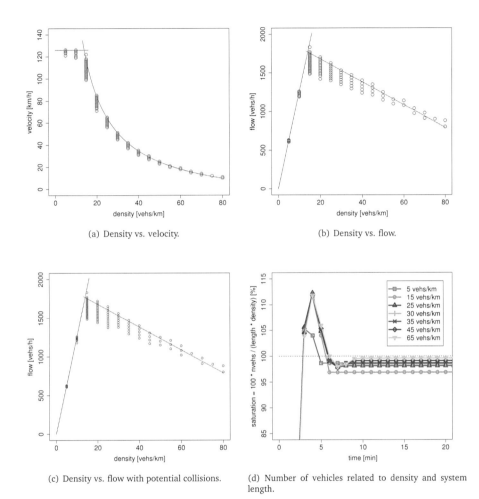

(a) Density vs. velocity.

(b) Density vs. flow.

(c) Density vs. flow with potential collisions.

(d) Number of vehicles related to density and system length.

Figure 5.3: Simulation runs of the Maude simulation engine with periodic boundary conditions. For each density, an almost identical traffic road has been instantiated. The traffic roads differ in the constant detector and the accumulator that the tuning elements use. The tuning elements try to assimilate the density measured by a global detector to the value given by the constant detector. The maximum simulation time 24000 s was not reached for the higher densities. The lines and curves in Figures 5.3(b), 5.3(a), and 5.3(c) show a possible but assumed continuous mean value. Figures 5.3(a) and 5.3(b) show the velocity and flow to density relations without collisions having occurred. Figure 5.3(c) shows the flow to density relationship with potential collisions. Figure 5.3(d) shows the number of vehicles in the system related to the density and the system length as saturation ratio in percent (captions and figures taken from [30, Fig. 7]).

and for specially crafted optimistic and pessimistic leader of the leader vehicles. For each velocity combination of the follower and the leader vehicle and the combination of their distance headway, a single update step has been evaluated. The combinations also included the leading vehicles of the leader

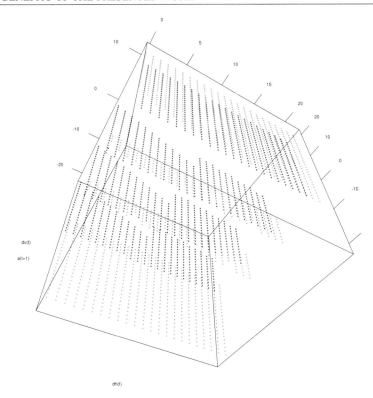

Figure 5.4: Acceleration in the deterministic traffic model by Pottmeier, i.e. without dawdling. For each velocity combination of the follower and the leader vehicle and the combination of their distance headway, a single update step has been evaluated. The combinations also included the leading vehicles of the leader that have been crafted specifically to match the first optimistic (which does not always yield into the optimistic case $\gamma_n^{(t)} = 0$) as well as the pessimistic case of Equation 3.40. Green points represent the optimistic case, red ones represent the pessimistic case. The $a_n^{(t+\Delta t)}$ values have been multiplied by ten and have been shifted by ± 1 cells/s² for a better visual distinction.

that have been crafted specifically, so that the first optimistic case matches with respect to $v_{n+2}^{(t)}$. Such a match does not always yield into the optimistic case $\gamma_n^{(t)} = 0$ as the latter has additional constraints. An additional vehicle that matches the pessimistic case of Equation 3.40 has also been crafted for each the combinations. Subsection 3.4.1.3 has also discussed the figure. The numerical solutions for the acceleration can serve as an invariant across implementations. When switching off dawdling, the invariant should show up even in successive vehicle coordinates, i.e. $\mathbf{k}_n^{(t)} \# \mathbf{q}^{(t)}$ and $\mathbf{k}_n^{(t)} \# \mathbf{q}^{(t-\Delta t)}$.

Even though plots might be essential for the simulationists, they give, in general only a first impression of the correctness of the implementation. The impression is only very vague as "programming errors in the computer code, deficiencies in the numerical algorithms, or inaccuracies in the numerical solution, for example, may cancel one another in specific validation calculations and give the illusion of an accurate representation of the experimental measurements" [154]. This is the reason why the plots 5.3(b) and 5.3(c) do not differ significantly (apart from the dot size). As both plots reproduce the fundamental diagrams, it is not possible to decide from the plots, whether the code still contains errors that might have an impact on the future simulation results in other scenarios. Additionally, the

```
fmod SIMULATION-NETWORK-EXAMPLE-CHECK is
  pr SIMULATION-NETWORK-EXAMPLE . inc SATISFACTION . inc MODEL-CHECKER . inc LTL-SIMPLIFIER .
  subsort SimulationNetwork < State .
  op __ : SimulationNetwork NzNat -> SimulationNetwork .
  ops sn30 sn35 : -> SimulationNetwork .
  op NoCrashedVehicles? : -> Prop .
  op checkRec : SimulationNetwork -> SimulationNetwork [iter] .
  ops checkRes checkRec : SimulationNetwork ModelCheckResult -> ModelCheckResult .
  var B : Bool . var N : NzNat . var P : Prop . var R : ModelCheckResult . var S : SimulationNetwork .
  eq S N = if N <= 0 then checkRec(S) else checkRec(Simulate^1(S)) (-1 + N) fi .
  eq sn30 = SimulationNetwork(Period(0, Seconds), List(rgp-11-30), EmptySwitchingAreas) .
  eq sn35 = SimulationNetwork(Period(0, Seconds), List(rgp-11-35), EmptySwitchingAreas) .
  eq S |= NoCrashedVehicles? = NoCrashedVehicles?(S) .
  eq checkRec(S) = checkRec(S, modelCheck(S, []NoCrashedVehicles?)) .
  ceq checkRec(S, B) = S if B .
  eq checkRec(S, R) = R [owise] .
  ceq checkRes(S, B) = B if B .
  eq checkRes(S, R) = R [owise] .
endfm
===============================================
reduce in SIMULATION-NETWORK-EXAMPLE-CHECK : checkRes(sn30 1200) .
rewrites: 260444217775 in 67494318ms cpu (67490103ms real) (3858757 rewrites/second)
result Bool: true
===============================================
reduce in SIMULATION-NETWORK-EXAMPLE-CHECK : checkRes(sn35 1200) .
rewrites: 389512323987 in 105011522ms cpu (105004958ms real) (3709234 rewrites/second)
result Bool: true
```

Listing 5.6: The `NoCrashedVehicles?` property checks a simulation network state between two update steps for possible vehicle collisions. The _ operator recursively applies the `checkRec` operation to verify that no collisions occured. The results were obtained for densities 30 and 35 vehs/km (source: Listing 10 [30]).

numerical results are only known for a fraction of the simulation scenarios. The expected numerical results for a specific scenario can, thus, only serve as a minimal standard for the simulation models and any potential implementation. At least for Figure 5.3(c) it turned out, that the code had serious discretisation errors that caused collisions to occur. The errors are described in the next section. Further verification can help to detect such kind of errors and is demonstrated also in the next section.

5.2.2 Collision freeness

Listing 5.5 demonstrates how to run the Maude simulation on a 10000 cells long single-lane with periodic boundary conditions. The NaSch [147] and the Knospe [118] microscopic traffic models are free of collisions due to their vehicles' unlimited braking capabilities. The collision-free property has to hold in the Maude implementation but also in the OLSIMv4 implementation as it employs the Knospe [118] model in addition to the model by Lee, Pottmeier, and Habel from Section 3.4. The latter is, however, probably not free of collisions according to Section 2.2.3. In the collision-prone model it is difficult but eventually still possible to verify the collision-free property as long as it is possible to distinguish between collisions that occur due to modelling intention and due to errors in the computer code. One possible error source for collisions are rounding errors, another one is improper discretisation.

Rounding errors in simulations with discrete time and space can occur when using a finer grained base discretisation to model a larger grained discretisation. As the traffic models of Section 3.4 rely on a discretisation of 1.5 m per cell and the NaSch model uses 7.5 m per cell (and for the vehicle), the author used the finer discretisation to model the larger one during the work on the Maude implementation. This is possible in principle, but requires special care in the implementation of the microscopic traffic model. Vehicle constellations with position values modulo five that do not equal zero represent invalid states in the simulation. When the microscopic traffic model tries to avoid the invalid states by simply rounding the position value to the next multiple of five, collisions can occur and happened during the work on the Maude article and "yielded to the fundamental diagram of Figure 5.3(c). Under certain circumstances, the velocities of the vehicles have been set to values that were not multiples of the 5 cells per second. As a result, collisions occurred.

The fundamental diagram with the corrected code is presented in Figure 5.3(b). It is not clear from the fundamental diagrams whether collisions occurred or not. However, in a scenario where the vehicle list container allows collisions, this may lead to very high densities that may be even greater than the maximum density. Thus, the collision freeness property of the NaSch model needs verification.

However, verification is not possible for all states because the model checker accepts only a finite state. Thus, a model checker offers only exhaustive search of simulation states for verification. Due to massive inserting of vehicles during the first few minutes collisions are very likely to appear in case they exist. Therefore, the verification focuses on the first few minutes.

The implementation for this property uses the binary operator `VehiclesCrashed?` from Listing 4.16. It has been extended successively for each sort up to sort `SimulationNetwork`. As the model checker accepts only finite states, the property `NoCrashedVehicles?` is checked recursively as shown in Listing 5.6" [30].

The Maude implementation might be useful for the investigation of the collisions that occur under certain circumstances in the model by Lee *et al.*, Pottmeier, and Habel. An inverse function to $\mathbf{F}_{drive}$ can help searching for the configurations that lead to collisions. Using the inverse function, a theorem prover such as those available for Maude could also try to prove that a given initial configuration can also result from established as sound configurations.

The following section describes additional ways to verify the microscopic traffic model as well the other components of the simulation model.

5.2.3 Further verification

The conceptual design of Chapter 3 offers now a variety of verification possibilities. Some have been verified as part of the Maude implementation and some still require activities. Those are described in the following.

The Maude implementation unveiled several deficiencies of the simulation model that have led to the enhancements in the conceptual design. With the multicomponent system specification and the network system specification, two variants of simulation models exist that have different semantics. Implementing both variants in addition to each other and letting the user choose among them at runtime should be possible, in principle. It may be even desirable to combine both simulation models in a single simulation scenario. The extremely long roadways such as the A3/A2 in NRW may require such a combination. The combination may be beneficial because it provides a way to split the occupation tables in several subtables with the same size and update them in parallel. Choosing both simulation models for implementation requires, however, to verify both. As the simulation model parts of the conceptual design operate concurrently and are intended to exploit CPU-level parallelism in the implementation, the concurrency model may also require the verification. Maude is well suited for this sort of tasks. The further increase of the parallelism in the simulation is possible according to Section 4.2.5 but has several approaches that need to be assessed first. The section described the so-called parallel update of the vehicles in an occupation table that can also be updated in concurrent order. Verification of the occupation table update involves showing the equivalence between the parallel update in sequential order and in concurrent order. As every state transition in a term rewrite system has a clearly defined semantics that is represented by a term, Maude is well suited to prove the equivalence between the two approaches.

The Maude implementation unveiled also that the random number generator had maintained a hidden state which turned out as a bottleneck in the productive implementation OLSIMv4. Subsections 4.2.5 and 4.2.10 discussed this in detail. Thus, the question emerges, where to store and maintain the state in a future implementation. The occupation tables and the traffic road elements are good candidates and the conceptual design chose the occupation tables to hold them. With the plus in concurrency, the state requires careful examination under the aspect of the number of random number generator updates and the state updates of the occupation tables. Maude is also well-suited for these kind of examinations. The occupation tables require also verification of the relation between the sections of the roadway and the vehicle lanes and positions. An occupation table model that wants to serve as a specification for a productive implementation might want to prove that the occupation tables do not contain vehicles on invalid lanes, positions, and velocities and that any vehicle operation of the occupation tables ensures this also. As the occupation tables have high efficiency requirements in the productive implementation, the question whether to merge the `VehiclesOnArea` and `SectionsOnArea` in a common data structure requires also investigation.

The conceptual design of the local detectors is almost complete. The traffic jam length detectors have not yet been explored. Additionally, the difficult parts in the local detector model include situations when vehicles take more than one time step to pass the detector and when the vehicles enter the lane on the detector position. The latter also applies to certain global detectors that cover not the whole cross-section of the roadway. Ensuring the robustness of the algorithms is thus the verification goal. The Maude implementation is well-suited for this task. The invariant detectors provide excellent facilities to explore the behaviour of the tuning elements. The latter require careful inspection and analysis of their behaviour as the conceptual tuning model is also not complete and requires concretisation first. With the concretisation, the verification of the strategy that tries to decrease the velocity by increasing the density becomes possible. The interesting question is, whether the velocity converges against the should-be value. Ensuring the effectiveness of this approach is the essential question during verification. When ensured, sensitivity analysis can provide valuable insights about the quantitative relationships between the input and output parameters. Sensitivity analysis for each relevant system component provides the additional benefit of advancing to the next class in the four-tier hierarchy. As part of a demonstration of verifying the tuning elements' convergence, Figure 5.3(d) has depicted the relative saturation ratio that shows the convergence of the actual to the should-be values in the simplified simulation scenario with periodic boundary conditions. Inverse simulation can also provide insights about the effectivity of inserting vehicles. Whenever a tuning element inserts a vehicle, an inverse simulation starts with the state that corresponds to the situation where the upstream tuning element could have inserted an additional vehicle. Rerunning the simulation from there on with the additional vehicle, will then show the correctness of the insertion strategy. Regarding the loop detector measurements from the on-ramps during the accumulate step may also enhance the accuracy further. The verification and validation of the tuning element models should address these questions. Which tool might be best suited is questionable. The Maude implementation as is can only process single-lane simulation scenarios which is at least a good starting point. Enhancing the efficiency of the Maude implementation and delaying more complex simulation scenarios into the productive implementation might be a way to go. Otherwise simulation tools such as SimuLink, Mathematica, and MatLab may be worth giving a try.

The vehicle structure from Equation 3.5 consists of several trajectory coordinates. The attributes or coordinates of the trajectory coordinates are not linearly independent as the position is derived from the velocity by integrating the latter over the single time step. For each pair of a vehicle's two succinctive coordinates $\mathbf{k}_n^{(t)}\#\mathbf{q}^{(t)}$ and $\mathbf{k}_n^{(t)}\#\mathbf{q}^{(t-\Delta t)}$, the difference, for example, of the position attributes $x^{(t)}$ and $x^{(t-\Delta t)}$ must hold the equation for position update $(v^{(t-\Delta t)} - D_{\max}\Delta t)\Delta t \leqslant x^{(t)} - x^{(t-\Delta t)} \leqslant (v^{(t-\Delta t)} + A_{\max}\Delta t)\Delta t$ with $v^{(t-\Delta t)}$ as the previous velocity, $A_{\max}$ as the maximum acceleration and $D_{\max}$ as the maximum deceleration. Similar constraints follow for the lane attributes and the velocities. Lowering the conditions for the equation to hold only to the state transition before and after the vehicle update of function $\mathbf{F}_{\text{drive}}$ and $\mathbf{F}_{\text{chlane}}$ from Sections 3.4.1 and 3.4.2 can be reasonable in the cases where, e.g., the tuning elements are allowed to replace vehicles with significant slower or faster vehicles.

The vehicle models, the tuning element models, and the traffic models proper functioning, enable then to explore the limitations of the simulation's road network capacity for various scenarios including the high traffic flows and the merging areas. The former focuses on the peaks of the traffic flows that the simulation still can handle and still can represent adequately. The OLSIMv4 traffic database enabled the research for the publication [114] in which traffic flows with more than 3000 vehs/h were found to be stable across several minutes. It has been discussed in Section 2.2.3. As it is known that the traffic model can't reproduce these high flows, the traffic model requires investigation. Additionally, the tuning elements behaviour requires a clear definition how to handle these extreme situations. The latter scenario, namely the limitation of the merging areas, focuses on the aspect of peaks of the number of

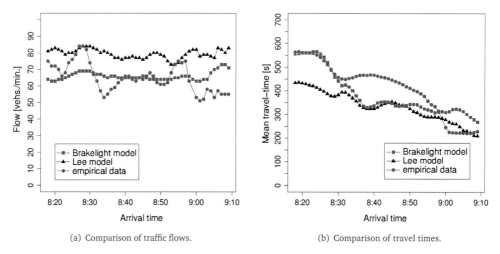

(a) Comparison of traffic flows.   (b) Comparison of travel times.

Figure 5.5: Stationary data from empirical observation and simulation results averaged over 5 minutes in an early test (source: [80, Fig. 6:1+3] in [31]).

lane changes from an on-ramp to the main roadway. Section 2.1.4.3 introduced the high traffic flows on some on-ramps to the network and provided two examples for their concurrencies in Tables 2.1.3 and 2.1.4. Depending on the type of simulation model, the behaviour of the tuning elements and the traffic model requires special treatment.

5.2.4   Travel-time validation

As any other microscopic traffic simulation system, OLSIMv4 also requires assessing of "how accurate the simulation results display the real world system" [30]. The assessment is widely known as validation. However, validation can only focus on individual aspects of the simulation results. Assessing several aspects of them simultaneously may be conflicting. When considering the tuning method, adapting the should-be velocity values by increasing the traffic flow may result in less accurate traffic flow values but increase also the accuracy of the travel times. To get a first impression of its accuracy for the further use in the calibration procedure, OLSIMv4 has been validated during an early "implementation phase of OLSIMv4 in [80] for travel times and traffic flow using a method based on section control [...]; on two subsequent bridges across a highway, vehicle travel times can be gained through taking photos in multi-shot mode using digital cameras. Later, the time-stamps of identical vehicles on photos from each bridge can be compared. Traffic flow can be gained with this method as well" [31]. As part of the validation results, "Figure 5.5 shows stationary traffic data of the 7 km section of A3 NRW highway, onto Breitscheid intersection, as recorded using section control on 25/11/2011 in [80]. The simulation results have been compared and are in good qualitative agreement – they reflect the drop of the empirical travel times after 8:25 a.m. even though the section includes an on- and off-ramp where the number of vehicles changes and impacts the traffic flow. Quantitative agreement is not expected since the used value of 108 km/h of the vehicles' maximum velocity, taken from literature, is too low for German highways" [31]. With the impression of the not really convincing quantitative agreement, the calibration process aimed at the increase of the quantitative agreement. As the result of the calibration process, the two parameters $v_{\mathrm{max},n}$ and t_{safe} of the traffic model from Section 3.4 were changed to $v_{\mathrm{max},n} = 22$ cells/s and $t_{\mathrm{safe}} = 4$ s. The calibration process was done by one person in significantly less than a full year which is not much compared to the efforts as discussed in Section 2.2.6. Later on, single vehicle data were used in another travel-times validation for the final assessment. As can be seen from the velocity related parts of the plots in Figure 5.2, OLSIMv4 calculated travel times are now in good agreement with the recorded ones. Only validating the section control travel-times again with the updated parameters would ensure the increase in accuracy. This should be done in the future.

5.2.5 Performance validation

As OLSIMv4 is a real-time simulation system, it has to fulfill the time requirements of one detector result in each minute. As has been mentioned afore, parallelisation forms the key to increase inform- ation accuracy and still meet the performance requirements. The increase can then be achieved by taking more traffic data into account. The parallelisation turned out as the difficult part of fulfilling the real-time requirements. The parallelisation gain lacked behind the expectations. Using profiling, the performance requirements could be met. However, the procedure unveiled certain disadvantages of the chosen Concurrent ML approach. As Scheme supports a `set!` procedure to manipulate storage, it is unclear in which part of the memory the storage cells exactly remain in. Although Bigloo supports thread-local storage, it requires explicit usage. A single hidden manipulation of a storage cell as well as the constructor and desctructor calls invoke the garbage collector which involves the operating system to send signals to the process. With the Bigloo profiler, the source code operations of large memory operations could be identified and fixed. However, the performance remains behind the expectations as not all threads work on a CPU load of 100%. The possible reasons are the garbage collector which might not play nicely with the multithreading as well as the random number generator state or any other hidden state. Sections 4.2.2.1 and 4.2.5 discussed this in more detail.

OLSIMv4 performs now, however, sufficiently well enough for handling the peaks of the vehicle traffic. Additionally, it regards an significantly increased amount of traffic information when compared to its predecessor. The computation of the traffic model is also much more complex and requires addi- tional computing power than those of the brakelight model. With the presented potential optimisation as depicted by Figure 5.4 and discussed in Subsection 5.2.1, a future implementation can probably take even more information into account. As OLSIMv4 also has a higher level of abstraction, supports multiple traffic models, several detector models, and tuning elements, the simulations complexity has increased drastically. OLSIMv4 has thus contributed to increasing the efficiency of the simulation.

5.3 Implementing the next version

Correctness, robustness, efficiency, code readability and maintainability form the keys of a successful implementation. With the conceptual design from Chapter 3, the implementation experiences from Chapter 4, and the concretised and verified models of the previous section, a well-explored specification for future implementations exists. The existence of the specification, the verified models, and the increase knowledge help in reducing the implementation effort for the next version significantly. Backed with all the details of the simulation system and the increased insights from the verification process, an experienced implementer should be able to implement the complete system in about a year. The reflections as presented in the previous sections and chapters allow already some first discussions about design decisions that evolved from the lessons learned. They are given in Section 5.3.1 together with potential future implementation issues thereafter in Section 5.3.2.

5.3.1 Design decisions

The parallelisation concept is maybe one of the longest reaching design decisions. As there are only a limited set of options, its choice manifests also the programming language. As the parallelisation concept depends on the simulation models and the occupation table updates to implement, they need to be selected first.

The previous section has already suggested the combined usage of the simulation models from Section 3.7 in favour of the extremely long roadways such as A3/A2 in NRW. Assuming the implementation of both variants of simulation models and as well as optionally choosing among several versions of occupation table updates, Haskell's nested data parallelism seems worth giving it a try although the author favours the ML language. In addition to the latter, the number of various parallelism concepts in Haskell grant a flexibility which might prove valuable with respect to the number of various parallel models that have not yet been implemented in a parallel environment. The degree of parallelism also impacts the type and implementation of the random number generator. A real parallel random number generator is probably favourable over the leap-frogging approach.

As a very rough estimation, Haskell programs range to be about two to three times slower than C or C++. With an appropriate benefit through proper parallelisation, the efficiency will be more than sufficient. The other advantages such as code readability and maintainability through minimal code size and better reasoning as well as the implementation duration will outperform imperative and object-oriented languages such as C and C++ easily. It is also questionable whether the NoSQL technologies will outperform the traditional relational database management systems approaches. As the latter have a strong plus in the mathematical foundations and the knowledge increased about the data model with this thesis and, thus, is probably not subject of frequent changes, they would still be preferable over the emerging technologies.

The Maude implementation employed already a physics library that used only exact rational numbers for the representation of the traffic observables. As the detector data employ only subsets of the natural numbers, a next version implementation should also prefer natural numbers, fixed point arithmetic, and rational numbers over floating point arithmetic. Additionally, a separate physics library provides better testing capabilities and encapsulates the difficulties in the development of appropriate unit systems such as MKS and CGS as well as transformations between the systems. The encapsulated system allows then better verification and examination of the intrinsic precision and uncertainty.

As new concept, the suspend and resume approach might be worth implementing. It will contribute in verifying the implementation. In the Maude implementation every system state has a defined semantic that corresponds to a finite term. An implementation can profit from this by dumping or restoring a system state that represents a term in the Maude format. The exchangeable term format allows then to verify computations across the systems.

5.3.2 Implementation issues

Figure 5.6 depicts the most relevant units and milestones for the implementation of a next version. For the implementation of the occupation tables, the inner most boxes in the figure, namely the vehicle types, the vehicles, the cells, and the sections represent the starting points. They are discussed first. Mapping the vehicle type onto the programming language's type system, might be a way to go. A context specific set of vehicle types might, however, be more realistic. Assertions for the coordinates as discussed in Section 5.2.3 might help in unveiling implementation errors of the vehicle model. Maybe the most important part for the implementation of the vehicle models affects the efficiency. As the vehicles are constructed several times in each time step, measurements of the construction times will give first hints about the best suited data structure. In contrast to the vehicles that require rapid construction, the cells and the sections for the roadways as part of the network model require rapid access to the contents. Measuring the access times will also indicate the best suited data structures.

The next milestones in Figure 5.6 are the `VehiclesOnArea` and `SectionsOnArea` procedures. With a vehicles and a sections data structure, measuring multiple vehicles as sets, lists, and vectors forms a next step in finding the best suited data containers. Due to the similarity given by the `VehiclesOnArea` and `SectionsOnArea` operations, the measurements should also include the sections and distinguish properly between access times of elements and construction times of large sets of elements in a purely functional language. The measurements should regard also the parallel environment and consider thread-local storage for the updates. The occupation tables support sorting out the neighbours and the topological analysis for any given vehicle. This operation is also done at least once for every vehicle and requires thus also careful examination under the efficiency aspect. As the data containers and the operations are not really difficult to implement, early measurements can help in establishing the fitness of the chosen approach.

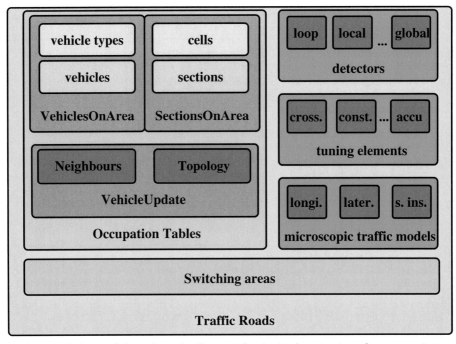

Figure 5.6: Some of the units and milestones for the implementation of a next version.

The choice among a set of the best suited data structures and containers should, however, not only rely on the efficiency aspect. The final efficiency is a mixture of several factors. The other keys for a successful implementation are also important. In case of premature optimisation, the code may suffer from readability and maintainability in favour of questionable performance gain. In the end, the performance efficiency gain is derived from analysing and identifying the critical parts in a profiler. For the OLSIMv4 implementation, these have been the vehicle updates of the occupation tables. The vehicle container and the topology container require, therefore, special attention. An experimental implementation should carefully examine whether it sufficiently supports exploiting CPU thread-level parallelism. Thereafter, decisions about the performance improvement can rely on the increased insights. The automatic construction of switching areas depending on two occupation tables and the switching area specification can follow thereafter.

Figure 5.6 depicts the determination of the neighbours and the topology for a vehicle as part of its update as the next milestone of the implementation. Once the automatic construction of a vehicle's neighbourhood and the topological analysis is available, the traffic models are ready for implementation. As the microscopic traffic models implement all the generic vehicle update procedures, Figure 5.6 displays them as a separate unit in a line with the detectors and the tuning elements. With the knowledge of the invariants for the deterministic case as presented in Section 5.4, the result verification for the simplified scenario is possible and should be done already in an early state. The invariant is not limited to the drive update case. Precalculating every single combination of relative distances in advance is possible also for the lane change case. Thereafter, iterating and refining of the vehicle update performance with continuous verification of the computation will yield a correct and highly optimised implementation of the vehicle model. The occupation tables can then use the latter for the update of their vehicles. This may already establish the earlier taken decisions or, if not, may require the refinement of the chosen data structures. In every single step, the components were ready to test afterwards which is an important part of the suggested implementation procedure. The suspend and resume feature might prove handy here. The occupation tables may also be subject of parallelisation in which case the latter is required exploring the performance gain and the CPU resource usage per thread. The nested data parallelism is probably the best approach for the implementation of the drive update. The lane change implementation is more difficult as it also requires considering alternative vehicles for more than two lanes in order to avoid collisions. Nested data parallelism might or might not improve performance.

The detector models come next in the implementation hierarchy. Figure 5.6 depicts them as a standalone unit. They require implementation in the programming language of the measuring method and of the accessing of the database. The latter, however, requires also implementation effort. The implementer has to decide whether to use a metamodel or not. It may help in structuring the implementation but may also prove superfluous. It mostly depends on the project dimensions. In case of OLSIM, it may be worth it. The database tuning and access time optimisation is a central aspect that requires later attention. As the importing of the traffic data may take a while, delaying the database throughput and access time optimisation to the end of the implementation cycle puts a major risk on the productive implementation. The delay occurred during the implementation of the first traffic database before OLSIMv4. In general, the detectors are not performance critical in the first place. Constructing fixed sequences of occupation tables and studying the measurements will then lead to the verification of the detectors implementation and enable also systematic testing.

The tuning element models work on top of the detector models and, thus, follow next. Figure 5.6 depicts them also as a standalone unit. Their implementation should focus on correctness as they are also not performance critical in the first place. The testing should start with simple simulation scenarios such as the one used in the context of Figure 5.3. As especially the aspect of the velocity decrease requires examination, invariant or loop detectors could provide specially crafted measurements where the tuning elements have to show up an expected behaviour. After the implementation and the verification of the tuning elements, the most part of the work will be to study the whole network of tuning elements and switch them on and off individually depending on the location and data quality.

The simulation models complete the implementation as probably the most interesting part of it. Their implementation starts with the definition of the traffic roads as the basic unit. In case of the multicomponent specification, the update of the traffic roads requires synchronisation after each vehicle update. In case of the network specification, a traffic road's list of accumulators for each of the traffic roads requires examination whether all detectors are part of the traffic road element or not. If not, the simulation engine must synchronise the accumulate operation for the affected traffic roads. The other variant provides more parallelism as no synchronisation is required. The different synchronisation requirements question what kind of parallel implementation should be used. As OLSIMv4 has already used a `parallel` map function successfully, this may be the first try. Otherwise nested data parallelism may also be worth giving a try. Apart from the synchronisation during one update step, the parallel update also has the following difficulties in an implementation. The occupation tables vary in length from $50\,km$ to $320\,km$ and also in size, i.e. the number of vehicles, significantly. The time to perform a single update step varies, thus, also significantly. A parallel map might lead to some "always" running threads and others that idle most of the time. Balancing the different computation loads across a set of CPUs involves, thus, exploring the possibilities and intensive testing.

The interfaces and components to external systems represent units that have been neglected almost completely during the implementation of OLSIMv4. The reason for this is simple. The scientific focus of the University lets the implementation of the interfaces and components appear secondary. In principle, this state needs more focus. The interfaces and tools to external systems have to grant data consistency and interoperability. In case where one is replaced by another, the simulation system should not be affected by the replacement. Furthermore, the implementation of the tools and the interfaces to the external systems should support later and easy adaption for the purpose of, for example, achieving a higher precision in mapping the data to and from a third party digital map. Also as part of a future work, the future implementation should also support placing higher level control algorithms on top of, for example, the tuning elements and the detectors. Adding parts of the road traffic network to the simulation dynamically comes also in mind as a further application domain of the higher level control algorithms. Any implementation of the interfaces and the tools to the external systems should be prepared for coupling with control algorithms.

Tools for the automatic setup of simulation scenarios have not yet been on the requirements list but are essential in setting up a simulation system. Also tools for the generation of the prognosis data are still missing. Backup and archive falls into the same category.

5.4 Summary

This chapter enumerated the benefits of the conceptual design and its implementations as well as the ones that result from reflecting over the implementation issues and the enhancement effort. They range from a significant progress in obtaining and distributing traffic information over improved verification and validation facilities to increased insights for implementing the next version of the simulation system.

The chapter started with Section 5.1 that details the progress in obtaining and distributing traffic information. OLSIMv4 and the simulation systems described in this thesis now support a variety of traffic information. The information includes road user and road operator traffic information, scientific and further traffic information. The growth in the variety denotes a significant advance over OLSIMv3. Subsection 5.1.1 desribed the road user traffic information. In addition to the common road user traffic information such as level-of-service, travel-time, density, and traffic flow, the systems now also support direct comparison of input and output detector data. The systems also support now the calculation of higher valued information that rely on the primitive traffic information as presented in this Subsection. Subsection 5.1.2 detailed the road operator traffic information. The possibility of combining traffic information with topological information and traffic messages facilitates searching for hybrid criteria. The enhanced research facilities eases exploring odd simulation results for plausibility and supports gaining advanced knowledge about the traffic network. The integrated combination of the data sources including the simulation and the real-world detectors supports also contrasting of the stationary detector data with the spatio-temporal traffic information. This is also a significant improvement over its predecessor as it supported only manual contrasting for a restricted subset of the traffic observables. The integration into a database supports also the generation of value added traffic information for advanced questions that Subsection 5.1.3 describes as part of the scientific traffic information. The detector concept of OLSIMv4 and this thesis supports also the integration of further detector types for more specialised scientific traffic information such as time headway distributions. The systems of this thesis also support advanced simulation setups that have not been possible in the context of OLSIMv3. Subsection 5.1.4 discusses these setups as part of the further traffic information.

The conceptual design facilitates better future verification and validation activities. Section 5.2 presented them as part of this chapter's benefits. Due to the lack of a proper specification and the minimal level of system knowledge provided by the chosen kind of system, not much was known about the functioning and the accuracy of the results. Due to the detailed specification, a future implementation can enter the next higher class of the four-tier hierarchy for the classification of systems with increasing verification degrees. Subsection 5.2.1 presented the numerical solution for a single lane scenario with periodic boundary conditions as part of some considerations about the numerical solution verification. The solution results in identical graphs and they do not depend on whether the code enables collisions or not. Subsection 5.2.2 demonstrated this and also showed a way to verify whether the simulation model is free of collisions or not. The collisions may occur due to rounding errors or due to errors in the update logic. More complex simulation scenarios may, however, show significant discrepancies in the results. Subsection 5.2.3 enumerates further verification possibilities that involve verification of the parallelisation model. This verification is mainly of interest for the multicomponent model as this verification for the network model will generally be much simpler. The further verification possibilities also include the equivalence verification of the presented models and a future development of the vehicle update with increased parallelism. Furthermore, they include the reproduction of well-known traffic characteristics such as breakdown due to high traffic flows, reproduction of the high traffic flow patterns, as well as corresponding reflection for each of the participating component models, namely the vehicle models, the tuning element models, and the traffic models. The verified simulation models can then participate in travel-time validation and performance validation. Subsection 5.2.4 presented the travel-time validation. The travel-times have been validated using a section control method in an early stage during the implementation phase of OLSIMv4. As part of the lessons learned, this thesis emphasizes to let them take place only after the successful verification of the system. The section closes with some considerations about performance validation take place in Section 5.2.5.

Section 5.3 presented the implementation considerations for the next version. The conceptual design in combination with the experiences from the implementation issues and the updated and verified component models ease the future development of a next version. Backed by this thesis, the future development can rely on a top-down implementation. The reliable design decisions that a future implementation can rely on is one of the reasons for the facilitated top-down implementation. The otherwise long-term learning process reduces to an implementation effort of about one person for one or two years for the future development of a next version. Subsection 5.3.2 explained the significantly reduced implementation issues. The orthogonal and in complexity significantly reduced models contribute to the reduction in the implementation effort.

Outlook

The results of this thesis equip a future implementation project of a microscopic traffic simulation with the insights and the knowledge to reduce the implementation effort significantly.

The thesis presented the simulation system that were developed based on sound theoretic foundations. The simulation system involved vehicle models, network models, microscopic traffic models, detector models, tuning models, and simulation system models. Defining such simulation models for microscopic traffic simulation systems appeared at hand. This thesis, however, showed that the models at hand have serious shortcomings such as a lack of specification, verification possibilities, parallelisation potential, minimality and orthogonality of the system components, superfluous dependencies, and undefined behaviour. To overcome the shortcomings, this thesis defined three simulation models as part of the conceptual design. Apart from the first model which concretises the discrete time system specification and that was already present in OLSIMv3, the models may participate in future versions of OLSIM. As part of the alternative models, this thesis defined microscopic traffic simulation models for the multicomponent specification and the network of systems specification. While the former depends on the state of connected components, the latter depends with its input only on the output of connected systems. The network of system specification provides, thus, the most system knowledge, the most degree of parallelisation possibilities, and the best verification possibilities. Defining a simulation model as such that the formalisms apply results, thus, in a significant gain. This approach is, of course, not limited to microscopic traffic simulations as the formalisms also apply to simulations in general. This thesis encourages, thus, to extend the approach onto other microscopic traffic simulations systems and even onto other kinds of simulation systems such as discrete event systems and differential equation systems.

The conceptual design of this thesis used semantic algebras to provide the (semi-)formal specifications for the simulation models. They consist of abstract data types suitable for an analogous implementation in functional programming languages. As the abstract data types in the specification correspond to those in a corresponding implementation, the specification form facilitates a straightforward implementation verification. The sets in the specification describe system states or component states in the implementation. The functions in the specifications describe system and component state transitions. The specifications split the systems into polymorphic components such as vehicles, network elements, detectors, tuning elements, and system states. The splitting into the components facilitates now component-wise verification, testing, and validation. It facilitates also implementing the state transition functions as generic functions. As this thesis also discussed the implementation issues that exist for each of the components and as there exists a direct correspondence between the specification and the implementation, the specification facilitates also specification verification by means of the implementation. Combining the specification method of using semantic algebras with the functional and declarative implementation methods yields increased verification possibilities which, in turn, makes any simulation systems more reliable, more robust, and more trustworthy. It is, thus, also a good choice for other microscopic traffic simulation systems and even other kinds of simulation systems as the ones mentioned above. The plus in verification possibilities and the therefrom resulting more trustworthy simulation systems smooth the way for the validation and the calibration process.

The simulation models in the presented form are not too distinct to implement each of it in addition to the others. The variety of the models enables the simulator to choose the best suited simulation model for any targeted simulation scenario among several models and even to combine them as part of a specific simulation scenario. The polymorphic components in combination with the generic models and the plurality of the microscopic traffic models facilitate the construction of simulation systems that can be adjusted to meet their specific requirements. The network model in combination with the generic microscopic traffic model ensure platform independency and enables the simulator to add further traffic models to the simulation system and apply them to any given network. The impact of the network model on the simulation model turned out to form a bottleneck. Analysing the impacts and freeing the simulation model of them turned out to be possible for the case of microscopic traffic simulations. The thesis also discussed the advantages of one simulation model over the other. Any future implementation can now rely on well-founded knowledge instead of speculation and experimentation. The plus in knowledge reduces the complexity significantly. It is now possible to use the definition for the multicomponent model in the optimised form for each scenario where a component state depends on the state of other components. In all the other cases the model following the network of systems specification is clearly preferred. It remains, however, an open question how exactly the simulation results differ across the simulation models with respect to the traffic information.

Acknowledgements

A computer scientist may wonder in the first place why the modelling of such a simulation system would require five versions to find an adequate representation. A computer scientist may also wonder why, apparently, it was not possible to write such a (semi-)formal specification earlier. The question is justified but so are the five versions. This section answers the question and explains the background.

At the time when the OLSIM project started, the domain of microscopic traffic simulations were a scientific domain only and reports of successful applications of microscopic traffic simulations to real-world networks were lacking. As a consequence, applying the microscopic traffic simulation to a real-world network with real-time requirements resulted in an experiment. The experimental nature required an iterative approach for facing and solving the arising problems–one after the other. The microscopic traffic models were subject of continuous development. The development phase of the simulation system up to the third version was dominated by the development of the traffic models. Without the significant progress that the former colleagues achieved within this development phase, this thesis would not have been possible. The development of the network model which has a significant impact on the simulation model and a proper representation of the network model were a difficult and long ongoing process. Several network models from various administrations and authorities had to be merged. Freely available digital maps and a complete tool chain such as OpenStreetMap were not available at the starting time. Introducing new technologies into administrations is a long ongoing process which means that a certain technology lasts for a long time once it is established. The technology includes wide-spread non-digital maps that were used in a catalogue for road-works, loop detectors, and road traffic properties such as speed limits, lane mergings and closings, and overtaking restrictions. Without the groundwork of the former colleagues, the creation of an electronic version as a graph would not have been possible. The same applies to the development prognosis methods, to the examinations of the loop detector data, and the third simulation model.

Another important point is the fact that the subject of microscopic traffic simulation is frequently underestimated. Possible reasons are the analogy to the participation in the daily road traffic. However, there exist subtle but relevant differences that complicate the subject of microscopic traffic simulation significantly. The underestimated complexity results in difficult project settings and worse software quality. Despite of all software quality deficiencies that is also present in OLSIMv4, without the groundworks of the OLSIMv3 simulation system this current work would not have been possible. The de facto iterative development process with four OLSIM versions, the Maude enhancement effort, and this thesis was, thus, justified.

Therefore, I thank Prof. Dr. Schreckenberg and his staff as well as all former colleagues. Also, without the financial support of the Project funder, namely the German Ministry of Economic Affairs, Technology, and Transportation of North-Rhine Westphalia (NRW) that initiated OLSIM in 1999 the presented work and this thesis would not have been possible. The Landesbetrieb Straßen.NRW also supported this work not only by providing loop detector data and the digital maps but also by answering patiently and kindly a lot of questions regarding the telematics and the NRW road traffic network. Prof. Dr. Schreckenberg also provided a great infrastructure and working environment to learn and to work in the field of transport and traffic. His patience with me, his expert knowledge, and his loyalty will always serve as good examples for me.

Prof. Dr. Luther showed me the way how to write the publications and this thesis. He provided many ideas to this thesis. He patiently read and annotated all my drafts to the publications and he responded always within a few days. He always found the time to discuss all my questions on this thesis. His patience with me, professional companionship, his enormous effort, and his immense expert knowledge will always serve as good examples for me.

Florian Knorr did an enormous work on this thesis. He read the most part of the thesis and corrected subject specific errors as well as language mistakes all in his spare time.

Peter Gilson and Stefan Maier helped with correcting language errors. They also did it in their spare time.

I also want to thank my family. My father facilitated my study in computer science when he no longer had to. My family always encouraged me to complete the thesis. My wife facilitated the work by giving me the time and taking some of my duties.

Last but not least, I thank God who equipped me with the necessary skills. He made it all happen.

Without any of this help this thesis would not have been possible. Thank you all.

References

[1] Hal Abelson, R. Kent Dybvig, Christopher T. Haynes, Guillermo J. Rozas, Norman I. IV Adams, Daniel P. Friedman, Eugene Kohlbecker, Guy Lewis Jr. Steele, David H. Bartley, Robert Halstead, Don Oxley, Gerald Jay Sussman, Gary Brooks, Chris Hanson, Kent M. Pitman and Mitchell Wand. 'Revised[5] Report on the Algorithmic Language Scheme'. In: *Higher-Order and Symbolic Computation* 11.1 (1998), pp. 7–105. issn: 1388-3690. doi: `10.1023/A:1010051815785`. url: `http://dx.doi.org/10.1023/A:1010051815785` (cit. on pp. 119, 125, 128, 135).

[2] PTV Planung Transport Verkehr AG. *PTV VISUM 12.5 USER MANUAL*. Ed. by PTV Planung Transport Verkehr AG. 12.5. PTV. 76131 Karlsruhe, Germany: PTV, July 2012 (cit. on p. 5).

[3] PTV Planung Transport Verkehr AG. *VISSIM 5.40 User Manual*. Ed. by PTV Planung Transport Verkehr AG. 5.40-03. PTV. 76131 Karlsruhe, Germany: PTV, May 2012 (cit. on pp. 5, 30, 33).

[4] Kayvan Aghabayk, Majid Sarvi, William Young and Lukas Kautzsch. 'A novel methodology for evolutionary calibration of VISSIM by multi-threading'. In: *Australasian Transport Research Forum 2013 Proceedings*. Oct. 2013. url: `http://www.atrf.info/papers/2013/2013_aghabayk_sarvi_young_kautzsch.pdf` (cit. on p. 35).

[5] Meaza Negash Akililu. 'Verification of Rural Traffic Simulator RuTSim 2'. Master Thesis. Norköping, Sweden: Linköping University – Department of Science and Technology, Sept. 2012. url: `http://www.diva-portal.se/smash/get/diva2:559245/FULLTEXT01.pdf` (cit. on p. 34).

[6] Jonathan Aldrich and Kevin Donnelly. 'Selective open recursion: Modular reasoning about components and inheritance'. In: (Oct. 2004). SIGSOFT2004/FSE-12 – 12th ACM SIGSOFT Symposium on the Foundations of Software Engineering, p. 26. url: `https://www.cs.ucf.edu/~leavens/SAVCBS/2004/savcbs04.pdf` (cit. on p. 125).

[7] Cécile Appert-Rolland. 'Experimental study of short-range interactions in vehicular traffic'. In: *Physical Review E* 80 (3 Sept. 2009), p. 036102. doi: `10.1103/PhysRevE.80.036102`. url: `http://link.aps.org/doi/10.1103/PhysRevE.80.036102` (cit. on p. 31).

[8] The Motor Industry Software Reliability Association. *MISRA-C:2004 — Guidelines for the use of the C language in critical systems*. Tech. rep. The Motor Industry Software Reliability Association, Oct. 2008. url: `http://www.misra-c.com` (cit. on p. 125).

[9] The Motor Industry Software Reliability Association. *MISRA-C++:2008 — Guidelines for the use of the C++ language in critical systems*. Tech. rep. The Motor Industry Software Reliability Association, June 2008. url: `http://www.misra-cpp.com` (cit. on p. 125).

[10] ASTM International. *Standard Specification for Telecommunications and Information Exchange Between Roadside and Vehicle Systems — 5 GHz Band Dedicated Short Range Communications (DSRC) Medium Access Control (MAC) and Physical Layer (PHY) Specifications*. Tech. rep. E2213. ASTM International, Mar. 2010. doi: `10.1520/E2213-03R10`. url: `http://dx.doi.org/10.1520/E2213-03R10` (cit. on p. 28).

[11] Paolo Atzeni, Christian S. Jensen, Giorgio Orsi, Sudha Ram, Letizia Tanca and Riccardo Tor-lone. 'The Relational Model is Dead, SQL is Dead, and I Don't Feel So Good Myself'. In: *SIG-MOD Special Interest Group on Management of Data Record* 42.2 (July 2013), pp. 64–68. issn: 0163-5808. doi: 10.1145/2503792.2503808. url: http://doi.acm.org/10.1145/2503792.2503808 (cit. on pp. 129, 130, 150).

[12] Ekatarina Auer, Wolfram Luther and Roger Cuypers. 'Process-oriented verification in biomech-anics'. In: *Safety, Reliability, Risk and Life-Cycle Performance of Structures and Infrastructures*. Ed. by George Deodatis, Bruce R. Ellingwood and Dan M. Frangopol. International Associ-ation for Structural Safety and Reliability. Columbia University, New York, NY, USA: CRC Press, June 2014, pp. 391–398. isbn: 978-1-138-00086-5. doi: 10.1201/b16387-59. url: http://dx.doi.org/10.1201/b16387-59 (cit. on pp. 34, 123, 197, 198).

[13] Ekaterina Auer and Wolfram Luther. 'Numerical Verification Assessment in Computational Bio-mechanics'. In: *Numerical Validation in Current Hardware Architectures*. Ed. by Annie Cuyt, Wal-ter Krämer, Wolfram Luther and Peter Markstein. Vol. 5492. Lecture Notes in Computer Science. Springer Berlin Heidelberg, 2009, pp. 145–160. isbn: 978-3-642-01590-8. doi: 10.1007/978-3-642-01591-5_9. url: http://dx.doi.org/10.1007/978-3-642-01591-5_9 (cit. on p. 198).

[14] John Backus. 'Can Programming Be Liberated from the Von Neumann Style?: A Functional Style and Its Algebra of Programs'. In: *Communications of the ACM* 21.8 (Aug. 1978), pp. 613–641. issn: 0001-0782. doi: 10.1145/359576.359579. url: http://doi.acm.org/10.1145/359576.359579 (cit. on pp. 124–127).

[15] Robert Barlovic, Torsten Huisinga, Andreas Schadschneider and Michael Schreckenberg. 'Open boundaries in a cellular automaton model for traffic flow with metastable states'. In: *Physical Review E* 66 (4 Oct. 2002), p. 046113. doi: 10.1103/PhysRevE.66.046113. url: http://link.aps.org/doi/10.1103/PhysRevE.66.046113 (cit. on pp. 33, 96).

[16] Robert Barlovic, Ludger Santen, Andreas Schadschneider and Michael Schreckenberg. 'Meta-stable states in cellular automata for traffic flow'. In: *The European Physical Journal B - Con-densed Matter and Complex Systems* 5.3 (1998), pp. 793–800. issn: 1434-6028. doi: 10.1007/s100510050504. url: http://epjb.epj.org/articles/epjb/abs/1998/19/b8242/b8242.html (cit. on p. 66).

[17] Rimon Barr, Zygmunt J. Haas and Robbert van Renesse. 'JiST: an efficient approach to simu-lation using virtual machines'. In: *Software: Practice and Experience* 35.6 (2005), pp. 539–576. issn: 1097-024X. doi: 10.1002/spe.647. url: http://onlinelibrary.wiley.com/doi/10.1002/spe.647/pdf (cit. on p. 26).

[18] Fernando J. Barros. 'Modeling Formalisms for Dynamic Structure Systems'. In: *ACM Transac-tions on Modeling and Computer Simulation* 7.4 (Oct. 1997), pp. 501–515. issn: 1049-3301. doi: 10.1145/268403.268423. url: http://doi.acm.org/10.1145/268403.268423 (cit. on pp. 104, 115).

[19] Michael Behrisch, Laura Bieker, Jakob Erdmann and Daniel Krajzewicz. 'SUMO - Simulation of Urban Mobility: An Overview'. In: *SIMUL 2011, 3rd International Conference on Advances in System Simulation*. Barcelona, Spain, Oct. 2011 (cit. on pp. 5, 17, 24, 32, 33, 56, 61, 77).

[20] Lars Bergstrom, Matthew Fluet, Mike Rainey, John Reppy, Stephen Rosen and Adam Shaw. 'Data-only Flattening for Nested Data Parallelism'. In: *ACM SIGPLAN Notices* 48.8 (Feb. 2013), pp. 81–92. issn: 0362-1340. doi: 10.1145/2517327.2442525. url: http://doi.acm.org/10.1145/2517327.2442525 (cit. on p. 141).

[21] Yves Bertot and Pierre Castéran. *Interactive theorem proving and program development: Coq'Art: the calculus of inductive constructions*. Springer Berlin Heidelberg, 2004. isbn: 3-540-20854-2 (cit. on pp. 36, 124).

[22] Aggelos Biboudis, Nick Palladinos and Yannis Smaragdakis. 'Clash of the Lambdas'. In: *Comput-ing Research Repository[cs.PL]* abs/1406.6631 (June 2014). url: http://arxiv.org/abs/1406.6631 (cit. on pp. 126, 127).

[23] Reinhold Bien. 'Gauß and Beyond: The Making of Easter Algorithms'. In: *Archive for History of Exact Sciences* 58.5 (2004), pp. 439–452. issn: 0003-9519. doi: 10.1007/s00407-004-0078-5. url: http://dx.doi.org/10.1007/s00407-004-0078-5 (cit. on p. 84).

[24] Barry W. Boehm. 'A spiral model of software development and enhancement'. In: *Computer* 21.5 (May 1988), pp. 61–72. issn: 0018-9162. doi: 10.1109/2.59 (cit. on pp. 121, 122).

[25] Sylvie Boldo. *Formal verification of tricky numerical computations*. SCAN 2014 Würzburg, Ger-many. Sept. 2014. url: http://www.scan2014.uni-wuerzburg.de/fileadmin/10030000/scan2014/talks/plenary_7.pdf (cit. on pp. 36, 124).

[26] Dietrich Braess, Anna Nagurney and Tina Wakolbinger. 'On a Paradox of Traffic Planning'. In: *Transportation Science* 39.4 (2005), pp. 446–450. doi: 10.1287/trsc.1050.0127. eprint: http://pubsonline.informs.org/doi/pdf/10.1287/trsc.1050.0127. url: http://homepage.rub.de/Dietrich.Braess/Paradox-BNW.pdf (cit. on p. 14).

[27] Elmar Brockfeld, Reinhart Kühne, Alexander Skabardonis and Peter Wagner. 'Toward Bench-marking of Microscopic Traffic Flow Models'. In: *Transportation Research Record: Journal of the Transportation Research Board* 1852 (Jan. 2003), pp. 124–129. issn: 0361-1981. doi: 10.3141/1852-16 (cit. on p. 30).

[28] Elmar Brockfeld and Peter Wagner. 'Calibration and Validation of Microscopic Traffic Flow Mod-els'. In: *Traffic and Granular Flow '03*. Ed. by Serge P. Hoogendoorn, Stefan Luding, Piet H.L. Bovy, Michael Schreckenberg and Dietrich E. Wolf. Springer Berlin Heidelberg, 2005, pp. 67–72. isbn: 978-3-540-25814-8. doi: 10.1007/3-540-28091-X_6 (cit. on p. 30).

[29] Johannes Brügmann. 'Scsh-make – a build system with the scheme shell (in German)'. Ger-man title: Scsh-make – Ein Build-System mit der Scheme Shell. Thesis. Tübingen, Germany: University Tübingen, Apr. 2005 (cit. on p. 128).

[30] Johannes Brügmann, Michael Schreckenberg and Wolfram Luther. 'A verifiable simula-tion model for real-world microscopic traffic simulations'. In: *Simulation Modelling Practice and Theory* 48C (Nov. 2014), pp. 58–92. issn: 1569-190X. doi: 10.1016/j.simpat.2014.07.002. url: http://www.sciencedirect.com/science/article/pii/S1569190X14001117 (cit. on pp. 6, 8, 27, 28, 32, 34, 39, 41, 45, 48, 49, 53, 56, 59, 87, 96, 97, 99, 119, 122, 123, 129, 130, 132–134, 141–149, 151–159, 161–168, 171, 172, 174, 176–181, 197, 199–201, 203, 204, 206).

[31] Johannes Brügmann, Michael Schreckenberg and Wolfram Luther. 'Real-Time Traffic Inform-ation System Using Microscopic Traffic Simulation'. In: *EUROSIM 2013 – 8th EUROSIM Con-gress on Modelling and Simulation*. Ed. by Khalid Al-Begain, David Al-Dabass, Alessandra Or-soni, Richard Cant and Richard Zobel. EUROSIM. Cardiff, Wales, UK: The Institute of Elec-trical and Electronics Engineers, Inc., Sept. 2013, pp. 448–453. isbn: 978-0-7695-5073-2. doi: 10.1109/EUROSIM.2013.83 (cit. on pp. 5, 13, 14, 20, 29, 34, 56, 77, 100, 119, 131, 143, 176, 206).

[32] Arthur W. Burks. 'A radically non-von-Neumann-architecture for learning and discovery'. In: *CONPAR 86*. Ed. by Wolfgang Händler, Dieter Haupt, Rolf Jeltsch, Wilfried Juling and Otto Lange. Vol. 237. Lecture Notes in Computer Science. Springer Berlin Heidelberg, 1986, pp. 1–17. isbn: 978-3-540-16811-9. doi: 10.1007/3-540-16811-7_148. url: http://dx.doi.org/10.1007/3-540-16811-7_148 (cit. on p. 126).

[33] Arthur W. Burks, Herman H. Goldstine and John von Neumann. 'Preliminary Discussion of the Logical Design of an Electronic Computing Instrument'. In: *The Origins of Digital Computers*. Ed. by Brian Randell. Texts and Monographs in Computer Science. Springer Berlin Heidelberg, 1982, pp. 399–413. isbn: 978-3-642-61814-7. doi: 10.1007/978-3-642-61812-3_32. url: http://dx.doi.org/10.1007/978-3-642-61812-3_32 (cit. on p. 125).

[34] Taylor R. Campbell. *Scheme-CML*. http://mumble.net/ campbell/darcs/scheme-cml/. Oct. 2010. url: `http://mumble.net/~campbell/darcs/scheme-cml/` (cit. on pp. 128, 138, 141).

[35] Jeffrey C. Carver. 'Software Engineering for Computational Science and Engineering'. In: *Computing in Science & Engineering* 14.2 (2012), pp. 8–11. doi: `10.1109/MCSE.2012.31`. url: `http://scitation.aip.org/content/aip/journal/cise/14/2/10.1109/MCSE.2012.31` (cit. on pp. 16, 121–123).

[36] Jeffrey C. Carver, Richard P. Kendall, Susan E. Squires and Douglass E. Post. 'Software Development Environments for Scientific and Engineering Software: A Series of Case Studies'. In: *Software Engineering, 2007. ICSE 2007. 29th International Conference on*. May 2007, pp. 550–559. doi: `10.1109/ICSE.2007.77` (cit. on pp. 16, 121–124, 127).

[37] Rick Cattell. 'Scalable SQL and NoSQL Data Stores'. In: *SIGMOD Special Interest Group on Management of Data Record* 39.4 (May 2011), pp. 12–27. issn: 0163-5808. doi: `10.1145/1978915.1978919`. url: `http://doi.acm.org/10.1145/1978915.1978919` (cit. on pp. 129, 130).

[38] Avik Chaudhuri. 'A Concurrent ML Library in Concurrent Haskell'. In: *ACM SIGPLAN Notices* 44.9 (Aug. 2009), pp. 269–280. issn: 0362-1340. doi: `10.1145/1631687.1596589`. url: `http://doi.acm.org/10.1145/1631687.1596589` (cit. on p. 141).

[39] Debashish Chowdhury, Ludger Santen and Andreas Schadschneider. 'Statistical physics of vehicular traffic and some related systems'. In: *Physics Reports* 329.4–6 (2000), pp. 199–329. issn: 0370-1573. doi: `10.1016/S0370-1573(99)00117-9`. url: `http://www.sciencedirect.com/science/article/pii/S0370157399001179` (cit. on pp. 28, 29, 31, 159).

[40] Roland Chrobok. 'Theory and Application of Advanced Traffic Forecast Methods'. PhD thesis. University Duisburg-Essen, 2005. url: `http://duepublico.uni-duisburg-essen.de/servlets/DerivateServlet/Derivate-5656/Chrobokdiss.pdf` (cit. on pp. 84, 86).

[41] Roland Chrobok, Sigurður Freyr Hafstein, Andreas Pottmeier and Michael Schreckenberg. 'OLSIM: Future Traffic Information'. In: *Proceedings of the IASTED International Conference, Applied Simulation and Modelling*. Ed. by M. H. Hamza. Applied Simulation and Modelling 2004 443-142. International Association of Science and Technology for Development (IASTED). Rhodes, Greece: IASTED/ACTA Press, June 2004, p. 592. url: `http://www.actapress.com/Abstract.aspx?paperId=18672` (cit. on pp. 5, 12, 18, 32).

[42] Roland Chrobok, Andreas Pottmeier, Sigurður Freyr Marinósson and Michael Schreckenberg. 'On-Line Simulation and Traffic Forecast: Applications and Results'. In: *6th IASTED International Conference, Internet and Multimedia Systems and Applications*. Ed. by M. H. Hamza. International Association of Science and Technology for Development (IASTED). Kaua'i, Hawaii, USA, 2002: IASTED/ACTA Press, 2002, pp. 113–118 (cit. on pp. 12, 18).

[43] Roland Chrobok, Joachim Wahle and Michael Schreckenberg. 'Traffic forecast using simulations of large scale networks'. In: *Intelligent Transportation Systems, 2001. Proceedings. 2001 IEEE*. Aug. 2001, pp. 434–439. doi: `10.1109/ITSC.2001.948696`. url: `http://ieeexplore.ieee.org/xpl/articleDetails.jsp?arnumber=948696` (cit. on p. 18).

[44] Manuel Clavel, Francisco Durán, Steven Eker, Patrick Lincoln, Narciso Martí-Oliet, José Meseguer and J.F. Quesada. 'Maude: specification and programming in rewriting logic'. In: *Theoretical Computer Science* 285.2 (2002). Rewriting Logic and its Applications, pp. 187–243. issn: 0304-3975. doi: `10.1016/S0304-3975(01)00359-0`. url: `http://www.sciencedirect.com/science/article/pii/S0304397501003590` (cit. on pp. 128, 129).

[45] Manuel Clavel, Francisco Durán, Steven Eker, Patrick Lincoln, Narciso Martí-Oliet, José Meseguer and Carolyn Talcott. *All About Maude - A High-Performance Logical Framework. How to Specify, Program and Verify Systems in Rewriting Logic*. Vol. 4350. Lecture Notes in Computer Science. Springer Berlin Heidelberg, 2007, p. 794. isbn: 978-3-540-71940-3. doi: 10.1007/978-3-540-71999-1 (cit. on pp. 119, 121, 128, 129).

[46] Manuel Clavel, Francisco Durán, Steven Eker, Patrick Lincoln, Narciso Martí-Oliet, José Meseguer and Carolyn Talcott. *Maude Manual (Version 2.5)*. June 2010. url: http://maude.cs.uiuc.edu/maude2-manual/html/maude-manual.html (cit. on pp. 154, 168).

[47] Manuel Clavel, Francisco Durán, Steven Eker, Patrick Lincoln, Narciso Martí-Oliet, José Meseguer and Carolyn Talcott. 'The Maude 2.0 System'. In: *Rewriting Techniques and Applications*. Ed. by Robert Nieuwenhuis. Vol. 2706. Lecture Notes in Computer Science. Springer Berlin Heidelberg, 2003, pp. 76–87. isbn: 978-3-540-40254-1. doi: 10.1007/3-540-44881-0_7. url: http://dx.doi.org/10.1007/3-540-44881-0_7 (cit. on pp. 119, 121).

[48] Randal E. Bryant, Klaus Sutner and Mark J. Stehlik. *Introductory Computer Science Education at Carnegie Mellon University: A Deans' Perspective*. SCS TECHNICAL REPORT COLLECTION. Pittsburgh PA 15213-3891: School of Computer Science, Carnegie Mellon University, Aug. 2010. url: http://reports-archive.adm.cs.cmu.edu/anon/2010/CMU-CS-10-140.pdf (cit. on p. 127).

[49] Edgar F. Codd. 'A Data Base Sublanguage Founded on the Relational Calculus'. In: *Proceedings of the 1971 ACM SIGFIDET (Now SIGMOD) Workshop on Data Description, Access and Control*. SIGFIDET '71. San Diego, California: Association for Computing Machinery, 1971, pp. 35–68. doi: 10.1145/1734714.1734718. url: http://doi.acm.org/10.1145/1734714.1734718 (cit. on p. 130).

[50] Edgar F. Codd. 'Relational Database: A Practical Foundation for Productivity'. In: *Communications of the ACM* 25.2 (Feb. 1982), pp. 109–117. issn: 0001-0782. doi: 10.1145/358396.358400. url: http://doi.acm.org/10.1145/358396.358400 (cit. on pp. 18, 129).

[51] Gordon Duncan. *Paramics Technical Report: Car-Following, Lane-Changing and Junction Modelling*. Tech. rep. Paramics, 1998. url: http://www.paramics-online.com/downloads/technicaldocs/Quadstone%20Paramics%20Car%20Follow.pdf (cit. on p. 30).

[52] Florian Mazur, Thomas Zaksek and Johannes Brügmann. *Dynamic Level-Of-Service boundaries (in German)*. German title: Erzeugung von dynamischen Verkehrszustandsgrenzen für Online-Verkehrssimulationen. 2008. url: http://www.dpg-verhandlungen.de/2008/berlin/dy17.pdf (cit. on p. 94).

[53] D. Eastlake 3rd, J. Schiller and S. Crocker. *Randomness Requirements for Security*. Tech. rep. 4086. Internet Engineering Task Force, June 2005. url: http://www.ietf.org/rfc/rfc4086.txt (cit. on p. 25).

[54] Sebastian Egner. *SRFI-27 Sources of Random Bits*. Apr. 2002. url: http://srfi.schemers.org/srfi-27/srfi-27.html (cit. on p. 142).

[55] Steven Eker. *Maude Higher-Order Functions*. Jan. 2006. url: http://lists.cs.uiuc.edu/pipermail/maude-help/2006-January/000154.html (cit. on p. 143).

[56] Steven Eker, José Meseguer and Ambarish Sridharanarayanan. 'The Maude LTL Model Checker'. In: *Electronic Notes in Theoretical Computer Science* 71 (2004). WRLA 2002, Rewriting Logic and Its Applications, pp. 162–187. issn: 1571-0661. doi: 10.1016/S1571-0661(05)82534-4. url: http://www.sciencedirect.com/science/article/pii/S1571066105825344 (cit. on pp. 36, 124, 129, 154, 168, 197).

[57] Heike Emmerich and Ernst Rank. 'An improved cellular automaton model for traffic flow simulation'. In: *Physica A: Statistical Mechanics and its Applications* 234.3–4 (1997), pp. 676–686. issn: 0378-4371. doi: http://dx.doi.org/10.1016/S0378-4371(96)00310-X. url: http://www.sciencedirect.com/science/article/pii/S037843719600310X (cit. on p. 26).

[58] Jörg Esser and Michael Schreckenberg. 'Microscopic Simulation of Urban Traffic Based on Cellular Automata'. In: *International Journal of Modern Physics C* 08.05 (1997), pp. 1025–1036. doi: 10.1142/S0129183197000904. eprint: http://www.worldscientific.com/doi/pdf/10.1142/S0129183197000904. url: http://www.worldscientific.com/doi/abs/10.1142/S0129183197000904 (cit. on p. 18).

[59] Constantinos Antoniou, Jaume Barceló, Mark Brackstone, Hilmi Celikoglu, Biagio Ciuffo, Vincenzo Punzo, Pete Sykes, Tomer Toledo, Peter Vortisch and Peter Wagner. *Traffic Simulation: Case for guidelines*. Tech. rep. Publications Office of the European Union, JRC Institute for Energy and Transport, 2014. doi: 10.2788/11382. url: http://publications.jrc.ec.europa.eu/repository/handle/JRC88526 (cit. on pp. 34, 35).

[60] Marc Feeley. *SRFI-18 Multithreading Support*. Mar. 2001. url: http://srfi.schemers.org/srfi-18/srfi-18.html (cit. on p. 141).

[61] Martin Fellendorf and Peter Vortisch. 'Microscopic Traffic Flow Simulator VISSIM'. In: *Fundamentals of Traffic Simulation*. Ed. by Jaume Barceló. Vol. 145. International Series in Operations Research & Management Science. Springer New York, 2010, pp. 63–93. isbn: 978-1-4419-6141-9. doi: 10.1007/978-1-4419-6142-6_2. url: http://dx.doi.org/10.1007/978-1-4419-6142-6_2 (cit. on pp. 31, 32, 59).

[62] Paul A. Fishwick. 'A taxonomy for simulation modeling based on programming language principles'. In: *IIE Transactions* 30.9 (1998), pp. 811–820. doi: 10.1080/07408179808966527. eprint: http://dx.doi.org/10.1080/07408179808966527. url: http://dx.doi.org/10.1080/07408179808966527 (cit. on pp. 25, 39, 41).

[63] Will Fitzgerald. *SRFI-19 Time Data Types and Procedures*. Aug. 2000. url: http://srfi.schemers.org/srfi-19/srfi-19.html (cit. on p. 142).

[64] Linjie Gao, Zhicai Juan and Peng Jing. 'The design and implement of parallel simulation algorithm of dynamic route solution for traffic network'. In: *System Simulation and Scientific Computing, 2008. ICSC 2008. Asia Simulation Conference - 7th International Conference on*. Oct. 2008, pp. 230–234. doi: 10.1109/ASC-ICSC.2008.4675360 (cit. on p. 32).

[65] Yu Gao. 'Calibration and Comparison of the VISSIM and INTEGRATION Microscopic Traffic Simulation'. Thesis. Blacksburg, Virginia: Virginia Polytechnic Institute and State University, Sept. 2008. url: http://scholar.lib.vt.edu/theses/available/etd-09102008-135101/unrestricted/Yu_Gao_Thesis.pdf (cit. on p. 35).

[66] Qiao Ge and Monica Menendez. 'Sensitivity Analysis for Calibrating VISSIM in Modeling the Zurich Network'. In: *Proceedings of ?* Monte Verità / Ascona: 12th Swiss Transport Research Conference (STRC 2012), May 2012. url: http://www.ivt.ethz.ch/svt/publications/papers/svtr31.pdf (cit. on pp. 35, 199).

[67] Jim Gray et al. 'The transaction concept: Virtues and limitations'. In: *Proceedings of the Very Large Database Conference* (1981), pp. 144–154 (cit. on p. 130).

[68] PostgreSQL Development Group. *PostgreSQL*. 2011. url: http://www.postgresql.org (cit. on pp. 120, 139).

[69] Lars Christian Habel. *Calibration and Tuning Times and Costs*. Personel communication. Dec. 2014 (cit. on pp. 191, 193).

[70] Lars Christian Habel. 'Modelling, implementation, and validation of route dependant driving behaviour and asymetric lane changing for the model of Lee, Barlovic, and Pottmeier (in German)'. German title: Modellierung, Implementierung und Validierung von streckenabhängigem Fahrverhalten und asymmetrischen Fahrstreifenwechseln für das Modell von Lee, Barlović und Pottmeier. Thesis. Duisburg, Germany: University Duisburg-Essen, Apr. 2011 (cit. on pp. 9, 26, 29, 31, 49, 61, 64, 65, 72, 123, 134, 136, 147, 149, 165, 167–169, 196).

[71]   Lars Christian Habel and Michael Schreckenberg. 'Asymmetric Lane Change Rules for a Microscopic Highway Traffic Model'. In: *Cellular Automata*. Ed. by Jarosław Wąs, GeorgiosCh. Sirakoulis and Stefania Bandini. Vol. 8751. Lecture Notes in Computer Science. Springer International Publishing, 2014, pp. 620–629. isbn: 978-3-319-11519-1. doi: 10.1007/978-3-319-11520-7_66. url: http://dx.doi.org/10.1007/978-3-319-11520-7_66 (cit. on pp. 9, 29, 31, 61, 72, 167–169, 200).

[72]   Theo Haerder and Andreas Reuter. 'Principles of Transaction-oriented Database Recovery'. In: *ACM Computing Surveys (CSUR)* 15.4 (Dec. 1983), pp. 287–317. issn: 0360-0300. doi: 10.1145/289.291. url: http://doi.acm.org/10.1145/289.291 (cit. on p. 130).

[73]   Thomas C. Hales. 'Formal Proof'. In: *Notices of the AMS* 55.11 (2008), pp. 1370–1380. url: http://www.ams.org/notices/200811/tx081101370p.pdf (cit. on pp. 34, 124, 197).

[74]   Fred Hall, Sarah Wakefield and Ahmed Al-Kaisy. 'Freeway Quality of Service: What Really Matters to Drivers and Passengers?' In: *Transportation Research Record: Journal of the Transportation Research Board, Journal of the* 1776.01-2286 (Feb. 2007), pp. 17–23. issn: 0361-1981. doi: 10.3141/1776-03. url: http://www.sciencedirect.com/science/article/pii/S1369847814001260 (cit. on p. 5).

[75]   Michael Hanus. 'Multi-paradigm Declarative Languages'. In: *Logic Programming*. Ed. by Véronica Dahl and Ilkka Niemelä. Vol. 4670. Lecture Notes in Computer Science. Springer Berlin Heidelberg, 2007, pp. 45–75. isbn: 978-3-540-74608-9. doi: 10.1007/978-3-540-74610-2_5. url: http://dx.doi.org/10.1007/978-3-540-74610-2_5 (cit. on p. 119).

[76]   Robert Harper. *Teaching FP to freshmen*. Blog Existential Type. Accessed 2014-08-02. Mar. 2011. url: http://existentialtype.wordpress.com/2011/03/15/teaching-fp-to-freshmen/ (cit. on p. 127).

[77]   Martin Hauschild. 'Weather and Road Surface Information and Hazard Warnings: Data content acquisition through advanced probe vehicle systems'. In: *12th World Congress on Intelligent Transport Systems 2005*. Ed. by Intelligent Transportation Society of America. Vol. 3. San Francisco, California: Curran Associates, Inc. (Mar 2007), Nov. 2005, pp. 1609–1619. isbn: 9 781604 236354. url: https://origin.bmwgroup.com/bmwgroup_prod/publikationen/d/2003/pdf/Weather_and_road_surface_information_and_hazard_warnings_paper_en_2003.pdf (cit. on p. 28).

[78]   Carl H. Hauser. *Concurrent ML as a Discrete Event Simulation Language*. Tech. rep. Washington State University, Oct. 2006. url: http://www.eecs.wsu.edu/~hauser/Publications/CMLSim-TechReport.pdf (cit. on p. 141).

[79]   Dirk Helbing. 'Traffic and related self-driven many-particle systems'. In: *Reviews of Modern Physics* 73 (4 Dec. 2001), pp. 1067–1141. doi: 10.1103/RevModPhys.73.1067. url: http://link.aps.org/doi/10.1103/RevModPhys.73.1067 (cit. on pp. 7, 28).

[80]   Peter Hemmerle. 'Validation of two cellular automaton models using empirical travel time data (in German)'. German title: Validierung zweier Zellularautomatenmodelle anhand empirischer Reisezeitendaten. Thesis. Duisburg, Germany: University Duisburg-Essen, Nov. 2012 (cit. on pp. 29, 176, 206).

[81]   Marko Hofmann. 'On the Complexity of Parameter Calibration in Simulation Models'. In: *The Journal of Defense Modeling and Simulation: Applications, Methodology, Technology* 2.4 (2005), pp. 217–226. doi: 10.1177/154851290500200405. eprint: http://dms.sagepub.com/content/2/4/217.full.pdf+html. url: http://dms.sagepub.com/content/2/4/217.abstract (cit. on p. 36).

[82]   Robert Hranac, Vijay Kovvali, Richard Margiotta and Vassili Alexiadis. *Next Generation Simulation (NGSIM) Task E.2: High-Level Verification and Validation Plan*. Tech. rep. FHWA-HOP-06-010. U.S. Department of Transportation–Federal Highway Administrataion, July 2004. url: http://ngsim-community.org/NGSIM%20Task%20E2.pdf (cit. on p. 34).

[83] Torsten Huisinga, Robert Barlovic, Wolfgang Knospe, Andreas Schadschneider and Michael Schreckenberg. 'A microscopic model for packet transport in the Internet'. In: *Physica A: Statistical Mechanics and its Applications* 294.1–2 (2001), pp. 249–256. issn: 0378-4371. doi: `10.1016/S0378-4371(01)00107-8`. url: `http://www.sciencedirect.com/science/article/pii/S0378437101001078` (cit. on p. 159).

[84] Bruce Ikenaga. *Divisibility and the Division Algorithm*. June 2008. url: `http://www.millersville.edu/~bikenaga/number-theory/divisibility/divisibility.pdf` (cit. on p. 70).

[85] ISO 14819-2:2003(E). *Event and information codes for Radio Data System – Traffic Message Channel (RDS-TMC)*. International Organization for Standardization. ISO copyright office, Case postale 56, CH-1211 Geneva 20, June 2003 (cit. on pp. 20, 193).

[86] ISO 14819-3:2004. *Location referencing for ALERT-C*. International Organization for Standardization. ISO copyright office, Case postale 56, CH-1211 Geneva 20, Aug. 2004 (cit. on pp. 16, 19–22, 32, 188, 192, 193).

[87] *ISO/IEC 14882:1998 Information technology — Programming languages — C++*. International Organization for Standardization. ISO copyright office, Case postale 56, CH-1211 Geneva 20, 1st Sept. 1998. url: `http://www.iso.org/iso/catalogue_detail.htm?csnumber=25845` (cit. on p. 125).

[88] *ISO/IEC 14882:2011 Information technology — Programming languages — C++*. International Organization for Standardization. ISO copyright office, Case postale 56, CH-1211 Geneva 20, 28th Feb. 2011. url: `http://www.iso.org/iso/catalogue_detail.htm?csnumber=50372` (cit. on pp. 125–127, 137).

[89] *ISO/IEC 14882:1999 Information technology — Programming languages — C*. International Organization for Standardization. ISO copyright office, Case postale 56, CH-1211 Geneva 20, 1st Dec. 1999. url: `http://www.iso.org/iso/home/store/catalogue_tc/catalogue_detail.htm?csnumber=29237` (cit. on p. 125).

[90] *ISO/IEC 14882:2011 Information technology — Programming languages — C*. International Organization for Standardization. ISO copyright office, Case postale 56, CH-1211 Geneva 20, 8th Dec. 2011. url: `http://www.iso.org/iso/home/store/catalogue_ics/catalogue_detail_ics.htm?csnumber=57853` (cit. on p. 125).

[91] Aubrey Jaffer. *SRFI-60 Integers as Bits*. Jan. 2005. url: `http://srfi.schemers.org/srfi-60/srfi-60.html` (cit. on p. 138).

[92] Leila Jalali, Carolyn Talcott, Nalini Venkatasubramanian and Sharad Mehrotra. 'Formal Specification of Multisimulations Using Maude'. In: *Proceedings of the 2012 Symposium on Theory of Modeling and Simulation - DEVS Integrative M&S Symposium*. TMS/DEVS '12. Orlando, Florida: Society for Computer Simulation International, 2012, 22:1–22:8. isbn: 978-1-61839-786-7. url: `http://dl.acm.org/citation.cfm?id=2346616.2346638` (cit. on p. 13).

[93] Panu Kalliokoski. *SRFI-69 Basic hash tables*. Sept. 2005. url: `http://srfi.schemers.org/srfi-69/srfi-69.html` (cit. on p. 139).

[94] Oliver Kaumann, Kai Froese, Roland Chrobok, Joachim Wahle, Lutz Neubert and Michael Schreckenberg. 'On-Line Simulation of the Freeway Network of North Rhine-Westphalia'. In: *Traffic and Granular Flow '99*. Ed. by Dirk Helbing, Hans J. Herrmann, Michael Schreckenberg and Dietrich E. Wolf. Springer Berlin Heidelberg, 2000, pp. 351–356. isbn: 978-3-642-64109-1. doi: `10.1007/978-3-642-59751-0_34` (cit. on pp. 33, 95, 97).

[95] Rupinder Kaur and Jyotsna Sengupta. 'Software Process Models and Analysis on Failure of Software Development Projects'. In: *International Journal of Scientific and Engineering Research* abs/1306.1068 (2011). url: `http://www.ijser.org/onlineResearchPaperViewer.aspx?Software_Process_Models_and_Analysis_on_Failure_of_Software_Development_Projects.pdf` (cit. on p. 121).

[96]   Richard Kelsey. *SRFI-9 Defining Record Types*. Sept. 1999. url: `http://srfi.schemers.org/srfi-9/srfi-9.html` (cit. on p. 135).

[97]   Richard Kelsey, William Clinger, Jonathan Rees, Robert Bruce Findler and Jacob Matthews. 'Revised[6] Report on the Algorithmic Language Scheme'. In: *Journal of Functional Programming* 19.S1 (26th Sept. 2007). Ed. by Michael Sperber, Kent R. Dybvig, Matthew Flatt and Anton van Straaten, pp. 1–301 (cit. on p. 125).

[98]   Richard Kelsey, William Clinger, Jonathan Rees, Robert Bruce Findler and Jacob Matthews. 'Revised[6] Report on the Algorithmic Language Scheme–Standard Library'. In: *Journal of Functional Programming* 19.S1 (26th Sept. 2007). Ed. by Michael Sperber, Kent R. Dybvig, Matthew Flatt and Anton van Straaten, pp. 1–301 (cit. on pp. 125, 136–138).

[99]   Richard Kelsey, Jonathan Rees, Mike Sperber, Marcus Crestani, Robert Ransom, Roderic Morris, Marcel Turino and Martin Gasbichler. *The Incomplete Scheme 48 Reference Manual for release 1.9.2*. reference version 1.8.0. Apr. 2014. url: `http://s48.org/1.9.2/manual/manual.html` (cit. on pp. 136, 137).

[100]   Boris S. Kerner. 'Criticism of generally accepted fundamentals and methodologies of traffic and transportation theory: A brief review'. In: *Physica A: Statistical Mechanics and its Applications* 392.21 (2013), pp. 5261–5282. issn: 0378-4371. doi: `10.1016/j.physa.2013.06.004`. url: `http://www.sciencedirect.com/science/article/pii/S0378437113004986` (cit. on pp. 28–30).

[101]   Boris S. Kerner. *Introduction to Modern Traffic Flow Theory and Control–The Long Road to Three-Phase Traffic Theory*. Springer Berlin Heidelberg, 2009, p. 265. isbn: 978-3-642-02605-8 (cit. on p. 193).

[102]   Boris S. Kerner, Hubert Rehborn, Mario Aleksic and Andreas Haug. 'Recognition and tracking of spatial–temporal congested traffic patterns on freeways'. In: *Transportation Research Part C: Emerging Technologies* 12.5 (2004), pp. 369–400. issn: 0968-090X. doi: `10.1016/j.trc.2004.07.015`. url: `http://www.sciencedirect.com/science/article/pii/S0968090X04000270` (cit. on pp. 5, 14).

[103]   Brian W. Kernighan and Dennis M. Ritchie. *The C Programming Language*. 2nd ed. Prentice-Hall, 1988, p. 318. isbn: 0-13-110362-8 (cit. on p. 124).

[104]   Oleg Kiselyov. 'Subclassing errors, OOP, and practically checkable rules to prevent them'. In: *Computing Research Repository* cs.PL/0301032 (2003). url: `http://arxiv.org/abs/cs.PL/0301032` (cit. on pp. 125, 126, 145).

[105]   Lawrence A. Klein, Milton K. Mills and David R.P. Gibson. *Traffic Detector Handbook: Third Edition—Volume I*. Tech. rep. FHWA-HRT-06-108. Turner-Fairbank Highway Research Center, 6300 Georgetown Pike, HRDO-04, Room No. T204, McLean, VA 22101-2296: U.S. Department of Transportation, Federal Highway Administration, Oct. 2006. url: `http://www.fhwa.dot.gov/publications/research/operations/its/06108/06108.pdf` (cit. on pp. 14, 27).

[106]   Lawrence A. Klein, Milton K. Mills and David R.P. Gibson. *Traffic Detector Handbook: Third Edition—Volume II*. Tech. rep. FHWA-HRT-06-139. Turner-Fairbank Highway Research Center, 6300 Georgetown Pike, HRDO-04, Room No. T204, McLean, VA 22101-2296: U.S. Department of Transportation, Federal Highway Administration, Oct. 2006. url: `http://www.fhwa.dot.gov/publications/research/operations/its/06139/06139.pdf` (cit. on pp. 14, 27).

[107]   Anthony Klug. 'Equivalence of Relational Algebra and Relational Calculus Query Languages Having Aggregate Functions'. In: *J. ACM* 29.3 (July 1982), pp. 699–717. issn: 0004-5411. doi: `10.1145/322326.322332`. url: `http://doi.acm.org/10.1145/322326.322332` (cit. on p. 130).

[108] Martina Klukas-Illen and Ralf E. Weiss. *Traffic control on German highways in Northrhine-Westphalia (in German)*. Tech. rep. German: Verkehrsbeeinflussung auf BAB in Nordrhein-Westfalen, Lastenheft Unterzentrale. Heusch/Boesefeldt GmbH Aachen, Aug. 2001 (cit. on pp. 93, 94).

[109] Florian Knorr. 'Applicability and Application of Microscopic Traffic Simulations'. PhD thesis. Duisburg, Germany: University Duisburg-Essen, Feb. 2013. url: http://duepublico.uni-duisburg-essen.de/servlets/DocumentServlet?id=31414 (cit. on pp. 7, 23, 29, 31, 61).

[110] Florian Knorr, Thorsten Chmura and Michael Schreckenberg. 'Route choice in the presence of a toll road: The role of pre-trip information and learning'. In: *Transportation Research Part F: Traffic Psychology and Behaviour* 27, Part A (2014), pp. 44–55. issn: 1369-8478. doi: 10.1016/j.trf.2014.09.003. url: http://www.sciencedirect.com/science/article/pii/S1369847814001260 (cit. on pp. 4, 5, 188).

[111] Florian Knorr and Michael Schreckenberg. 'Counting the corners of a random walk and its application to traffic flow'. In: *Journal of Physics A: Mathematical and General* 45.31 (2012), p. 315001. doi: doi:10.1088/1751-8113/45/31/315001. url: http://stacks.iop.org/1751-8121/45/i=31/a=315001 (cit. on p. 28).

[112] Florian Knorr and Michael Schreckenberg. 'On the reproducibility of spatiotemporal traffic dynamics with microscopic traffic models'. In: *Journal of Statistical Mechanics: Theory and Experiment* 2012.10 (2012), P10018. url: http://stacks.iop.org/1742-5468/2012/i=10/a=P10018 (cit. on pp. 21, 23, 26, 29–31, 33).

[113] Florian Knorr and Michael Schreckenberg. 'The comfortable driving model revisited: traffic phases and phase transitions'. In: *Journal of Statistical Mechanics: Theory and Experiment* 2013.07 (2013), P07002. doi: 10.1088/1742-5468/2013/07/P07002. url: http://stacks.iop.org/1742-5468/2013/i=07/a=P07002 (cit. on pp. 23, 29).

[114] Florian Knorr, Thomas Zaksek, Johannes Brügmann and Michael Schreckenberg. 'Statistical Analysis of High-Flow Traffic States'. In: *Traffic and Granular Flow '13*. Ed. by Mohcine Chraibi, Maik Boltes, Andreas Schadschneider and Armin Seyfried. Forschungszentrum Jülich. 52425 Jülich, Germany: Springer International Publishing Switzerland, Sept. 2013, pp. 557–562. isbn: 978-3-319-10628-1. doi: 10.1007/978-3-319-10629-8_62. url: http://my.arxiv.org/arxiv/FilterServlet/pdf/1312.3589v1 (cit. on pp. 27, 30, 130, 173, 205).

[115] Wolfgang Knospe, Ludger Santen, Andreas Schadschneider and Michael Schreckenberg. 'A realistic two-lane traffic model for highway traffic'. In: *Journal of Physics A: Mathematical and General* 35.15 (2002), p. 3369. doi: 10.1088/0305-4470/35/15/302. url: http://stacks.iop.org/0305-4470/35/i=15/a=302 (cit. on pp. 21, 25, 29, 31, 43, 45, 60, 73, 143).

[116] Wolfgang Knospe, Ludger Santen, Andreas Schadschneider and Michael Schreckenberg. 'Disorder effects in cellular automata for two-lane traffic'. In: *Physica A: Statistical Mechanics and its Applications* 265.3–4 (1999), pp. 614–633. issn: 0378-4371. doi: 10.1016/S0378-4371(98)00565-2. url: http://www.sciencedirect.com/science/article/pii/S0378437198005652 (cit. on p. 26).

[117] Wolfgang Knospe, Ludger Santen, Andreas Schadschneider and Michael Schreckenberg. 'Empirical test for cellular automaton models of traffic flow'. In: *Physical Review E* 70 (1 July 2004), p. 016115. doi: 10.1103/PhysRevE.70.016115. url: http://link.aps.org/doi/10.1103/PhysRevE.70.016115 (cit. on pp. 23, 26, 28, 29).

[118] Wolfgang Knospe, Ludger Santen, Andreas Schadschneider and Michael Schreckenberg. 'Towards a realistic microscopic description of highway traffic'. In: *Journal of Physics A: Mathematical and General* 33.48 (2000), p. L477. url: http://stacks.iop.org/0305-4470/33/i=48/a=103 (cit. on pp. 25, 29, 45, 60, 61, 65, 203).

[119] Daniel Krajzewicz, Jakob Erdmann, Michael Behrisch and Laura Bieker. *SUMO User Document-ation*. Ed. by Institute of Transportation Systems at the German Aerospace Center. 2014. url: http://sumo-sim.org/userdoc/index.html (cit. on pp. 48, 77, 100).

[120] Daniel Krajzewicz, Georg Hertkorn, Julia Ringel and Peter Wagner. 'Preparation of Digital Maps for Traffic Simulation; Part 1: Approach and Algorithms'. In: *3rd Industrial Simulation Confer-ence 2005*. Ed. by J. Krüger, A. Lisounkin and G. Schreck. EUROSIS-ETI, June 2005, pp. 285–290. url: http://elib.dlr.de/21013/ (cit. on p. 32).

[121] Axel van Lamsweerde. 'Formal Specification: A Roadmap'. In: *Proceedings of the Conference on The Future of Software Engineering*. ICSE '00. Limerick, Ireland: Association for Computing Machinery, 2000, pp. 147–159. isbn: 1-58113-253-0. doi: 10.1145/336512.336546. url: http://doi.acm.org/10.1145/336512.336546 (cit. on p. 122).

[122] Pierre L'Ecuyer and Serge Côté. 'Implementing a Random Number Package with Splitting Fa-cilities'. In: *ACM Transactions on Mathematical Software* 17.1 (Mar. 1991), pp. 98–111. issn: 0098-3500. doi: 10.1145/103147.103158. url: http://doi.acm.org/10.1145/103147.103158 (cit. on p. 142).

[123] Pierre L'Ecuyer, Boris Oreshkin and Richard Simard. 'Random Numbers for Parallel Computers: Requirements and Methods'. In: *Mathematics and Computers in Simulation* (2014). url: http://www.iro.umontreal.ca/~lecuyer/myftp/papers/parallel-rng-imacs.pdf (cit. on pp. 56, 142, 154).

[124] Hyun Keun Lee, Robert Barlovic, Michael Schreckenberg and Doochul Kim. 'Mechanical Re-striction versus Human Overreaction Triggering Congested Traffic States'. In: *Physical Review Letters* 92 (23 June 2004), p. 238702. doi: 10.1103/PhysRevLett.92.238702. url: http://link.aps.org/doi/10.1103/PhysRevLett.92.238702 (cit. on pp. 9, 29–31, 36, 61–64, 66, 157, 167–169, 199).

[125] Jong-Keun Lee, Ye-Hwan Lim and Sung-Do Chi. 'Hierarchical Modeling and Simulation Envir-onment for Intelligent Transportation Systems'. In: *SIMULATION* 80.2 (2004), pp. 61–76. doi: 10.1177/0037549704042860. eprint: http://sim.sagepub.com/content/80/2/61.full.pdf+html. url: http://sim.sagepub.com/content/80/2/61.abstract (cit. on p. 26).

[126] Ken Ka-Yin Lee, Wai-Choi Tang and Kup-Sze Choi. 'Alternatives to relational database: Com-parison of NoSQL and XML approaches for clinical data storage'. In: *Computer Methods and Programs in Biomedicine* 110.1 (2013), pp. 99–109. issn: 0169-2607. doi: 10.1016/j.cmpb.2012.10.018. url: http://www.sciencedirect.com/science/article/pii/S0169260712002805 (cit. on p. 130).

[127] Hong Liu, Ping Li, Bojian Li and Yu Wen. 'Lane-Based Network Data Model for Micro-Traffic Flow Simulation'. In: *Intelligent Control and Automation, 2006. WCICA 2006. The Sixth World Congress on*. Vol. 2. 2006, pp. 8635–8639. doi: 10.1109/WCICA.2006.1713666 (cit. on p. 32).

[128] John W. Lloyd. 'Practical Advtanages of Declarative Programming'. In: *1994 Joint Conference on Declarative Programming, GULP-PRODE'94*. Ed. by María Alpuente, Roberto Barbuti and Isidro Ramos. Vol. 1. Peñiscola, Spain, Sept. 1994, pp. 18–30 (cit. on pp. 124, 126, 127).

[129] Emmanuel López-Neri, Antonio Ramírez-Treviño and Ernesto López-Mellado. 'A modeling framework for urban traffic systems microscopic simulation'. In: *Simulation Modelling Prac-tice and Theory* 18.8 (2010). Modeling and Simulation for Complex System Development, pp. 1145–1161. issn: 1569-190X. doi: 10.1016/j.simpat.2009.09.007. url: http://www.sciencedirect.com/science/article/pii/S1569190X09001312 (cit. on p. 26).

[130] Stefan Maier. *Update optimization*. Personel communication. 2012 (cit. on p. 145).

[131] Marek Małowidzki, Michał Mazur, Tomasz Dalecki and Przemysław Bereziński. 'Traffic Routes for Emergency Services'. In: *EUROSIM 2013 – 8th EUROSIM Congress on Modelling and Simulation*. Ed. by Khalid Al-Begain, David Al-Dabass, Alessandra Orsoni, Richard Cant and Richard Zobel. EUROSIM. Cardiff, Wales, UK: The Institute of Electrical and Electronics Engineers, Inc., Sept. 2013, pp. 466–471. isbn: 978-0-7695-5073-2. doi: 10.1109/EUROSIM.2013.119 (cit. on pp. 13, 44, 166, 168, 193).

[132] Linda Mannila and Michael de Raadt. 'An Objective Comparison of Languages for Teaching Introductory Programming'. In: *Proceedings of the 6th Baltic Sea Conference on Computing Education Research: Koli Calling 2006*. Baltic Sea '06. Uppsala, Sweden: Association for Computing Machinery, 2006, pp. 32–37. doi: 10.1145/1315803.1315811. url: http://doi.acm.org/10.1145/1315803.1315811 (cit. on p. 124).

[133] Sigurður Freyr Marinósson, Roland Chrobok, Andreas Pottmeier, Joachim Wahle and Michael Schreckenberg. 'Simulation Framework for the Autobahn Traffic in North Rhine-Westphalia'. In: *Cellular Automata*. Ed. by Stefania Bandini, Bastien Chopard and Marco Tomassini. Vol. 2493. Lecture Notes in Computer Science. Springer Berlin Heidelberg, 2002, pp. 315–324. isbn: 978-3-540-44304-9. doi: 10.1007/3-540-45830-1_30 (cit. on p. 18).

[134] George Marsaglia. *The Marsaglia Random Number CDROM including the Diehard Battery of Tests of Randomness*. last visited 2014-08-14. url: http://www.stat.fsu.edu/pub/diehard/ (cit. on p. 142).

[135] Narciso Martí-Oliet, Miguel Palomino and Alberto Verdejo. 'A Tutorial on Specifying Data Structures in Maude'. In: *Electronic Notes in Theoretical Computer Science* 137.1 (2005), pp. 105–132. issn: 1571-0661. doi: http://dx.doi.org/10.1016/j.entcs.2005.01.041. url: http://www.sciencedirect.com/science/article/pii/S157106610505084X (cit. on pp. 119, 121).

[136] Akhilesh Kumar Maurya and Partha Chakroborty. 'Microscopic Model for Simulation of Uninterrupted Mixed Traffic Streams Without Lane Discipline'. In: *International conference on Best Practices to Relieve Congestion on Mixed-Traffic Urban Streets in Developing Countries*. IIT Madras, Chennai, India, Sept. 2008, pp. 165–175 (cit. on pp. 72, 146, 168, 193).

[137] Adolf Darlington May. *Traffic Flow Fundamentals*. Prentice-Hall, 1990, p. 464. isbn: 0 13 926072 2 (cit. on p. 31).

[138] Florian Mazur, Roland Chrobok, Sigudur F. Hafstein, Andreas Pottmeier and Michael Schreckenberg. 'Future of Traffic Information – Online-Simululation of a Large Scale Freeway Network'. In: *IADIS International Conference WWW/Internet 2004*. Vol. 1. Madrid, Spain, 2004, pp. 665–672 (cit. on p. 29).

[139] Florian Mazur, Roland Chrobok, Sigurður Freyr Hafstein, Daniel Pottmeier Andreas Weber and Michael Schreckenberg. 'Modelling of Entries and Lane-Blocks for Up-to-Date Traffic Information Systems'. In: *Modelling, Simulation, and Optimization (MSO 2005)*. Ed. by G. Tonella. Vol. 1. Oranjestad, Aruba, Aug. 2005, p. 333. url: http://www.actapress.com/Abstract.aspx?paperId=21575 (cit. on pp. 12, 15, 18).

[140] Linda McIver and Damian Conway. 'Seven Deadly Sins of Introductory Programming Language Design'. In: *Proceedings of the 1996 International Conference on Software Engineering: Education and Practice (SE:EP '96)*. SEEP '96. Washington, DC, USA: The Institute of Electrical and Electronics Engineers, Inc., 1996, pp. 309–316. isbn: 0-8186-7379-6. url: http://dl.acm.org/citation.cfm?id=829500.829930 (cit. on p. 124).

[141] Doug Merritt. *Von Neumann Bottleneck*. Sept. 2009. url: http://c2.com/cgi/wiki?VonNeumannBottleneck (cit. on p. 126).

[142] Matthew Might, Yannis Smaragdakis and David Van Horn. 'Resolving and Exploiting the k-CFA Paradox: Illuminating Functional vs. Object-oriented Program Analysis'. In: *SIGPLAN Not.* 45.6 (June 2010), pp. 305–315. issn: 0362-1340. doi: 10.1145/1809028.1806631. url: http://arxiv.org/pdf/1311.4231 (cit. on p. 126).

[143]   Scott G. Miller. *SISC — Second Interpreter of Scheme Code*. Feb. 2007. url: `http://sisc-scheme.org` (cit. on p. 123).

[144]   Iain Milne and Glenn Rowe. 'Difficulties in Learning and Teaching Programming—Views of Students and Tutors'. In: *Education and Information Technologies* 7.1 (2002), pp. 55–66. issn: 1360-2357. doi: `10.1023/A:1015362608943`. url: `http://dx.doi.org/10.1023/A:1015362608943` (cit. on p. 124).

[145]   Ministerium für Bauen, Wohnen, Stadtentwicklung und Verkehr des Landes Nordrhein-Westfalen. *Mobility in North-Rhine Westphalia (in German)*. Tech. rep. Mobilität in Nordrhein-Westfalen: Daten und Fakten 2011. Jürgensplatz 1, 40219 Düsseldorf, Germany: Ministerium für Bauen, Wohnen, Stadtentwicklung und Verkehr des Landes Nordrhein-Westfalen, 2011. url: `http://www.mbwsv.nrw.de/verkehr/strasse/Strassenverkehr/Daten_und_Fakten/Daten_und_Fakten_2011.pdf` (cit. on pp. 4, 13).

[146]   Ministerium für Wirtschaft, Energie, Bauen, Wohnen und Verkehr des Landes Nordrhein-Westfalen and University Duisburg-Essen. *Traffic information for highways in North-Rhine Westphalia*. Ed. by Ministerium für Wirtschaft, Energie, Bauen, Wohnen und Verkehr des Landes Nordrhein-Westfalen. 2014. url: `http://www.autobahn.nrw.de` (cit. on pp. 4, 12, 187, 188).

[147]   Kai Nagel and Michael Schreckenberg. 'A Cellular Automaton Model for Freeway Traffic'. In: *Journal de Physique I France* 2.12 (Dec. 1992), pp. 2221–2229. doi: `10.1051/jp1:1992277`. url: `https://www.uni-due.de/imperia/md/content/ptt/paper/1992_origca.pdf` (cit. on pp. 25, 29, 43, 65, 167, 173, 203).

[148]   Kai Nagel, Peter Wagner and Richard Woesler. 'Still Flowing: Approaches to Traffic Flow and Traffic Jam Modeling'. In: *Operations Research* 51.5 (2003), pp. 681–710. doi: `10.1287/opre.51.5.681.16755`. eprint: `http://pubsonline.informs.org/doi/pdf/10.1287/opre.51.5.681.16755`. url: `http://pubsonline.informs.org/doi/abs/10.1287/opre.51.5.681.16755` (cit. on pp. 28, 31).

[149]   Kai Nagel, Dietrich E. Wolf, Peter Wagner and Patrice Simon. 'Two-lane traffic rules for cellular automata: A systematic approach'. In: *Physical Review E* 58 (Aug. 1998), pp. 1425–1437. doi: `10.1103/PhysRevE.58.1425`. eprint: `cond-mat/9712196` (cit. on pp. 29, 167).

[150]   Lutz Neubert. 'Statistische Analyse von Verkehrsdaten und die Modellierung von Verkehrsfluss mittels zellularer Automaten'. PhD thesis. Duisburg, Germany: University Duisburg-Essen, 2000. url: `https://portal.d-nb.de/opac.htm?method=showFullRecord&currentResultId=Statistische+Analyse+von+Verkehrsdaten+und+die+Modellierung+von+Verkehrsfluss+mittels+zellularer+Automaten+&any&currentPosition=0` (cit. on p. 173).

[151]   Lutz Neubert, Ludger Santen, Andreas Schadschneider and Michael Schreckenberg. 'Single-vehicle data of highway traffic: A statistical analysis'. In: *Physical Review E* 60 (6 Dec. 1999), pp. 6480–6490. doi: `10.1103/PhysRevE.60.6480`. url: `http://link.aps.org/doi/10.1103/PhysRevE.60.6480` (cit. on p. 31).

[152]   Richard S. Norman. *Massively-parallel processor array with outputs from individual processors directly to an external device without involving other processors or a common physical carrier*. US Patent 5,801,715. Sept. 1998. url: `http://www.google.com/patents/US5801715` (cit. on p. 126).

[153]   William Louis Oberkampf and Timothy Guy Trucano. *Verification and validation benchmarks*. Tech. rep. SAND2007-0853. Sandia National Laboratories, Feb. 2007. doi: `10.2172/901974`. url: `http://www.osti.gov/scitech/servlets/purl/901974` (cit. on pp. 34, 35, 123, 131).

[154] William Louis Oberkampf and Timothy Guy Trucano. 'Verification and validation benchmarks.' In: *Nuclear Engineering and Design* 238.3 (2008). Benchmarking of CFD Codes for Application to Nuclear Reactor Safety, pp. 716–743. issn: 0029-5493. doi: `10.1016/j.nucengdes.2007.02.032`. url: `http://www.sciencedirect.com/science/article/pii/S0029549307003548` (cit. on pp. 34, 35, 123, 131, 197, 202).

[155] William Louis Oberkampf, Timothy Guy Trucano and Charles Hirsch. 'Verification, validation, and predictive capability in computational engineering and physics, SAND2003-3769'. In: *Applied Mechanics Reviews* 57.SAND2003-3769 (Dec. 2004), pp. 345–384. doi: `10.1115/1.1767847` (cit. on p. 34).

[156] Johan Janson Olstam and Andreas Tapani. *Comparison of Car-following models*. Tech. rep. E2213. Development of the VTI rural traffic simulation model and Simulated traffic for the VTI driving simulator. SE-581 95 Linköping Sweden: VTI meddelande 960A, 2004, 36+2 Appendices. url: `http://www.vti.se/en/publications/pdf/comparison-of-car-following-models.pdf` (cit. on p. 30).

[157] Kunle Olukotun and Lance Hammond. 'The Future of Microprocessors'. In: *Queue* 3.7 (Sept. 2005), pp. 26–29. issn: 1542-7730. doi: `10.1145/1095408.1095418`. url: `http://doi.acm.org/10.1145/1095408.1095418` (cit. on p. 126).

[158] Peter Csaba Ölveczky. 'Semantics, Simulation, and Formal Analysis of Modeling Languages for Embedded Systems in Real-Time Maude'. In: *Formal Modeling: Actors, Open Systems, Biological Systems*. Ed. by Gul Agha, Olivier Danvy and José Meseguer. Vol. 7000. Lecture Notes in Computer Science. Springer Berlin Heidelberg, 2011, pp. 368–402. isbn: 978-3-642-24932-7. doi: `10.1007/978-3-642-24933-4_19`. url: `http://dx.doi.org/10.1007/978-3-642-24933-4_19` (cit. on pp. 36, 124).

[159] Pahio. *Empty sum*. 2009. url: `http://planetmath.org/emptysum` (cit. on p. 71).

[160] David A. Patterson and John L. Hennessy. *Computer Architecture – A Quantitative Approach*. 2nd ed. San Francisco, California: Morgan Kaufmann Publishers, Inc., 1996. isbn: 1-55860-329-8 (cit. on pp. 125, 126).

[161] Simon Peyton Jones, Roman Leshchinskiy, Gabriele Keller and Manuel M. T. Chakravarty. 'Harnessing the Multicores: Nested Data Parallelism in Haskell'. In: *IARCS Annual Conference on Foundations of Software Technology and Theoretical Computer Science*. Ed. by Ramesh Hariharan, Madhavan Mukund and V Vinay. Vol. 2. Leibniz International Proceedings in Informatics (LIPIcs). Dagstuhl, Germany: Schloss Dagstuhl–Leibniz-Zentrum fuer Informatik, 2008, pp. 383–414. isbn: 978-3-939897-08-8. doi: `10.4230/LIPIcs.FSTTCS.2008.1769`. url: `http://drops.dagstuhl.de/opus/volltexte/2008/1769` (cit. on p. 141).

[162] Fedor Pikus. *Safe Labels in C++*. online resource. Oct. 2007. url: `http://www.artima.com/cppsource/safelabels.html` (cit. on p. 138).

[163] Zoltán Porkoláb, József Mihalicza and Ádám Sipos. 'Debugging C++ Template Metaprograms'. In: *Proceedings of the 5th International Conference on Generative Programming and Component Engineering*. GPCE '06. Portland, Oregon, USA: Association for Computing Machinery, 2006, pp. 255–264. isbn: 1-59593-237-2. doi: `10.1145/1173706.1173746`. url: `http://doi.acm.org/10.1145/1173706.1173746` (cit. on p. 126).

[164] Andreas Pottmeier. 'Realistic Cellular Automaton Model for Synchronized Two-Lane Traffic'. PhD thesis. Duisburg, Germany: University Duisburg-Essen, 2007. url: `http://d-nb.info/987587420` (cit. on pp. 26–29, 31, 61, 64, 72, 89, 91, 145, 159, 167–169, 173, 200).

[165] Andreas Pottmeier, Roland Chrobok, Sigurður Freyr Marinósson, Florian Mazur and Michael Schreckenberg. 'OLSIM: Up-to-date traffic information on the web.' In: *Communications, Internet, and Information Technology*. Ed. by M. H. Hamza. International Association of Science and Technology for Development (IASTED). IASTED/ACTA Press, 2004, pp. 572–577. url: `http://dblp.uni-trier.de/db/conf/ciit/ciit2004.html#PottmeierCHMS04` (cit. on p. 12).

[166] Andreas Pottmeier, Roland Chrobok, Joachim Wahle and Michael Schreckenberg. 'On-Line Sim-
       ulation of Large Scale Networks'. In: *Selected Papers of the International Conference on Operations
       Research (OR 2001)*. Ed. by Peter Chamoni, Rainer Leisten, Alexander Martin, Joachim Min-
       nemann and Hartmut Stadtler. Vol. XVI. Operations Research Proceedings, Conference 2001.
       German Operations Research Society (GOR). Duisburg, Germany: Springer Berlin Heidelberg,
       Sept. 2001, p. 501 (cit. on p. 12).

[167] Andreas Pottmeier, Christian Thiemann, Andreas Schadschneider and Michael Schrecken-
       berg. 'Mechanical Restriction Versus Human Overreaction: Accident Avoidance and Two-Lane
       Traffic Simulations'. In: *Traffic and Granular Flow'05*. Ed. by Andreas Schadschneider, Thorsten
       Pöschel, Reinhart Kühne, Michael Schreckenberg and Dietrich E. Wolf. Springer Berlin Heidel-
       berg, 2007, pp. 503–508. isbn: 978-3-540-47640-5. doi: 10.1007/978-3-540-47641-
       2_46 (cit. on p. 31).

[168] ProofWiki. *Division Theorem*. Feb. 2014. url: http://www.proofwiki.org/wiki/
       Division_Theorem (cit. on p. 70).

[169] Richard Dowling, Alexander Skabardonis and Vassili Alexiadis. *Traffic Analysis Toolbox Volume
       III: Guidelines for Applying Traffic Microsimulation Modeling Software*. Ed. by CA 94612 Dowl-
       ing Associates Inc. 180 Grand Avenue Suite 250 Oakland. FHWA COTR: John Halkias, Office of
       Transportation Management. U.S. Department of Transportation, Federal Highway Administra-
       tion. Research, Development, and Technology, Turner-Fairbank Highway Research Center, 6300
       Georgetown Pike, McLean, VA 22101-2296: National Technical Information Service Springfield
       VA 22161, July 2004. url: http://ops.fhwa.dot.gov/trafficanalysistools/tat_
       vol3/Vol3_Guidelines.pdf (cit. on pp. 34, 35, 199).

[170] Atanas Radenski, Jeff Furlong and Vladimir Zanev. 'The Java 5 generics compromise ortho-
       gonality to keep compatibility'. In: *Journal of Systems and Software* 81.11 (2008), pp. 2069–
       2078. issn: 0164-1212. doi: 10.1016/j.jss.2008.04.008. url: http://www.
       sciencedirect.com/science/article/pii/S0164121208000848 (cit. on pp. 124,
       137).

[171] Nikolaus Rajewsky, Ludger Santen, Andreas Schadschneider and Michael Schreckenberg. 'The
       asymmetric exclusion process: Comparison of Update Procedures'. In: *Journal of Statistical Phys-
       ics* 92 (1–2 1998), pp. 151–194. issn: 0022-4715. doi: 10.1023/A:1023047703307 (cit. on
       pp. 26, 35, 159).

[172] Nikolaus Rajewsky and Michael Schreckenberg. 'Exact results for one-dimensional cellu-
       lar automata with different types of updates'. In: *Physica A: Statistical Mechanics and its
       Applications* 245.1–2 (1997), pp. 139–144. issn: 0378-4371. doi: 10.1016/S0378-
       4371(97)00010-1. url: http://www.sciencedirect.com/science/article/
       pii/S0378437197000101 (cit. on pp. 26, 35).

[173] Jens Christian Refsgaard and Hans Jørgen Henriksen. 'Modelling guidelines—terminology and
       guiding principles'. In: *Advances in Water Resources* 27.1 (2004), pp. 71–82. issn: 0309-1708.
       doi: 10.1016/j.advwatres.2003.08.006. url: http://www.sciencedirect.com/
       science/article/pii/S0309170803001489 (cit. on p. 35).

[174] John H. Reppy. 'CML: A Higher Concurrent Language'. In: *ACM SIGPLAN Notices* 26.6 (May
       1991), pp. 293–305. issn: 0362-1340. doi: 10.1145/113446.113470. url: http://doi.
       acm.org/10.1145/113446.113470 (cit. on pp. 128, 141).

[175] John H. Reppy. *Concurrent Programming in ML*. ISBN-13: 9780521480895. Cambridge Univer-
       sity Press, Aug. 1999, p. 324. isbn: 0521480892 (cit. on p. 141).

[176] John Reppy, Claudio V. Russo and Yingqi Xiao. 'Parallel Concurrent ML'. In: *ACM SIGPLAN
       Notices* 44.9 (Aug. 2009), pp. 257–268. issn: 0362-1340. doi: 10.1145/1631687.1596588.
       url: http://doi.acm.org/10.1145/1631687.1596588 (cit. on p. 141).

[177] Christian Roszak. *Uniform data center software for traffic data centers (in German)*. Ed. by Geschäftsstelle NERZ e. V. Einheitliche Rechnerzentralensoftware für Verkehrsrechnerzentralen. Feb. 2014. url: `http://www.nerz-ev.de/www.nerz-ev.de/home.php` (cit. on pp. 15, 16, 18, 20, 33, 123, 186, 187).

[178] Damian Rouson, Jim Xia and Xiaofeng Xu. *Scientific Software Design: The Object-Oriented Way*. 32 Avenue of the Americas, New York, NY 10013-2473, USA: Cambridge University Press, June 2011, p. 406. isbn: 978-0-521-88813-4. url: `http://www.cambridge.org/9780521888134` (cit. on p. 123).

[179] Amr Sabry. 'What is a purely functional language?' In: *Journal of Functional Programming* 8 (01 Jan. 1998), pp. 1–22. issn: 1469-7653. url: `http://journals.cambridge.org/article_S0956796897002943` (cit. on p. 120).

[180] Robert G. Sargent. 'Verification and validation of Simulation Models'. In: *Simulation Conference (WSC), Proceedings of the 2010 Winter*. Dec. 2010, pp. 166–183. doi: `10.1109/WSC.2010.5679166` (cit. on pp. 6, 7, 34, 41, 197).

[181] Hayssam Sbayti and David Roden. *Best Practices in the Use of Micro Simulation Models*. Tech. rep. 3101 Wilson Boulevard, Arlington, VA 22201: AECOM, Mar. 2010. url: `http://jdpckc7ap7dq.az.pl/download/WARSZTATY/SIMULATION/259_NCHRP-08-36-90.pdf` (cit. on pp. 34, 36).

[182] Andreas Schadschneider. 'Cellular Automaton Approach to Highway Traffic: What do we Know?' In: *Traffic and Granular Flow '07*. Ed. by Cécile Appert-Rolland, François Chevoir, Philippe Gondret, Sylvain Lassarre, Jean-Patrick Lebacque and Michael Schreckenberg. Springer Berlin Heidelberg, 2009, pp. 19–34. isbn: 978-3-540-77073-2. doi: `10.1007/978-3-540-77074-9_2` (cit. on pp. 30, 31).

[183] Andreas Schadschneider, Debashish Chowdhury and Katsuhiro Nishinari. *Stochastic Transport in Complex Systems – From Molecules to Vehicles*. 1st. Elsevier, Oct. 2010, p. 582. isbn: 978-0-444-52853-7 (cit. on pp. 25, 26, 28, 30, 31, 90).

[184] Andreas Schadschneider and Michael Schreckenberg. 'Garden of Eden states in traffic models'. In: *Journal of Physics A: Mathematical and General* 31.11 (1998), p. L225. doi: `doi:10.1088/0305-4470/31/11/003`. url: `http://stacks.iop.org/0305-4470/31/i=11/a=003` (cit. on p. 159).

[185] David A. Schmidt. *Denotational Semantics: A Methodology for Language Development*. McGraw-Hill Professional, 1987. isbn: 0-205-08974-7. url: `https://www.scss.tcd.ie/Andrew.Butterfield/Teaching/CS4003/DenSem-full-book.pdf` (cit. on p. 41).

[186] Michael Schreckenberg, Roland Chrobok, Sigurður Freyr Hafstein and Andreas Pottmeier. 'OLSIM - Traffic Forecast and Planning using Simulations'. In: *SCS/ASIM - 17. Symposium in Magdeburg, Simulationstechnik*. Ed. by R. Hohmann. SCS - Society for Modeling and Simulation International/ASIM - Arbeitsgemeinschaft Simulation. Magdeburg, Germany, 2003, pp. 11–18. url: `https://www.uni-due.de/imperia/md/content/ptt/paper/2003_asim2003.pdf` (cit. on pp. 12, 18, 19).

[187] Robert W. Sebesta. *Concepts of programming languages*. 10th ed. Pearson Education, Inc., publishing as Addison-Wesley, 2012. isbn: 978-0-13-139531-2 (cit. on p. 124).

[188] Reinhard Selten, Thorsten Chmura, Thomas Pitz, Sebastian Kube and Michael Schreckenberg. 'Commuters route choice behaviour'. In: *Games and Economic Behavior* 58.2 (2007), pp. 394–406. issn: 0899-8256. doi: `10.1016/j.geb.2006.03.012`. url: `http://www.sciencedirect.com/science/article/pii/S089982560600039X` (cit. on pp. 4, 5).

[189] Manuel Serrano. *Bigloo–A practical Scheme compiler, User manual for version 3.8c*. Inria Sophia-Antipolis. Cedex, France, 2011. url: `http://www-sop.inria.fr/indes/fp/Bigloo/doc/bigloo.html` (cit. on pp. 119, 120, 135).

[190]   Olin Shivers. *SRFI-1 List Library*. Oct. 1999. url: `http://srfi.schemers.org/srfi-1/srfi-1.html` (cit. on pp. 135, 143).

[191]   Special Working Group Digital Map and TMC Locations. *CentroMap Technical Specifications*. Tech. rep. 1.2.5. CENTRICO, Act. Dom. 2, Traffic Centers, 2001. url: `http://www.centromap.org/ts_1_2_5.pdf` (cit. on pp. 20, 21, 32, 33, 48).

[192]   Michael Sperber and Marcus Crestani. 'Form over Function: Teaching Beginners How to Construct Programs'. In: *Proceedings of the 2012 Annual Workshop on Scheme and Functional Programming*. Scheme '12. Copenhagen, Denmark: Association for Computing Machinery, 2012, pp. 81–89. isbn: 978-1-4503-1895-2. doi: `10.1145/2661103.2661113`. url: `http://doi.acm.org/10.1145/2661103.2661113` (cit. on pp. 125, 127).

[193]   *Information technology - Portable Operating System Interface (POSIX) Operating System Interface (POSIX)*. International Organization for Standardization, The International Electrotechnical Comission, The Institute of Electrical and Inc. Electronics Engineers. ISO copyright office, Case postale 56, CH-1211 Geneva 20, Sept. 2009 (cit. on p. 141).

[194]   Bundesanstalt für (BASt) Straßenwesen. *Traffic signs that apply in Germany*. 2013. url: `http://www.bast.de/cln_030/nn_169964/DE/Aufgaben/abteilung-v/referat-v1/v1-verkehrszeichen/vz-download.html` (cit. on pp. 44, 51, 152).

[195]   Bjarne Stroustrup. 'The Real Stroustrup Interview'. In: *Computer* 31.6 (June 1998), pp. 110–114. issn: 0018-9162. doi: `10.1109/MC.1998.683014`. url: `http://dx.doi.org/10.1109/MC.1998.683014` (cit. on p. 126).

[196]   Transport Simulation Systems. *Aimsun 7 Dynamic Simulators User's Manual*. 2012 (cit. on pp. 32, 33, 56, 100).

[197]   Transport Simulation Systems. *Aimsun 7 User's Manual*. 2012 (cit. on pp. 5, 30, 33).

[198]   Daimin Tang, Xiang Li and Yijuan Jiang. 'Microscopic traffic simulation oriented road network data model'. In: *Future Computer and Communication (ICFCC), 2010 2nd International Conference on*. Vol. 2. May 2010, pages. doi: `10.1109/ICFCC.2010.5497347` (cit. on p. 32).

[199]   Bundesanstalt für (BASt) Straßenwesen. *Technical delivery conditions specification for route stations (in German)*. Tech. rep. Technische Lieferbedingungen für Streckenstationen. Invalidenstraße 44, 10115 Berlin: Bundesministerium für Verkehr, Bau und Stadtentwicklung, 1998 (cit. on pp. 14, 15, 27).

[200]   Bundesanstalt für (BASt) Straßenwesen. *Technical delivery conditions specification for route stations (in German)*. Tech. rep. Technische Lieferbedingungen für Streckenstationen. Invalidenstraße 44, 10115 Berlin: Bundesministerium für Verkehr, Bau und Stadtentwicklung, 2002 (cit. on pp. 14, 15, 27).

[201]   Bundesanstalt für (BASt) Straßenwesen. *Technical delivery conditions specification for route stations (in German)*. Tech. rep. Technische Lieferbedingungen für Streckenstationen. Invalidenstraße 44, 10115 Berlin: Bundesministerium für Verkehr, Bau und Stadtentwicklung, 2012. url: `http://www.bast.de/cln_030/nn_42742/DE/Publikationen/Regelwerke/tls/tls-2012,templateId=raw,property=publicationFile.pdf/tls-2012.pdf` (cit. on pp. 14, 15, 27, 44).

[202]   Oregon Department of Transportation, DKS Associates and PTV America. *Protocol for VISSIM Simulation*. Tech. rep. Oregon Department of Transportation, June 2011. url: `http://www.oregon.egov.com/ODOT/TD/TP/APM/AddC.pdf` (cit. on pp. 34, 35).

[203]   Martin Treiber and Dirk Helbing. 'Reconstructing the spatio-temporal traffic dynamics from stationary detector data'. In: *Cooper@tive Tr@nsport@tion Dyn@mics* 1 (May 2002), pp. 3.1–3.24. url: `http://vwitme011.vkw.tu-dresden.de/TrafficForum/journalArticles/ctd02051401.pdf` (cit. on pp. 5, 14).

[204] Martin Treiber, Ansgar Hennecke and Dirk Helbing. 'Congested traffic states in empirical obser-
       vations and microscopic simulations'. In: *Physical Review E* 62 (2 Aug. 2000), pp. 1805–1824.
       doi: 10.1103/PhysRevE.62.1805. url: http://link.aps.org/doi/10.1103/
       PhysRevE.62.1805 (cit. on p. 29).

[205] Patrick O'Neil Meredith, Mark Hills and Grigore Roşu. *A Formal Rewriting Logic Semantic Defin-
       ition of Scheme*. Tech. rep. Computer Science Research and Tech Reports. University of Illinois,
       July 2007. url: http://hdl.handle.net/2142/11368 (cit. on p. 128).

[206] University of Illinois, Department of Computer Science, Formal Methods and Declarative Lan-
       guages Laboratory. *The Maude System*. 2014. url: http://maude.cs.uiuc.edu/ (cit. on
       p. 129).

[207] Angel Urbina, Sankaran Mahadevan and Thomas L. Paez. 'Quantification of margins and un-
       certainties of complex systems in the presence of aleatoric and epistemic uncertainty'. In: *Re-
       liability Engineering & System Safety* 96.9 (2011). Quantification of Margins and Uncertain-
       ties, pp. 1114–1125. issn: 0951-8320. doi: 10.1016/j.ress.2010.08.010. url: http:
       //www.sciencedirect.com/science/article/pii/S0951832011000640 (cit. on
       p. 36).

[208] Peter Wagner, Kai Nagel and Dietrich E. Wolf. 'Realistic multi-lane traffic rules for cellular
       automata'. In: *Physica A: Statistical Mechanics and its Applications* 234.3–4 (1997), pp. 687–
       698. issn: 0378-4371. doi: 10.1016/S0378-4371(96)00308-1. url: http://www.
       sciencedirect.com/science/article/pii/S0378437196003081 (cit. on p. 31).

[209] Joachim Wahle, Roland Chrobok, Andreas Pottmeier and Michael Schreckenberg. 'A Micro-
       scopic Simulator for Freeway Traffic'. In: *Networks and Spatial Economics* 2.4 (Dec. 2002),
       pp. 371–386. url: http://www.ingentaconnect.com/content/klu/nets/2002/
       00000002/00000004/05100309 (cit. on p. 18).

[210] Siri Krishan Wasan, Vasudha Bhatnagar and Harleen Kaur. 'The impact of data mining tech-
       niques on medical diagnostics'. In: *Data Science Journal* 5 (2006), pp. 119–126. doi: 10.2481/
       dsj.5.119 (cit. on p. 130).

[211] Stefan Wick. *Road information data base of North-Rhine Westphalia (in German)*. Ed. by Landes-
       betrieb Straßenbau NRW Fachcenter Vermessung/Straßeninformationssysteme. Straßeninform-
       ationsbank Nordrhein-Westfalen (NWSIB). Deutz-Kalker-Straße 18-26, 50679 Köln, 2014. url:
       http://www.nwsib-online.nrw.de/ (cit. on pp. 19, 21).

[212] David Wilkie, Jason Sewall and Ming C. Lin. 'Transforming GIS Data into Functional Road
       Models for Large-Scale Traffic Simulation'. In: *IEEE Transactions on Visualization and Computer
       Graphics* 18.6 (June 2012), pp. 890–901. issn: 1077-2626. doi: 10.1109/TVCG.2011.116
       (cit. on p. 32).

[213] Victor Winter. 'Bricklayer: An Authentic Introduction to the Functional Programming Language
       SML'. In: *Proceedings of the 3rd International Workshop on Trends in Functional Programming
       in Education*. Soesterberg, Netherlands, May 2014. url: http://wiki.science.ru.nl/
       tfpie/images/5/5e/Tfpie2014_submission_1.pdf (cit. on pp. 126, 127).

[214] Project BEST-ZEIT von Institut der deutschen Wirtschaft. *History of the working time flexib-
       ilisation (in German)*. Geschichte der Arbeitszeitflexibilisierung. 2002. url: http://www.
       flexible-arbeitszeiten.de/Kompakt/Geschichte1.htm (cit. on p. 4).

[215] Ludwig Wittgenstein. *Tractatus Logico-Philosophicus*. version 0.41 (February 11, 2014). London:
       Kegan Paul, 1922, p. 122. url: http://people.umass.edu/klement/tlp/tlp.pdf (cit.
       on p. 124).

[216] Bernard P. Zeigler, Tag Gon Kim and Herbert Praehofer. *Theory of Modeling and Simulation –
       Integrating Discrete Event and Complex Continuous Dynamic Systems*. 2nd. Academic Press, Jan.
       2000, p. 510. isbn: 9-780127-784557 (cit. on pp. 7, 24–26, 80, 99–101, 104, 105, 111, 114,
       198).